Health Economics

Health Economics

CHARLES E. PHELPS

University of Rochester

HarperCollins*Publishers*

To Dale—Our time has just begun.

Sponsoring Editor: John Greenman
Project Editor: Nora Helfgott
Design Supervisor: Jan Kessner
Cover Design: Jared Schneidman Design Inc.
Director of Production: Kewal K. Sharma
Production Assistant: James Spillane
Compositor: TCSystems, Inc.
Printer and Binder: R.R. Donnelley & Sons Company
Cover Printer: New England Book Components, Inc.

Health Economics

Library of Congress Cataloging-in-Publication Data
Phelps, Charles E.
   Health economics/Charles E. Phelps.
     p. cm.
   Includes bibliographical references and index.
   ISBN 0-673-38746-1
   1. Medical economics. I. Title.
   [DNLM: 1. Delivery of Health Care—economics. 2. Economics,
Medical. W 74 P538ha]
RA410.P48 1992
338.4'33621—dc20
DNLM/DLC
for Library of Congress

91-20910
CIP

95 9 8 7 6

Contents

Preface

All those who wish to develop a deeper understanding of health economics—including undergraduate and graduate students in business, public health, social sciences, or public policy—should find *Health Economics* helpful. It is written to serve a variety of users, and the content should be accessible to any student who has taken an introductory economics course.

One of the first major economics journal articles on medical care (by Kenneth J. Arrow, who later received a Nobel Prize in economics) emphasized the importance of uncertainty, and it is my hope that *Health Economics* will carry forward and extend that tradition. This text stresses the effects of uncertainty and incomplete information, both on the market for medical care and on the sustenance of health. The importance of uncertainty in generating the demand for health insurance is obvious, and every analyst of the medical care and insurance markets has encountered this uncertainty.

We understand intuitively that treatments don't always produce the desired results, so the production of health is itself a substantially uncertain process. As numerous studies have demonstrated, the medical profession as a whole harbors considerable uncertainty about the *average* effectiveness and desirability of using many medical interventions, ranging from coronary bypass surgery to the hospitalization of a child for a middle-ear infection. Recently, questions of information have become more prominent, particularly when asymmetric information exists—in other words, when the patient and the health care provider have differing information.

Uncertainty also pervades the structure of health insurance, the contractual arrangements that exist between patients and doctors, and the relationships between doctors and hospitals. A wide range of regulatory interventions in the health market (e.g., licensure and new drug testing) are directly linked to uncertainty, and the presence of other regulations (e.g., laws limiting hospital construction or cost control regulations) can be indirectly linked to uncertainty as a social response to the "side effects" of health insurance. *Health Economics* examines all of these topics in detail.

Wherever I introduce concepts that might not have been covered thoroughly in an introductory economics course, I include boxed material to elaborate on the ideas. Chapters dealing mostly with theoretical concepts are followed immediately by companion chapters that develop applications of those concepts.

The text avoids the use of calculus, but I have included in appendixes to several chapters additional material that depends in part on calculus and statistics. I have also included optional topics that lie beyond the scope of some courses for which this text might be appropriate. The chapter on international comparisons of health care systems, for example, might prove particularly useful to students of international health or to students seeking to examine U.S. health care spending in a comparative context.

The referenced readings are intended to expand students' understanding of the text material. Any student contemplating further study in health economics should become accustomed to dealing with original source materials such as journal articles and reports. Empirical studies in this field are burgeoning so rapidly that the only way to "keep up" is to read original sources.

Finally, the set of problems and questions at the end of each chapter are designed to help students assimilate text material and to stimulate further thinking about some of the ideas presented. I have provided prototype answers to nearly half of these questions at the back of the book.

I would like to thank the following individuals who read parts or all of this book's manuscript: David Dranove, Northwestern University, Kellog School of Management; Randall P. Ellis, Boston University; Richard W. Foster, University of Colorado; Richard G. Frank, Johns Hopkins University; John Godderris, Michigan State University; Mark Schaefer, Georgia State University; Michael Grossman, National Bureau of Economic Research; Roger T. Kaufman, Smith College; Robert H. Lee, University of North Carolina, Chapel Hill; Willard Manning, University of Michigan; Michael A.

Morrisey, University of Alabama, Birmingham; Joseph P. New-house, Harvard University; Mark V. Pauly, University of Pennsylvania; John Rapoport, Mount Holyoke College; Jody L. Sindelar, Yale University; Frank A. Sloan, Vanderbilt University; Stephanie So, University of Rochester; Doug Staiger, Stanford University; Marticia Wade, Carnegie-Mellon University; Gerard J. Wedig, University of Pennsylvania; and Barbara L. Wolfe, University of Wisconsin. These people provided valuable suggestions for improvements, and I appreciate their assistance on my behalf. I would also like to thank the staff at HarperCollins, as well as George Lobel, a former Scott Foresman editor, who encouraged me to undertake this book.

Charles E. Phelps

Chapter

1

Why Health Economics?

*H*ealth care represents a collection of services, products, institutions, regulations, and people that now accounts for over 12.5 percent of our gross national product (GNP), or $1 out of every $8 spent in our country. Personal spending on medical care—which occurs only rarely for most individuals—is exceeded on average only by food and housing, which "happen" every day. Estimates for 1991 show that aggregate spending on medical care has surpassed $750 billion, of which $670 billion represents "personal" health care expenditures, with the remainder arising from research, construction, and other administrative expenses. Personal health care expenses amount to nearly $2700 annually for each of the 250 million persons living in this country. This alone makes the study of health care a topic of potential importance.[1] Yet, many other aspects of health care economics make the topic fascinating.

Almost every person has confronted the health care system at some point, often in situations of considerable importance or concern to the individual. Even the most casual contact with this part of the economy confirms that something is indeed quite different about health care. Indeed, the differences are often so large that one wonders whether anything we have learned about economic systems and markets from other areas of the economy will apply, even partly, in the study of health care. Put most simply, does anybody behave as a "rational economic actor" in the health care market?

---

[1] The most detailed accounting of national health expenditures available at this writing appears in *Health Care Financing Review* 11:4 (1990), which contains detailed data through 1988. Extrapolations to 1991 provide a more contemporary context for the reader. Such estimates appear yearly in this journal, published monthly by the Health Care Financing Administration, the U.S. government organization responsible for administering Medicare, the federal program for health insurance for the elderly.

IMPORTANT (IF NOT UNIQUE) ASPECTS OF HEALTH CARE ECONOMICS

While the health care sector shares many *individual* characteristics with other areas of the economy, the *collection* of unusual economic features that appears in health care markets seems particularly large. The unusual features include (1) the extent of government involvement; (2) the dominant presence of uncertainty at all levels of health care, ranging from the randomness of individuals' illnesses to the understanding of how well medical treatments work, and for whom; (3) the large difference in knowledge between doctors (and other providers) and their patients, the consumers of health care; and (4) externalities—behavior by individuals that imposes costs or creates benefits for others. Each of these is present in other areas of the economy as well, but seldom so much as in health care, and never in such broad combination. A brief discussion of each issue follows.

As background to each of these ideas, and indeed for the entire book, the student of health economics will be served by the following notion: Uncertainty looms everywhere. Uncertain events guide individual behavior in health care. This major uncertainty leads to the development of health insurance, which in turn controls and guides the use of resources throughout the economy. The presence of various forms of uncertainty also accounts for much of the role of government in health care. Thus, if all else fails, search for the role of uncertainty in understanding health care. Such a search will often prove fruitful, and will lead to a better understanding of why the health economy works the way it does and why the institutions in those markets exist.

Government Intervention

The government intrudes into many markets, but seldom as commonly or extensively as in health care. Licensure of health professionals, of course, is common. Many other professionals also require a license before they may practice, including barbers, beauticians, airplane pilots, attorneys, scuba divers, bicycle racers, and (ubiquitously) automobile drivers. But almost every specialist in health care has a formal certification process to pass before practicing, including physicians, nurses, technicians, pharmacists, opticians, dentists, dental hygienists, and a host of others. The certification processes include not only government licensure, but often private certification of competence as well. Why does our society so rigorously examine the competence of health care professionals?

The government also intrudes into health care markets in ways unheard of in other areas. For example, federal and state programs provide insurance or financial aid against health expenses for an extremely diverse set of people, including *all* elderly persons, the poor, military veterans, children with birth defects, persons with kidney disease, persons permanently disabled, the blind, migrant workers, families of military personnel, and schoolchildren of all stripes. In addition, a broad majority of people living in the United States can walk into a county hospital and claim the right to receive care for free, if they have no obvious way of paying for the care. Probably only in public education do various levels of government touch as many individuals at any given time as in health care. Over a life cycle, nothing besides education comes close: Since Medicare has mandatory enrollment at age 65, every person who lives to that age will become affected by an important government health care program. By contrast, many individuals go through private schools, and never see the light of day in a public school. Why does the government involve itself so much in the financing of health care?

The government also controls the direct economic behavior of health care providers such as hospitals, nursing homes, doctors, and other providers far more than in other sectors of the economy. We have seen economywide price controls sporadically in our country's history, and considerable regulation in various sectors such as petroleum, banking, and (by local governments) housing rental rates. After OPEC raised the price of oil fourfold in 1973, petroleum regulations became a national phenomenon for several years, with such unintended consequences as gasoline shortages and hours-long queues to buy a tank of gas. However, such intervention pales in comparison to government involvement in prices in the health sector. At least in some form, the government has been controlling prices in the health care industry continually since 1971, and these controls have become more rigorous and binding over time. During the same time, the government decontrolled prices in a broad array of industries, including airlines, trucking, telephones, and petroleum. Why do we spend so much effort controlling health care prices, in contrast to those in other industries?

For decades, the United States has also seen direct controls on the simple decision to enter the business of providing health care.

Even ignoring licensure of professionals as an entry control, we have seen a broad set of regulations requiring such things as a "certificate of need" before a hospital can add so much as a single bed to its capacity. Similar laws control the purchase of expensive pieces of equipment such as diagnostic scanning devices. The reverse process also attracts considerable attention: If a hospital wants to close its doors, political chaos may ensue. What leads the government to intensively monitor and control the simple process of firms entering and exiting an industry?

Quite separately, both federal and state governments have commonly provided special assistance for providing education to people entering the health care field, through direct financial aid to professional schools and generous scholarships to students in those schools. This financial aid often directly benefits a group of persons (e.g., medical students) who will enter among the highest-paying professions in our society. For what reasons do governments proffer this support for the medical education process?

Government research is also prominent in the health care sector. While the government accounts for considerable research in other areas, most notably those involving national security (such as aircraft design, electronics, computers, and the like), its concern with research in health care is unique. The campus of the National Institutes of Health (NIH) in Bethesda, Maryland (on the outskirts of Washington, D.C.), surpasses that of almost every major university in the country in health-related research and education. In no other nonmilitary area does the government directly undertake research at such a scale. How did biomedical research reach such a level of prominence?

Before any new drug reaches the market, it must undergo a rigorous series of hurdles—research requirements imposed by the government on pharmaceutical firms, drug testing, and reporting of potential side effects. New medical devices confront similar regulations now. By contrast, for example, no regulations exist to control the study of health economics—perhaps, you might argue, fortuitously for this textbook! Why does the government concern itself so much with the drugs we stuff into our mouths, and so little with the knowledge we stuff into our heads?

Other apparently minor aspects of the government can dramatically affect our lives through the health sector. A simple provision of the tax code makes employer-paid health insurance exempt from income taxation. Another tax provision (Section 501.c.3 of the Internal Revenue Code) grants corporate immunity from income taxes to most hospitals and to insurance plans that account for about half of the private health insurance in the United States. In turn, most states adopt this same tax treatment for state income and sales taxes, and most local governments exclude the same organizations from paying property taxes. Why does the health care sector receive these favored tax treatments, and how much has its size and shape been changed thereby?

These ideas only touch briefly the extent of government involvement in health care, and indeed much of the remainder of this book refers continually to the presence and effects of the government. Astonishingly, though, the government's role in the health sector in the United States is *much less* than in almost all other countries. A later chapter describes how other nations have made choices to involve the government even more than in this country, and attempts to understand the consequences of such choices.

Uncertainty

Uncertainty lurks in every corner of the health care field. Many decisions to use health care begin because of seemingly random events—a broken arm, an inflamed appendix, an auto accident, or a heart attack. Most other medical events are initiated because an individual is concerned about the possibility of some illness—"Do I have cancer, Doctor? Am I crazy, Doctor?" "Why am I so tired, Doctor?"

Uncertainty may begin with the consumer-patient in health care, but it certainly doesn't end there. Providers also confront large uncertainty, although often they don't appear to recognize it individually. Yet, in similar situations, doctors often recommend treatment at vastly different rates, and often diverge greatly on which treatment they recommend. Therapies of choice change through time, often with little or no scientific basis for the decision. How can such medical confusion persist in a modern, scientific society?

The contrast between our approach to uncertainty in some areas of medical care (e.g., new drugs) and to similarly large uncertainty in other areas (e.g., the efficacy of a new surgical technique) also commands attention. In one case, we regulate the market intensively. In the other, we license the providers broadly, and then entrust them to make appropriate decisions. Thus, new therapies may sweep through the country with not so much as a *single* case-control study, let alone a true randomized controlled trial such as would be required for a new drug. Why do we behave so differently in these areas of uncertainty?

Asymmetric Knowledge

Symmetry exists when two objects are identical in size, shape, or power. When two people bargain in an economic exchange, and one holds far more relevant information than the other, the issues of asymmetric information arise. "Knowledge is power," so goes the old saying. This holds equally in international arms control negotiations and discussions between a doctor and patient. In the former case, however, both sides have similar opportunity and (presumably) similar skill in evaluating the positions and claims of the other. In health care, just the opposite holds true: One party (the doctor) generally has a considerably, possibly massively greater level of knowledge than the other about the issues at hand, namely, the diagnosis and treatment of disease. Not only that, the incentives to reveal information differ. In the arms control case, the two parties hold similar incentives to reveal or hide information. In the doctor-patient case, the patient clearly wishes to reveal information to the doctor, but the doctor may be in a different position. Professional duty, ethics, and personal responsibility make the doctor want to be open and honest. Conflicting with this, however, the simple profit motive can lead the doctor into different choices. Put most simply, if so desired, the doctor might be able to deceive the patient, and make more money doing so. In addition, the patient could have no way of telling when this was happening (if at all). Patients, after all, decide to consult with doctors because they want the doctors' advice.

As with many other facets of health care economics, this situation is not unique to health care. Most adults have confronted a similar circumstance in the most common setting imaginable— auto repair. There, the auto mechanic is in a position to do the same thing, that is, deceive the customer into believing that repairs are needed, and then possibly not even undertake them, since nothing really needed to be fixed.

We have evolved a variety of mechanisms to protect untutored consumers-patients in setting like these, some applying with more force in health care than others. As Kenneth Arrow (1963) has discussed so well, one of the important reasons for "professions" to evolve, with a code of ethics and, commonly, professional licensure, is to provide an institutional mechanism to help balance transactions like these.[2]

Arrow also emphasizes the importance of trust in ongoing relationships, and recent developments in the study of health care and other similar markets have formalized these ideas more completely: When two parties know that they will deal with one another for a long time—the classic "doctor-patient relationship"—their behavior can differ considerably from that in a one-time transaction.

The logic of this idea is quite simple: If a mechanic at a cross-country-route service station tells you that you need new shock absorbers in your car, pointing ominously to some oil dripping from one of your shock absorbers, the chances are a lot higher that the mechanic squirted the oil moments before, than that you really need new shocks. However, local mechanics have both less opportunity and less reason to try such stunts. First, they can't keep selling you new shocks each week! Second, they know that if you catch them once at attempted fraud, your relationship will end, and you will also tell your friends to avoid them. Mid-desert service stations will never see you again, and thus have no such constraints on their behavior.

Consumers can protect themselves against fraud by learning more about the activity at hand. With auto repair, many people can learn to be effective mechanics on their own, and hence less likely subjects to fraud. With the purchase of stereo equipment, we at least have the defense of listening to the quality of the sound. With these and many other activities, we also have the defense that the product can be returned to the seller, or the device to the repair person, if it doesn't function correctly. You can always go back to your mechanic and say, "Do it again until you get it right!" Not only that, mechanics probably have incentives to try, if they value the lasting relationship with you.

With medical care, as with other areas where "professionals" dominate the supply of the activity, things seem at least qualitatively different. First, the disparity of knowledge between the

---

[2] Every student of health care economics should read Kenneth Arrow's classic "Uncertainty and the Welfare Economics of Medical Care." Read it now, and also after you have completed your work with this book; you will get a lot more out of it the second time through. As with all citations in this book, the full reference for Arrow's essay appears in the bibliography at the end of the book.

doctor and the patient is larger than that between the customer and the auto mechanic. A reasonably intelligent person can learn quite a bit about auto repair in a relatively short period of time. The frequency of do-it-yourself (DIY) books on this topic and the number of auto parts stores attests to the commonality of this practice. The Yellow Pages in Rochester, N.Y., for example, list over 200 auto parts stores. By contrast, only 3 establishments offer to sell surgical instruments, suggesting that DIY surgery seems unpopular relative to DIY auto repair.

Perhaps more important, it may prove difficult to trade in a "service" when it doesn't work properly. By its very nature, a "service" involves the participation of the patient's body. If a surgical mistake is made, trade-ins may be hard to come by. Obviously, many medical mistakes are self-correcting, and many others can be restored with further medical intervention, but it seems reasonable to state that, on average, mistakes are harder to correct with services than with goods, also having less recourse to trade-ins as an ultimate fallback strategy.

The ability of individual customers to learn about the activity they purchase from others places a constraint on the amount of fraud one might expect. Alas, there are so many things to learn about in this world that we can't learn enough to protect ourselves on every possible front. Adam Smith pointed this out several hundred years ago, when he noted that "division of labor is limited by the extent of the market." In frontier communities, we would all operate as "Jacks (and Jills) of all trades"; in a larger society, we all specialize. Because we specialize, we must depend upon (and trust) others, leading to the possibility of fraud. Some of this fraud is not worth confronting. There is, in the words of one study, an "optimal amount of fraud" that we should learn to live with (Darby and Karni, 1973).

Fortunately, we are protected somewhat by our friends and neighbors who take the time to learn about auto repair (for example). They can help steer us away from clearly fraudulent mechanics toward those whom they trust. This process of acquiring information about the quality of mechanics (or doctors and dentists) proves important in the functioning of health markets, a topic to which we will return in a later chapter.

Externalities

Another area importantly separating health care from many other (but not all) economic activities is the common presence of "externalities," both positive and negative. External benefits and costs arise when one person's actions create benefits for or impose costs on others, and when those benefits and costs are not privately accounted for in individuals' decisions. Many early successes in medicine dealt with communicable diseases, probably the purest form of an event with externalities. When people get sick with a communicable disease such as polio or the flu, they not only bear their own illness, but they also increase the risk that their relatives, friends, and neighbors will contract the same illness. When they take steps to avoid such diseases, they confer a benefit not only on themselves, but on those around them as well. For example, the social benefit of getting a flu shot exceeds the private benefit. If people balance the costs of flu shots (including monetary costs, time, inconvenience, pain, and the risk of an adverse reaction) with their private benefits (the reduced risk of contracting the flu for a season), they will underinvest in flu shots from a societal perspective.

Many health care activities have little or no external benefit or cost, but surprisingly many such activities do. Most of the major health care activities with significant externalities have become such a part of the background of our society that we seldom recognize their presence or consequences. Sewage control, mosquito abatement, quarantine rules for certain diseases, and massive inoculation programs for infectious diseases often pass unnoticed by the average person.

Other apparently private activities also create external costs. For example, every time a patient receives an antibiotic injection, the odds go up slightly that a drug-resistant strain of the bacterium will emerge, immune to the current antibiotic. In relatively closed communities like nursing homes, this can become a serious problem (Phelps, 1989).

A number of other private actions affect other people's health and safety, but the health care system deals with them only at the end of the process. Most notable are individuals' decisions to drink and drive. Half of the vehicle fatalities in this country involve at least one driver who has consumed alcohol, and the number of "external" deaths caused by drunk drivers staggers the imagination. Every two years, for example, drunk drivers cause the deaths of more people on American roads than *all* deaths among U.S. soldiers in the entire Vietnam War. While these issues typically are not considered as "health care economics," the death and injury associated with such events may be more important than most diseases (and their cures) in our society.

As with everything else discussed in this section, the issue of externalities is not confined to the health care sector. Such simple local activities as fire and police protection have at least some element of externality (or "public good") about them, and on a grander scale, national defense and the formation of alliances such as NATO create the same issues. Air and water pollution, obnoxious "boom boxes" at the beach, and cars with noisy mufflers provide other examples of externalities outside of the area of health. Thus, while externalities may be an important part of some medical activities, they are not unique to health care markets.

HOW TO THINK ABOUT HEALTH AND HEALTH CARE (OR . . . HOW HEALTH ECONOMICS?)

After establishing that health care has some important differences from other markets, we should now turn to the more important question—how *can* we think about health care and health from an economic perspective? Do any of the normal tools of economics apply? If so, how must we change and modify their normal use to make them most fruitful?

Health as a Durable Good

To begin, think of the most fundamental of all building blocks of consumer demand theory, the "good" that increases a person's utility. Is anybody prepared to believe that having a dentist drill into a molar is a "good" in the traditional sense? Which of us really enjoys getting weekly allergy injections, or for that matter, once-in-a-lifetime yellow fever shots? These services—the actual events delivered and paid for in health care markets—cannot meaningfully be thought of as "goods" in the traditional sense. They do not augment utility directly. They hurt. They cause anxiety. Sometimes they have bad side effects. These sound more like "bads" than "goods"!

We are better served by backing up a step, and asking what really does create more "utility" for an individual. The most helpful answer is also the most sensible: Health itself creates happiness. We can begin to think about a reservoir of "health" that people have, and ask how medical care fits into this picture. Michael Grossman has most prominently explored the idea of "health" as an economic good, and showed how a rational economic person would have a demand curve for medical services that "derived" from the underlying demand for health itself.[3]

The ideas flowing from the simple concept of a stock of health permeate modern health economics. While we explore these ideas more fully in Chapters 3 and 4 on the demand for medical care, the basic idea deserves an early airing. You can think about health as a durable good, much like an automobile, a home, or a child's education. We all come into the world with some inherent "stock" of health, some more than others. A normal healthy baby has a relatively high stock of health. An infant born prematurely, with lung disease, the risk of brain damage, and possible blindness, has a very low initial stock of health. Almost every action we take for the rest of our lives affects this stock of health.

If we think of a bundle of other goods as X, and a stock of health (unobservable) as H, then we can say that a person's utility function is of the form

Utility = $U(X, H)$

Technically, we should think of the flow of services produced by the stock of health that creates utility, just as the transportation services from a car produce utility. However, to keep the wording from becoming too clumsy, we can continue to say that the "stock of health" creates utility rather than the more technically accurate expression that "the flow of services from the stock of health creates a flow of utility."

In the usual fashion for "goods," we would say that "more is better," so that more health creates more utility. It also seems plausible that the pleasure of other goods and services, which we can designate as the ubiquitous X, might increase with health. It's more fun to go to the zoo when you don't have a headache, for example. Thus, as Figures 1.1a and 1.1b show, both X and H produce more utility as the consumption of each expands. Figure 1.1a contains a series of plots showing how utility grows with X, each having a different level of H associated with it (i.e., H is held constant at a specific value on each line in Figure 1.1a). Figure 1.1b shows how H increases utility for given X. We can combine these

---

[3] Highly recommended for the serious student, Michael Grossman's *The Demand for Health* (1972) first explored the ideas of a stock of health and the derived demand for medical care. This work stems from a broader set of considerations of "household production" of economic goods, arising from the work of Gary Becker (1965) and Kevin Lancaster (1966).

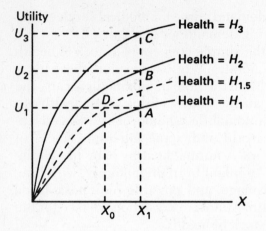

Figure 1.1a Expanding utility as a function of expanding goods.

two figures into one, for example, by picking some specific value such as $X = X_1$, and finding the level of utility associated with various values of H (H_1, H_2, H_3, etc.), at the points labeled A, B, and C in Figure 1.1a.

These same points appear in Figure 1.2, on a map showing combinations of X and H that produce the same level of utility. For example, in Figure 1.2, the combinations of X and H at points A and D both create the level of happiness U_1. This being the case, the

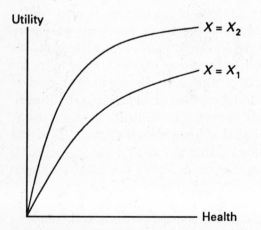

Figure 1.1b Expanding utility as a function of expanding stock of health.

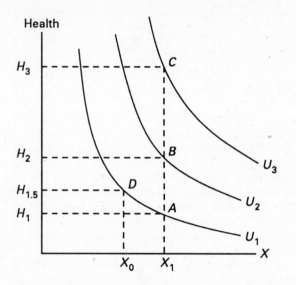

Figure 1.2 The same level of utility produced by different combinations of X and H.

consumer would be wholly *indifferent* between having a bundle like A or a bundle like D. The point D has more H and less X than point A, but both create the same level of happiness. The same point D appears in Figure 1.1a, showing the same relationship to point A (with more H and less X). The only difference between the two is that Figure 1.2 identifies the consumption points in combinations that create the same level of utility. We call curves like the one labeled $U = U_1$ *indifference curves*. (Another name for them might be *iso-utility* curves, stemming from the Greek *iso-*, meaning "the same.")

Maps like these provide a particularly powerful way to describe many of the economic forces driving health care (and other) behavior, so we will use them frequently.[4] You see maps like this commonly in everyday life, so they need not intimidate you. See Box 1.1 for some familiar examples.

---

[4] The appendix to this chapter develops some particular characteristics about indifference maps (and other ideas) in more detail, using mathematical concepts from the calculus. Those who wish to learn the concepts in this book at the more precise level allowed by formal mathematical structures can use the appendices. The main text, however, will remain free of such formalization, except in occasional footnotes.

Box 1.1 **READING MAPS**

Figure 1.2 in the text shows an "indifference map" but it is quite similar to something more familiar. Anybody who has ever used topographical maps should recognize Figure A, showing lines of equal altitude. (One might call them "iso-altitude lines." Similarly, one might call the indifference curves in Figure 1.2 "iso-attitude lines" if one were prone to that sort of verbal tomfoolery.) This particular topographical map shows the elevations in 100-foot contour lines. If one walks along the path depicted by such a contour line, one's altitude never changes, although one's location in east-west and north-south dimensions changes. (The location depicted on this map was featured in the science fiction movie *Close Encounters of the Third Kind.)*

Weather maps appearing in the daily newspaper show lines of constant temperature, perhaps called "iso-thermal maps." Figure B shows the average number of hailstorms annually in the United States. It reads just like a normal weather map, except that it shows different information. Perhaps we might call these "iso-ice contours." Notice the benign weather conditions prevailing in upstate New York.

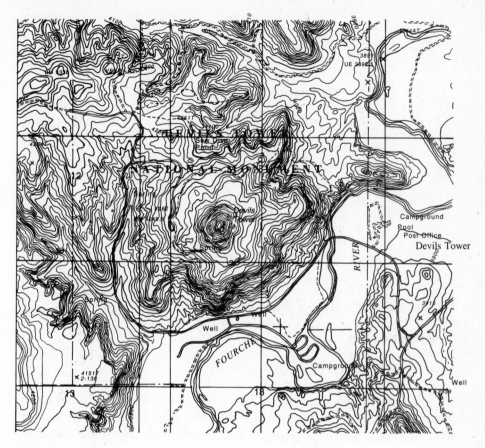

Figure A

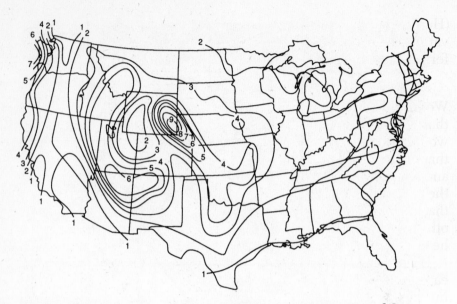

Figure B *Source:* Courtesy of Weather Bureau, U.S. Department of Commerce.

THE PRODUCTION OF HEALTH

Where does health come from? In part, it seems clear that we can produce health, or at least restore part of it after an illness, by using what we call "medical care," a set of activities designed specifically to restore or augment the stock of health.

In the usual economic terms, an auto company can produce automobiles using steel, plastic, labor, tires, wires, and so on as inputs. The process of transforming medical care into health can be thought of as a standard production function. The demand for the final product (such as automobiles) leads in turn to a *derived demand* for the productive inputs (such as steel or auto workers), or sometimes for subassemblies (such as motors). The same is true for health and medical care: Our underlying desire for health itself leads us to desire medical care to help produce health.

We can think about the process of transforming medical care (m) into health (H) in a fashion similar to the transformation of meat, energy (heat), buns, and mustard into hamburgers. In economics, we define such a process as a production function, the relationships that transform inputs (such as medical care) into outputs (such as health). We can call this function anything we wish

(Harry, Martha, or g), showing the functional relationship between various levels of m and H. "g" is more compact, making it a preferred name to Harry or Martha. Thus

$$H = g(m; D)$$

We would normally presume that more m produces more H, that is, that *the marginal productivity of medical care is positive.* We would also presume, in common with other economic phenomena, that *the incremental effect of m on H diminishes as more m is used,* and after a while, may even become negative. This would occur if the negative side effects of a drug or treatment occurred so often that they swamped any good provided. (The Greeks provide another word for us here: *iatrogenesis,* combining the words for healer [*iatros*] and origin [*genesis*].)

Figure 1.3 shows production functions for three different disease processes. Disease I doesn't make the individual terribly sick initially (without medical intervention), and medical care offers some help in healing, eventually reaching a near plateau. Allergies or asthma offer a useful example to think about. Disease II starts the individual out at worse health, but here, doctors have more to offer, and they can return the person to a higher level of health finally, although it may take more medical care. A broken arm provides a useful example. Finally, Disease III doesn't start the person out very sick, but doctors have almost no ability to help.

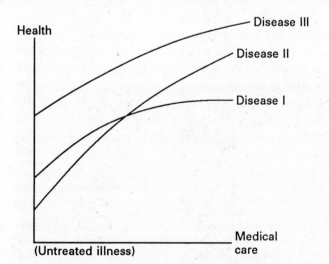

Figure 1.3 Health production functions for three diseases.

Thus, the person's health will not vary much with m. A common cold provides a classic example, following the old maxim that it takes a week to get over a cold if you don't see a doctor, but seven days if you do. In cases like these, the *productivity* of medical care is very small or zero over all ranges of use.

Several ideas deserve careful mention here. First, for almost every possible medical intervention, there reaches some point where the *incremental productivity* ("marginal productivity") of medical care will fall very low, or possibly even become negative. However, the *average* productivity can be quite high. The production process for Disease II in Figure 1.3 represents a good case— on average, medical care had done a lot of good, but it is possible to expand the use of m to the point where its marginal product also falls to zero—that is, the plot of health versus medical care flattens out. We should not confuse *average* and *marginal* effects. In later chapters, we will discuss the effects of changes *on the margin* in medical care in further detail, but it is important to remember that saying something has no *marginal productivity* doesn't mean that it is a worthless endeavor.

Second, it should be quite clear that the notation describing "medical care" as a homogenous activity m is hugely simplistic. There are literally thousands of identifiable medical procedures, and similar numbers of identified diseases and injuries.[5] Thus, "the medical production function" must really be thought of as a collection of various medical interventions, each applying to specific diseases and injuries.

Third, many medical interventions do not change the eventual level of health a person returns to, but they can considerably speed up the process of a "cure." The process of healing a cut hand provides an example. If you do nothing with a large cut, it can keep breaking open and may even get infected, but it usually heals itself eventually. If you use bandages and antibacterial ointments, you can speed up the healing process, but you get to the same point in

---

[5] The catalogs of diseases and procedures became more standardized as insurance became more important in health care. Insurers wanted ways to classify the procedures they pay for, and a number of competing systems emerged to provide that classification. The most commonly used system now is the Current Procedural Terminology system, now in version 4, called CPT4 almost universally in the health care system (American Medical Association, 1990). Similarly, as hospitals, doctors and insurers sought common systems to classify illnesses, separate coding systems emerged. Most of these follow from the International Classification of Diseases methods. The most recent version in common use is the International Classification of Diseases, Version 9, known more cryptically as ICD-9 (Department of Health and Human Services, 1980).

either case, because the body's own healing mechanisms work so well. This same idea carries at least some truth in phenomena ranging from intestinal upsets to seriously injured backs, with both of which people often return to the same level of health no matter what medical interventions are used. The same holds for many important diseases for which there is no cure, but with which medical intervention can at best slow the process of decline to death. Prominent diseases in this category include (at this writing) AIDS, Alzheimer's disease, and some forms of cancer.

Finally, we should always remember that medical care does not stand alone in affecting health. The production process contains a lot more than "medical care," including, most prominently, our own life-style. Also, the relationships between medical care and health may appear quite fuzzy, making it difficult to tell just how effective a medical intervention really is. This occurs because some people "get better" in ways unpredicted by doctors, and sometimes, they get worse or die despite the very best medical treatment, or sometimes because of side effects of that treatment. Later chapters contain more about these ideas. At this point, we should try to retain the basic ideas: Medical care produces health; health produces utility.

HEALTH THROUGH THE LIFE CYCLE

Like any durable good, our stock of health wears out over time. We call this process aging. As the stock of health falls low enough, we lose our ability to function, and eventually die. Again, in economic terms, our stock of health depreciates. "Normal aging" as measured in our society represents the average rate at which this depreciation takes place, but we should recognize that there is nothing biologically intrinsic about this process. Life expectancy has increased dramatically during this century, for example, implying that the depreciation rate on people's stock of health has slowed down through time. Public-health efforts (such as sanitation, vaccination against communicable disease, etc.) and individual medical care all serve to slow down the rate of depreciation of health, or to restore health to (or near) its original level after all illness or injury. Thus, if we plotted a typical person's health stock through time, it would look something like Figure 1.4, with a steady increase during childhood, and then a gradual decline from "aging," all the while punctuated by random events of illness and injury that can cause precipitous declines in one's health. At some

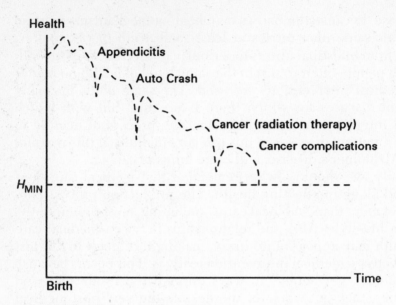

Figure 1.4 Time path of health stock.

critical H_{min} the person dies. Medical care forms an important part of the process of restoring health after such events, unless H_{min} has been reached. For example, the appendicitis attack shown in Figure 1.4 would have continued in a rapid slide downward to death if the person's appendix ruptured and no medical care was obtained. In addition, Table 1.1 shows aggregate mortality rates by age inter-

Table 1.1   OVERALL DEATH RATES BY AGE
            (Accelerated Depreciation of "Health" Stock in Humans)

Age	Aggregate annual death rate per 100,000 persons in age group
1–4	52
5–14	26
15–24	102
25–44	167
45–64	875
65–74	2,848
75–84	6,399
Over 85	15,224

Source: U.S. Department of Health and Human Services, National Center for Health Statistics, *Vital and Health Statistics of the United States,* 1986.

val for U.S. citizens, dramatically portraying the decrease in the stock of health associated with aging.

LIFE-STYLE AND HEALTH

In addition to the "random" events of health care, many of the other things we do and consume during our lives affect both the rate of aging (the slope of the smooth trend-line in Figure 1.4) and the frequency and severity of the "spikes." Our own life-styles can greatly contribute to our health. Stepping behind the broad portrait provided earlier, the bundle of goods and services we have called X can take on many characteristics, some of which augment and others of which dramatically reduce our stock of health. New medical research shows increasingly that the old adage "you are what you eat" is at least partly correct. Perhaps a better notion comes from the Bible: "As Ye sow, so shall ye reap." Prominent among such life-style choices are the decisions to smoke tobacco, consume alcohol, or use drugs (legal and illegal); the composition of diet; the nature of sexual activities; and the amount of exercise one undertakes. Therein lies the rub: Many of the things we enjoy (goods in the composite bundle X) cause life itself to trickle away. Not only can X and H substitute for one another in producing utility (see the discussion above on indifference curves) but X also affects H in a production sense as well. We should probably think about categories of X that have different effects on H. There are "good" types of X (X_G) that augment health. Moderate exercise provides a good example. There are clearly "bad" types of X (X_B) in terms of their effects on health, such as alcohol consumption and high-cholesterol foods. Still other neutral goods, such as books or saxophone jazz concerts, will have no apparent effect on health, aside from possibly increasing or reducing tranquility for the person involved. Thus, we can expand the production function idea even further, to include these types of X, where the signs over each element of the production function indicate the direction of effect that these activities should have on the health stock:

$$H = g(\overset{-}{X}_B, \overset{+}{X}_G, \overset{+}{m})$$

We would be on very thin ice indeed to argue that people "shouldn't" consume "bad" items. Presumably, people's goals are to maximize utility as much as possible within their budget constraint, and X_B presumably increases utility. However, we should also understand the role of these choices in affecting health. In

Table 1.2   FATALITIES BY CAUSE OF DEATH FOR PERSONS AGED 15–24

Cause of death	Annual deaths per 100,000 persons, 1984	
Vehicle crash	39.0	
Other accidents	12.2	
Homicide	14.2	
Suicide	13.1	
Subtotal: "Violent deaths"		78.5
Cancer	5.4	
Heart disease	2.8	
All other nonviolent causes	15.6	
Subtotal: All nonviolent causes		23.8
Total all deaths		102.3
Percent of all deaths due to violence	76.7%	

Source: U.S. Department of Health and Human Services, National Center for Health Statistics, *Vital and Health Statistics in the United States,* 1986.

many ways, these behavioral choices dominate a person's health far more than the medical care system.

To see the importance of life-style events most clearly, we can turn to the most clear-cut of all measures of health, namely, whether people live or die. In particular, age-adjusted death rates, by cause of death, illuminate best the role of life-style most clearly.

First consider the role of life-style in young adults. Table 1.2 provides the age-specific fatality rates for persons ages 15–24 in the United States. *All* of the really important single causes of death in this age group relate to "violence" in some way or another. The leading killer is vehicle crashes (called "accidents" in official statistics), and next on the list are other "accidents." Homicide and suicide follow close behind. These four categories alone account for over three-quarters of all deaths in this age bracket.[6] Most of these events lie outside the medical care system. While doctors and hospitals (most probably emergency departments) can try to patch up the consequences of these events, it is clear that the role of medical care is quite small, relative to that of "life" itself.

In older adults, a different pattern emerges, but the story re-

---

[6] For young adult black males, the fatality rates from violence just stagger the imagination: In the 15–24 age category, the death rate is 148 per 100,000. For blacks males aged 25–34, the violent death rate rises to 208, and persists at 180 for those aged 35–44. The comparable rates for white males aged 25–34 are 102.5, for white females, 25.9, and for black females, 44.8. Eventually, about 1 out of every 20 black males dies a violent death in this country.

Table 1.3   CAUSES OF DEATH FOR PERSONS AGE 65 AND OVER

Cause of death	Annual deaths per 100,000 persons, 1984		
	65–74	75–84	Over 85
Heart disease	1,103	2,749	7,251
Cancer	835	1,273	1,604
Cerebrovascular disease (stroke)	177	626	1,884
Chronic obstructive lung disease	141	270	331
Diabetes mellitus	59	126	217
Influenza and pneumonia	54	216	883
Accidents	50	107	257
All causes	2,848	6,399	15,233

Source: U.S. Department of Health and Human Services, National Center for Health Statistics, *Vital and Health Statistics of the United States,* 1986.

mains the same—"As Ye sow. . . ." Here, the major causes of death shift to the more familiar killers of heart disease, cancer, and stroke, all illnesses commonly associated with life-style choices. (Chronic obstructive lung disease is also heavily associated with tobacco consumption.) Table 1.3 shows death rates by cause of death for adults over age 65, and Table 1.4 shows the same data for persons at intermediate age 45–64. The leading causes of death in this intermediate group are a blending of those dominating the

Table 1.4   CAUSES OF DEATH FOR PERSONS AGED 45–64

Cause of death	Rate per 100,000 in 1984
Cancer	303
Heart disease	283
Strokes	37
Accidents	33
Chronic obstructive lung disease	28
Chronic liver diseases and cirrhosis	26
Diabetes mellitus	18
Suicide	17
Pneumonia and influenza	13
Homicide	7
All other causes	110
All causes	875

Source: U.S. Department of Health and Human Services, National Center for Health Statistics, *Vital and Health Statistics of the United States,* 1986.

BOX 1.2 A TALE OF TWO STATES

Victor Fuchs (1974) has portrayed the effects of life-style on health by comparing the causes of death in two adjacent western states in the United States,
comparable in climate, income, and medical resources per capita. One of these
two states (Utah) has a long history of avoiding such commodities as alcohol and
tobacco, primarily due to the predominance of members of the Church of Jesus
Christ of Latter-day Saints (Mormons) in that state. Tax levels, state law, tradition, and personal beliefs of many residents of this state maintain a low rate of
alcohol and tobacco use (as well as other commodities, such as coffee).

The neighboring state in Fuchs' comparison is Nevada, a region that has
certainly established a different reputation, deserved or not. For whatever
reasons, people in Nevada consume alcohol and tobacco at a markedly higher
rate than those in Utah. While aggregate statistics are somewhat misleading
(because of the heavy volume of tourists in Las Vegas and Reno, whose alcohol
and tobacco consumption would distort any measures of per capita use), it
seems clear that important differences in life-style exist between these two
states.

Fuchs shows the age-specific "excess" death rates in Nevada compared
with Utah:

AGE	MALES	FEMALES
<1	42%	35%
1–19	16%	26%
20–39	44%	42%
40–49	54%	69%
50–59	38%	28%
60–69	26%	17%
70–79	20%	6%

He punctuates his message with the cause-specific, age-specific death
rates in the two states. For every life-style–associated illness or event, the
age-specific death rates in Nevada substantially exceed those in Utah. These
data show the excess death rates for liver cirrhosis and lung cancer, one
associated with heavy drinking, and the other with tobacco smoking:

AGE	MALES	FEMALES
30–39	590%	443%
40–49	111%	296%
50–59	206%	205%
60–69	117%	227%

We should be cautious not to condemn or applaud one or the other of these
life-styles. However, we should not ignore the health effects of one or the other
life-styles when trying to understand the use of medical resources. To paraphrase an old idea in economics, there is no such thing as a free cigarette, even
if you "bum" it from a friend.

beginning and ending periods of adult life, as Table 1.4 shows, although by this age, heart disease and strokes have become the prominent killers.

The role of life-style choices in the causes of death for young adults seems clear. Yet, the same holds true for the major causes of death for the elderly: The risk of each of these events is strongly affected by life-style choices people make. Epidemiological data show systematically increased risks for most of these causes of death due to life-style choices. For heart attacks, to extend this point, consider the relative risks of a person with and without a particular life-style choice. Persons who smoke 1 or more packs of cigarettes daily have 2.5 times the risk of nonsmokers for fatal heart attacks. High blood pressure creates a 2.1-fold risk. Persons with elevated serum cholesterol have 2.4 times the heart attack risk of those with low serum cholesterol, and most doctors now believe that dietary intake can notably affect serum cholesterol. Finally, persons who lead a sedentary life have a 2.1 risk factor for heart attacks, compared with persons who exercise three times a week for at least 20 minutes. See Box 1.2.

SUMMARY

This chapter develops two main ideas. First, the study of health care and health contains important elements that demand careful attention for good economic analysis. Many of these issues pervade other areas of the economy as well, but seldom are they combined so much, and seldom with such importance, as in the economics of health care. These issues include:

- Government intervention
    health practitioners
    drugs and products
    price controls
    capital construction, entry, and exit
    provision of insurance
    research and development
    professional education
    favored tax treatment
- Uncertainty
    random illness striking individuals
    random outcomes from medical intervention
    professional uncertainty about efficacy of treatment

· Asymmetric knowledge
  health professionals know more than patient about healing
    processes
  consumers know more about their own health conditions
    than insurers
· Externalities
  communicable diseases
  reckless life-styles (e.g., drunk driving)
  production of knowledge

This chapter also explores a particular way of thinking about the relationship between health care and health itself, using the idea of a productive process. This formulation begins with the idea that utility is produced by health (H) and other goods (X). In turn, medical care systematically augments health. The relationships between health and other goods are more complicated than this simple model suggests, however. Some of the things we enjoy directly (part of the bundle of goods called X) add to our health, and others reduce it. Exercise and proper dietary composition add to health. Consumption of cigarettes, alcohol, and other drugs, as well as certain types of foods reduce our health. Other life-style choices dominate health outcomes, particularly for younger persons, such as the combination of drinking and driving. The risks for most of the primary causes of death at any age are strongly affected by our own life-style choices.

PROBLEMS

1. Give at least four separate areas of medical care where uncertainty creates an important, if not overriding, consideration in the analysis of health economics.
2. What are the leading causes of death for young people? How much does the medical care system have to do with their prevention?
3. What are the leading causes of death for older people? Which affects these more—medical intervention or individual life-style choices?
4. "Medical care is so special that normal economic forces don't apply." Comment.
5. "Medical care is obviously not a 'good' in the normal economic sense, so we can't use normal economic tools to think about and analyze the use of medical care." Comment.
6. What features most distinguish the study of demand for health from the study of demand for other assets (durable goods) that create utility (such as housing, stereo systems, etc.)?

APPENDIX

This appendix develops a formal model of a utility-maximizing individual who gains utility from both health (H) and other goods (X), and develops briefly the ideas of a production function for health.

Begin with a person who has a utility function of the form $U = U(X,H)$. If both are normal goods, then the *marginal utility* of both X and H is positive. Using the notation of calculus, $U_X = \partial U/\partial X > 0$ and $U_H = \partial U/\partial H > 0$. An indifference curve (iso-utility curve) as pictured in Figure 1.2 has slope $-U_X/U_H$. To see this, take the full derivative of the utility function, and then hold the *change in utility* equal to zero, while allowing X and H to vary. This will show how X and H can vary while U does not—in other words, the set of combinations of X and H producing the same level of utility. The total change in utility is given by

$$dU = dX\,U_X + dH\,U_H$$

and if we hold $dU = 0$, while still allowing dX and dH to be non-zero, we can assess the trade-offs between the two that still keep total utility the same. Thus, set $dU = 0$ and solve for the ratio dX/dH, which gives *the slope of the isoquant* in Figure 1.2. Simple algebra gives

$$dX/dH = -U_H/U_X$$

This gives the usual economics notion of trade-off, called the "marginal rate of substitution" between X and H in producing utility.

We can model the production of health in a similar fashion. Begin with the simple production function $H = g(m)$. Then the marginal product of m is the derivative $dH/dm = g'(m)$, the slope of any of the various production functions shown in Figure 1.3.

Since we can convert m into H, we can embed that relationship in the utility function, so that

$$U = U(X, H) \rightarrow U[X, g(m)]$$

Using the chain rule, we can easily find the relationship between medical care and utility:

$$\partial U/\partial m = (\partial U/\partial H)(dH/dm) = U_H g'(m)$$

We will use these ideas more completely in Chapter 3, where we expand on the transformation of medical care into health, and Chapter 4, where we develop the demand for medical care more formally.

Chapter
2

An Overview of How Markets Interrelate in Medical Care and Health Insurance

*T*he study of the economic and public policy issues in health care can take many paths, and we need a plan to organize our thinking. Chapter 1 set forth some broad conceptual ideas, including some notions of why health care issues both differ from and resemble other economic models, and a general idea of how to view medical care and the utility of health. The next step to gain a better understanding of health care economics should provide a framework upon which more detailed information can be assembled. This chapter provides that framework; it spells out the relationships between health care and health insurance, and establishes the major forces affecting supply and demand in each market. Later chapters focus on each of these subjects in greater detail; our purpose here is to set the stage, and to show more in outline form how these various actors relate to one another.

Our first analysis of health markets will be "static," assuming initially that the world stands still for a while. In this analysis, we seek to learn toward what sort of equilibrium the health care markets would naturally move. Next, we consider dynamic issues, particularly those arising from development of new knowledge about medical care and health, and those arising from broad economic events (such as persistent economic growth). After we have this structure in hand, we will go back to the important elements in this world—supply and demand for insurance, supply and demand for medical care, and technical change—and study each in more detail.

28

MEDICAL CARE MARKETS WITH FIXED TECHNOLOGY

One fruitful way to analyze medical markets links together the supply and demand for health insurance with the supply and demand for medical care. In each separate market (as in any market), supply and demand interact to create the *observed quantity demanded* and the *observed price,* the product of which is actual *spending* in the market. Except for this direct interaction, the analysis of competitive markets assumes that supply and demand in a market are independent of one another. Put differently, consumers shouldn't care about the cost of the inputs for a good, just its output price. Similarly, producers (at least in a competitive market) don't need to know the incomes of consumers in order to decide how to price their products. In noncompetitive markets, different ways of analyzing behaviors of sellers and buyers are needed, as we shall see in later chapters.

Health care differs considerably from most other markets in the following way: Many consumers face a different price to buy the product than sellers receive. This happens because of health insurance, which *lowers the price* of medical care to the consumer at the time of purchase. Of course, the premium charged by the insurance company must eventually recover all of the costs of that insurance, including medical care purchased through the plan, but the net result of insurance still leads to a lower *relevant* price for decision making by the consumer. Chapter 4 discusses these ideas in detail, but for now, we need merely to create a link between insurance and medical markets: We cannot talk meaningfully about medical markets or health insurance markets separately.

Figure 2.1 shows how these (and related) markets interact. Each box in the diagram characterizes either a supply or demand side of one of the relevant markets. Dotted "circles" show the phenomena we *observe* as these markets interact, such as prices and quantities consumed.

With this basic structure in mind, we now turn to a basic description of each component in these markets.

Demand for Medical Care

Illness Events   As we saw in Chapter 1, people's demand for medical care flows from their underlying demand for health. Thus, we might expect that shocks to people's health should cause their

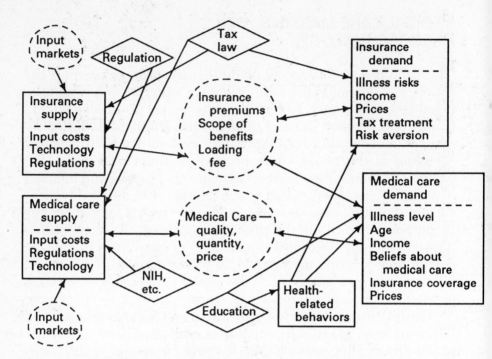

Figure 2.1 The interaction of medical, health insurance, and other markets.

medical care demands to change, and this is indeed true. Think of some person at the beginning of a year with some level of health H_0. If nothing happens to her during the year, her health stock will degrade slightly due to aging, but nothing else has happened. However, if she confronts a serious illness or injury, her health stock falls, say by the amount l (for "loss"). Any medical care (m) she buys will offset that by $g(m)$, so that her net amount of health at the end of the period will be

$$H = H_0 - l + g(m)$$

In general, the bigger the loss l, the more the person will want to try to restore health—that is, buying more medical care. Thus, medical care demand should vary directly with illness severity, as long as medical care has some ability to heal the patient.

As Chapter 4 discusses in more detail, illness events in many ways dominate the patterns of care through time for individuals. Averaging over larger groups washes out the financial importance of individual illness events, but for the individual, the random illness event takes primacy. Within the group, however, we can identify various subgroups prone to certain *systematic* illnesses or patterns of medical care use that deserve special attention.

Systematic Factors   Beyond individual illness events, the rate at which a person's health depreciates varies systematically with a number of things—most notably, age and sex. As we age, our bodies not only wear out, but do so at an apparently increasing rate. Thus, age should affect demand for medical care. Similarly, some diseases have strong genetic predisposition to occur. The most obvious factor is a person's sex: Some diseases (such as cervical cancer) can occur only in females, and others (such as enlarged prostate gland) only in males. None of the medical events surrounding a pregnancy can happen to a male. For example, childbirth is one of the most common reasons for hospitalization, dwarfing most other single events or illnesses. Systematic differences in medical use emerge because of differences like these. Other more arcane differences also abound in the medical literature, where certain genetically related groups face higher risks of disease.

Beliefs   Demand for care will also vary with a person's beliefs about the efficacy of care. That is, while doctors or scientists may have one view of the production function $g(m)$ transforming medical care into health for a particular disease or injury, patients may have a separate belief. Their beliefs may diverge in either direction: Many people continue to use (and to profess belief in) healing methods that "science" debunks, while at the same time others may distrust and refuse to seek care that scientists and doctors would describe as efficacious.

Advice from Providers   As with other complicated goods and services, consumers depend on providers of medical care for advice about when to use care, which treatment is most desirable, and how much is needed. Patients don't always follow this advice, but it seems clear that the advice received by patients can affect their choices. Cumulatively, such advice (and the effects of receiving recommended treatment) help form people's beliefs about the efficacy of care.

We should proceed with caution before characterizing "medicine" or "science" as holding uniform views on the efficacy of various medical interventions. The advice provided by physicians to patients differs greatly across the country and around the world for apparently similar patients. Chapter 3 discusses in more detail the variability in use of various medical procedures in different regions of this and other countries.

Income As with any other economic good, a person's demand for medical care may depend on income. If health care is a "normal" good, then more income will create more demand for care. Here, of course, we must be quite careful in our definitions. Statements like this should always carry the warning that *all other things are held equal.* In this case, for example, we want to be sure to hold constant the level of illness. If low-income people (for a variety of reasons) contract illnesses more frequently than high-income people, then the quantity of care they are observed using may exceed that of high-income people. The "pure" effect of income becomes contaminated in such cases, and it often is helpful to separate the various effects of income on demand. Another important difference is that of health insurance coverage, which increases directly with income on average. This would also distort simple comparisons of medical care use by income group.

Money Price Economic logic says that people living within a budget constraint (as we all must do) will buy less of any good as its price rises, and, conversely, more of the good as its price falls.[1] While empirical studies are needed to show the importance of price in affecting people's use of medical care, logic tells us that the possibility exists. Other things being equal, people's demand for care should decline as price rises.

Time Price Services like medical care require that people attend the proceedings with their own bodies. One cannot hire a "stand-in" to take a physical examination or have abdominal surgery. Medical care is a participatory phenomenon. Following the adage that "time is money," we can expect that people facing higher time prices for medical care will use less of it, other things being equal. Time price can rise or fall for two separate reasons: the actual time involved (e.g., travel time to the doctor) or the value of time to the person. Both elements of time cost can affect demand for care.

---

[1] One important anomaly occurs when the good consumes a large share of people's budgets, and yet it is an "inferior" good, that is, one where they would choose to buy less of it as their income increased. This is almost certainly not relevant in the study of health care. The classic study of these "Giffen goods" focused on demand for potatoes in Ireland during a famine.

The Supply of Medical Care

Medical care supply depends primarily on economic factors—the costs of inputs, the price received for the final product, and the constraints imposed by available technology. Nothing said here distinguishes medical care supply from that of any other good or service. Medical care supply does have some important differences, but the common ideas of supply from standard economics remain intact. The primary concepts include (1) the idea of a production function, (2) the difference between input costs and final product price, (3) the legal and regulatory structure in which production takes place, (4) the types of organization providing health care, and, finally (5) the way firms interact in the final product market. We can look at each of these in more detail.

Changing Inputs into Outputs: A Production Function   Any good or service transforms a set of inputs into a final product. In medical care, the inputs include capital (beds, diagnostic devices, operating rooms), supplies (ranging from bed sheets to sophisticated drugs), a wide variety of workers (ranging from nurses and doctors to janitors and clerks), and patients. (Remember—health care is a participatory event.)

Health care organizations differ from assembly lines because each patient is unique. Hospitals and doctor's offices are *job shops,* producing a diverse array of products, each tailored specifically to the individual patient. Many other organizations have similar production patterns, including barber shops, auto repair, electronic repair, and grocery stores. What does differ in the health care setting, particularly the hospital, is the management structure. The persons deciding which activities to undertake (physicians) are often not employees of the organization. Chapters 6 (for physician-firms) and 8 (for hospitals) analyze these issues more fully. We should also recall here, although discussing how medical care produces health, that many other activities (personal life-style choices) also affect health.

Input Costs and Final Product Price   In almost every imaginable market structure economists have ever analyzed, higher prices for inputs lead to higher cost (and hence higher price) for the final product. This holds true in health care markets, no matter whether the firms involved are not-for-profit hospitals or profit-maximizing drugstores.

Laws and Regulation   In any economic endeavor, legal structures importantly shape the way firms organize and conduct their business. The most fundamental laws relate to the rights to own, use, and trade property. Without property rights, much of our economic system, including health care, would differ radically from its current structures. Many of the same functions of private firms can be carried out by government agencies. For example, the British National Health Service operates almost the entire health care system of England and Wales without any role for private enterprise, except as suppliers of some inputs. In the United States, however, private production dominates the health care system and must be understood in the context of a system of private property rights. More so in health care than in other sectors of the economy, the government intervenes to limit or modify what would normally be held as standard property rights. A section in Chapter 1 touched on these issues, and Chapter 16 focuses more intensively on them. Here, in this overview, we need remember only that a good understanding of the production of health care must include a careful assessment of the regulations and laws governing the activity.

Organizational Structures   Many health care organizations differ from those in most industries because of the presence of unusual ownership and management structures. Most important, many firms, including most hospitals and many large health insurers, operate as *not-for-profit* firms. This does not preclude such firms' making a profit, but rather it eliminates the shareholder as the "residual claimant" of such profits. Thus, not-for-profit organizations must decide how to disperse such profits (including the possibility of lowering price to consumers), and these choices can affect composition and amount of output, the price, and even the mode of production. Chapter 8 details these issues more fully.

Final Product Price   In competitive markets, with each firm taking the market price as given, the role of price in affecting output is clear and simple: Higher prices bring forth more production. In less-simple market structures, life gets more complicated. Particularly in regulated markets and in markets with imperfect or limited information, the relationships among costs, prices, and final product output can get quite complex. Indeed, in many health care markets, the idea of "a" price seems difficult to define or support. Rather, we must learn to think about distributions of prices in some health markets. For example, doctors with apparently the same quality and credentials in the same city and under similar conditions may charge very different prices for "a simple office visit." Actual data often show two- or threefold variation (from highest to lowest) in the prices charged in single markets by doctors or dentists for the same procedure. However, for many health care questions, an assumption of competition probably will assist in clarifying thinking about such issues as health insurance, retail pharmacies, vision care, and (with some modification) nursing homes, hospitals, doctors, dentists, and the like.

The Demand for Health Insurance

Financial Risk   Illness and injuries befall people quite randomly, creating a health risk that in turn creates a financial risk as people seek medical care to alleviate the effects of the illness. Most people prefer to avoid financial risks, and will seek insurance against them. This provides the primary motivation for purchasing health insurance—the avoidance of risks. Obviously, people differ in the amount of risk they prefer to confront, or more important, on the amount they are willing to reduce or eliminate financial risk. Thus, attitudes toward risk importantly affect demand for insurance.

A special warning should be issued here: Traditional approaches in economics to the study of decision making with uncertainty, including the decisions to purchase insurance, have proven less adept in elucidating people's actual behavior than we might hope. Broad disagreements exist within economics, spreading into other disciplines such as psychology, about the correct ways to model people's behavior in uncertain environments. *Caveat emptor!* Chapter 10 provides a more extended discussion of these issues.

Price of Insurance   As with any other economic good, prices alter people's desire to insure against risks. The price of insurance is not the same thing as the *premium* paid to the insurance company. Since (at least on average) an insurance policy will return money to the insured person as *benefits* of the policy, it is best to think about the price of insurance in relation to premium costs that exceed the (statistically) expected benefits. If insurance premiums were to cost the same amount annually as the expected benefit payments, then the provision of the insurance itself would be free—that is, at a zero price. However, insurers must set their premiums to pay for the costs of operating the insurance company, and may add any additional payment for risk bearing that the market will sustain. This "loading" on top of expected benefits constitutes the price of insurance. In general, the higher the loading fee, the lower the demand for insurance.

Tax Laws   Through a variety of mechanisms, the U.S. government (with almost all state governments following suit) has provided favorable tax treatment for health insurance purchases through time. In the past, taxpayers could deduct part of any premium payments they made personally, providing a subsidy to insurance proportional to the marginal tax rate of the person. While that particular deduction has nearly evaporated through time, a more important tax incentive remains intact. Any payments made by an employer to buy health insurance for an employee (or family of an employee) are not taxable income. Again, this provides a subsidy to insurance proportional to the taxpayer's marginal tax rate. We can think of such a subsidy as reducing the price of insurance, with predictable effects on demand for insurance. While a variety of tax reform proposals have considered changing these laws, the favorable tax treatment of health insurance has persisted. Chapter 10 devotes considerable attention to these issues.

The Supply of Health Insurance

Consumers can find health insurance in a dizzying array of plans, but ironically, most Americans eventually select from one of only a few plans offered to them. This occurs because most private insurance in this country is sold through groups, rather than individually, and the groups typically offer only limited choice, if any, to individuals within the group. The very large majority of such groups arise through the place where the individual is employed —the work group. This mechanism dominates the market for two reasons, discussed in more detail in later chapters. First and foremost, the U.S. tax code makes payments by employers toward the cost of the insurance exempt from income taxes, making this a very low-cost way to buy insurance. However, even if this tax break did not exist, employer group insurance would probably still persist, because the work group does something else valuable for insurers:

It brings together a collection of individuals for some other purpose than buying insurance, thus providing the insurer with a strong confidence that their actuarial projections of cost will remain valid. However, if the group assembled itself for the purpose of buying insurance (such as a "Health Insurance Buyers Club"), such a group would largely attract only sickly individuals who expected high medical costs, and the insurance company would nearly face the same problem as selling insurance to one person at a time.

The problem that groups "solve" for insurers remains with individual insurance sales; persons buying the insurance inherently know more about their own health conditions than the insurer. Thus, persons who know they have contracted a serious illness, or who anticipate surgery or having children, or who know of family propensities to get certain diseases all have enhanced incentives to buy insurance. Insurers cannot know such details about individuals well. They can ask questions and even get physical exams, to sort out such "high-risk" patients, but they always confront the possibility that some people will slip through any filter they design. The concern for this, plus the costs of administering the filtering system, present two reasons for individual policies to cost more than group policies for the same level of coverage. This problem, known as "adverse selection," complicates the ability of insurers to sell insurance policies without losing money, and potentially even destroys the ability of an insurance market to function.

The cost of insurance is really just the administrative and risk-bearing cost of operating the plan. The "premium" that the subscriber pays is the sum of the average benefits ("expected" benefits in the statistical sense) plus any charge for administration, sales, and risk bearing. This extra charge is called the "loading fee" or more simply, "the load." One convenient way to describe the loading fee expresses it as a fraction of the expected benefits. In this structure,

Premium = (1 + Loading Fee) × (Expected Benefits)

Thus, if the Loading Fee $(L) = 0$, the premium just matches the expected benefits, and the insurance itself would be "free."

Of course, insurance companies cannot operate without using resources. Thus, all insurance has a positive loading fee, at least in equilibrium. (If insurers have an unlucky year or make mistakes in setting the policy premium, they might end up paying out more benefits than they collected in premiums, but this cannot persist for long or the company will go broke.)

Group health insurance policies in the United States have an average loading fee of about 15 to 20 percent. That is, for each $100 in expected benefits, about $15 to $20 more is added for administration, sales, and risk bearing. In very large groups, the loading fee can fall much lower, perhaps even to 5 percent or so, because the sales costs per enrollee are very low, and the administration is simplified by having so many people enrolled in the same plan.

Nongroup policies, by contrast, have loading fees of 60 to 80 percent and often more. Many such policies have loading fees near 100 percent. This means that only half of the premium paid eventually returns to enrollees as benefits, on average. This distinction in costs is one reason why employer group insurance remains attractive, even with the limited choice of coverage. Chapter 11 expands on the discussion of the supply of insurance.

Interaction of Insurance and Medical Markets

Most insurance in other areas of life (e.g., life insurance, auto insurance, and home insurance) provides a specific dollar amount paid to the consumer in the event a loss occurs. Often, a claims adjuster determines the amount of the loss after reviewing claims information, but when all is completed, the insurance company sends a check to the insured person for the amount of the loss, no matter what the person decides to do with that money. For example, if you get a crumpled fender on your auto, an amount is paid that would suffice to replace it (sometimes based on several bids from repair companies), but *you do not have to repair the auto in order to claim the insurance payment.* The "loss" is the collision, and the payment does not depend on actions you take subsequently. You always reserve the right to drive around with a crumpled fender, and spend the insurance payment on something else.

Health insurance functions quite differently. "Events" in health matters are more difficult to define in advance than "crumpled fender" or "building burned down." Getting estimates to restore the damage is harder. Probably for this reason more than others, health insurance does not pay on the basis of health events, but rather on the basis of medical care events. In the old "Star Trek" series, Dr. McCoy could hold up a little box to the patient that would go "Wrmblwmrblwmrbl" and provide a readout of the patient's condition. By contrast, real doctors cannot commonly define the illness event clearly, and have even greater difficulty providing a precise estimate of the "costs of repair." A major reason for this is that the "repair" process contains a lot of uncertainty. Thus, writing a health insurance policy on the basis of "health events" wouldn't provide the kind of financial protection that people seek.

Health insurance pays only when people buy medical care, and then only for types of care specified in advance in the insurance contract. The insurance policy will pay for part or all of any medical care the enrollee selects. This has the effect of reducing the price that people pay for medical care at the time they purchase it. In exchange for this, they pay a lump sum of money each year (the insurance premium). In economic terms, we can think of this as a simultaneous exchange of lower income for a lower price of medical care. *Health insurance acts to subsidize medical care* at the time of purchase. This feature of health insurance makes it different from almost all other types of insurance, and also dominates our analysis of the effects of health insurance.

Reaching Closure

Health insurance and medical care interact with one another in important ways. People's *anticipations* of their medical care events drive their purchases of health insurance, when they have a meaningful choice. Of course, the risks of ill health (and attendant financial events) create a serious financial threat, also leading people to purchase insurance. In turn, people's health insurance coverage increases their use of health care, by reducing the price of care at the time it is purchased. Together, health care demand and insurance coverage demand create an intricate pattern of behavior. The forces affecting all of these have led our health care system to quadruple in size over the past several decades. We now explore how this intricate system has evolved over time.

DYNAMIC ISSUES—CHANGES THROUGH TIME

In any economic system, changes in one part of the system trigger corresponding changes elsewhere, often with feedback to the original variables. Thus, while economic analysis often focuses on "long-run" equilibria, we might characterize the system we actually see and measure as a series of transitional short-run equilibria,

---

Box 2.1   **ADJUSTING FOR INFLATION**

When comparing economic data through time, the most useful comparisons adjust the raw data to show the "real" buying power people have in any year. In its most simple form, if all prices and incomes double from one period to the next, real buying power would remain unchanged, but the observed "nominal" income would appear to be twice as high.

The correct way to adjust for general inflation is to "deflate" time trends with the appropriate measure of inflation. A common such measure is the Consumer Price Index (CPI), which measures the prices of a particular "basket" of goods and services, each receiving a particular weight in the total, where the weight corresponds (one hopes!) to overall spending patterns. Thus, for example, changes in the prices of housing and food receive more weight than changes in the prices of tobacco or shoes.

It doesn't matter in such calculations which year one takes as the "base." The CPI usually has some specific year with a base index of 100, but that base changes periodically. At present, most reports of the CPI use the price level of 1967 as 100 and show relative prices in current years as a multiple of 100. Thus, for example, when 1967 prices are "100," 1987 prices were 338.

To put 1967 and 1987 incomes in the same metric, we either need to multiply 1967 incomes by 3.38 (putting everything into 1987 dollars) or divide 1987 incomes by 3.38 (putting everything into 1967 dollars). The same sort of calculation is made on data for each year, once the base year is selected. Using a recent year as the base has the advantage that the level of purchasing power represented by a particular dollar amount is easier for people to understand. Nothing logically prevents us from measuring all incomes in terms of prices in the year 1900, except that it is quite hard for us to envision what kind of buying power that really implies.

---

always interacting and changing. Nevertheless, a series of important and persistent patterns appears in health care markets that bear discussion. These include (1) changes in the overall economy; (2) demographic changes, particularly the aging of our society; and (3) changes in medical knowledge, brought about mostly by biomedical research.

Economywide Income Growth

Like almost all commodities and services, health care seems to be a "normal good," so that people's desires to use the good increase as their income increases. Studies of the effects of income on demand for care offer sometimes puzzling contradictions (see Chapters 5 and 17), but we should at least have in mind the overall change in per capita income in order to study the growth in health care spending.

Table 2.1   PER CAPITA INCOME AND INCOME GROWTH SINCE 1950 IN THE
UNITED STATES

Year	Per capita GNP income (1982 dollars)	Annual growth rate over five-year interval
1950	5,476	1.8%
1955	5,940	1.64%
1960	6,352	1.35%
1965	7,512	3.41%
1970	9,063	2.10%
1975	10,200	1.98%
1980	11,427	2.23%
1985	12,451	1.73%
1990	13,200	1.2%

Sources: U.S. Department of Commerce, *Statistical Abstracts of the United States,* 1989; Table 673 for per capita income, and Table 738 for consumer price index. Data in column 1 calculated from these sources; data in column 2 calculated from column 1 data, plus corresponding data from 1945.

Beginning after World War II (when modern health insurance began to flourish seriously), per capita income has increased with a steady if somewhat erratic pace. Overall price levels have also grown during this period (considerably!), so any meaningful comparison of per capita spending should adjust for inflation, to give "constant buying power." (Box 2.1 describes how this is done.) Table 2.1 shows the growth in per capita income over the years from 1950 to the present (the last year shown is a projection). With some obvious ups and downs in the growth rate, the U.S. economy has grown steadily since World War II, at an average annual rate of about 1.75 percent in real purchasing power. While that might not sound like much, when compounded over 40 years, it more than doubles the per capita real income.

Demographics

At the same time that our society has grown richer, it has grown older. After the famous postwar baby boom in the 1950s and 1960s, the population has settled in with a mere 7.4 percent under the age of 5. At the same time, the proportion of our population aged 65 or over has grown slowly but steadily from 8.1 percent in 1950 to 11.3 percent in 1980 and reached nearly 13 percent by 1990. The important implication from this aging society stems directly from

Table 2.2   AGE DISTRIBUTION OF THE U.S. POPULATION

Year	Percent over 65 years old	Percent under 5 years old
1950	8.1%	10.8%
1955	8.7%	11.0%
1960	9.2%	11.3%
1965	9.5%	9.7%
1970	9.8%	8.4%
1975	10.5%	7.4%
1980	11.3%	7.2%
1985	11.9%	7.5%
1990	12.6%	7.4%
2000	13.0%	6.3%

Source: U.S. Department of Commerce, Statistical Abstracts of the United States, 1989, Tables 13 and 17.

the underlying biological phenomenon of aging: As we get older, our health stock deteriorates faster, and we use more medical care. Table 2.2 shows these patterns through time.

R&D and Technical Change

Still another important phenomenon in the dynamic health economy is technical change. As science and medicine progress, we learn how to do things we could not do previously. The most prominent examples in people's minds include spectacular surgical interventions involving artificial hearts, organ transplants, and the like, but the phenomenon of technical change pervades every level of the health care system. Complex cancer therapies involving surgery, drugs, and radiation have increased tremendously. Genetic manipulation opens new realms of diagnosis and treatment. Complicated diagnostic devices like the CT and MRI scanners depend on massive computing power, unobtainable even in the 1960s, and now nearly portable in size.[2] Casts for broken arms can now be made lighter and waterproof. Even the humble Band-Aid differs significantly from its 1950s ancestor.

---

[2] The acronym CT stands for computed tomography. The Greeks gave us the word *tomography* as well as many others in medicine. *Tomos* means a slice or cut. Computed tomography literally creates pictures of cross-sectional slices of the body.

Much of this technical change has been fueled by biomedical research, supported by both public and private funds. The medical interventions deriving from this research have found ready financial support in a market with considerable (and growing) health insurance coverage.

Much of the private biomedical research has been funded by pharmaceutical companies, many in the United States, but many others elsewhere around the world. Almost all new drugs created (and tested) anywhere in the world find applications in most nations, so to refer to "U.S." private research, especially in pharmaceuticals, would provide a narrow and misleading picture. In addition, accurate and systematic data on research and development (R&D) spending in health care itself is almost impossible to obtain. Indeed, much of the research with health care implications begins in other sectors of the economy, ranging from biology and chemistry to lasers and computers.

In government-funded biomedical research, however, the U.S. role in worldwide research is prominent and provides a reasonable portrayal of overall research activity. Table 2.3 shows the pattern of U.S. health research funded by various arms of the Department of Health and Human Services (formerly the Department of Health, Education and Welfare), primarily through the National Institutes of Health. The growth in research funding, particularly between 1955 and 1965, was spectacular by any standards, with a subsequent flattening out of real spending through time in the 1980s.

Table 2.3   U.S. GOVERNMENT SPENDING ON BIOMEDICAL RESEARCH

Year	Total medical research via DHEW/DHHS budgets (millions of 1984 dollars)
1950	173
1955	271
1960	1259
1965	2893
1970	3239
1975	4628
1980	4697
1985	5400

Source: Arnett, H. R., McKusick, D. R., Sonnefeld, S. T., and Cowell, C. S., "Projections of Health Care Spending to 1990," *Health Care Financing Review,* Spring 1986, 7(3):1–36.

Price and Spending Patterns

These and other forces in the economy have led to a persistent growth in "real" medical spending in every year, at least since World War II. Indeed, this pattern of apparently ever-increasing spending growth has formed the basis of considerable private and public concern, provided the impetus for massive government regulation in health care, and created a large industry of "cost-conscious" health care plans in the United States over the past decade, all in attempts to stem the flow of dollars before the patient bleeds to death financially. Our next goal here is to examine the temporal patterns of medical care prices and spending, and health insurance coverage and cost. These data form the background tapestry for the more detailed study of the health markets that follows.

Let us begin by examining the overall trends in health care spending. As before, all of these data occur in a world where general inflation has taken place, and we really want to study a world "as if" no *general* inflation had occurred. Thus, all of these data are converted to "constant dollars," in this case, using 1986 as the base year.

The Growth in Medical Prices

Medical care prices are difficult to measure meaningfully, because the nature of the service changes. To understand this requires some understanding of how the Bureau of Labor Statistics constructs the CPI and its component measures. For the overall CPI, they select a "basket" of goods that represents a typical pattern of purchases for urban consumers, including food, clothing, housing, transportation, medical care, entertainment, and other goods and services. Within each of these categories a specific set of commodities and services is chosen. In medical care, the index includes a measure of physicians' fees, dentists' fees, a hospital room and board charge, and a measure of prescription drugs and over-the-counter (nonprescription) drugs. Each of these requires still further specification. The "prescription drug" indicator is a small sample of the many thousands of drugs actually available by prescription. It includes drugs such as a tranquilizer (Valium), an antibiotic (penicillin), and so on. Similarly, "a physician office visit" can embrace a wide range of choices, including the type of specialist, the complexity of the visit, and even whether it is a first visit or a revisit to the doctor. In 1986, for example, family practitioners charged an average of $31 for a first office visit, and $25 for a revisit. Nonsurgical specialists charged $46 and $28 for comparable visits, and surgical specialists $46 and $26. Within the surgical domain, a threefold variation exists in office visit fees between the lowest (general surgeons) and highest (neurosurgeon) fees.

The CPI measures changes in this bundle of good and services, but even with the components in the bundle held constant, the CPI can misstate the "inflation" in some sense. The fly in the CPI ointment is technical change. As the quality of a good or service changes, its price will likely change as well, but if its "name" does not change, the CPI ignores the change in quality.

Several common examples will illuminate this problem. First, consider the most important element in the medical CPI, the "hospital room and board" charge, the basic fee charged for a semiprivate hospital room. Compared (say) with 1950, the services that come with that basic room have changed considerably through time: Standby emergency equipment pervades the hospital; the skill level of the nursing staff has increased; the bed's position is now controlled electronically by the patient, rather than manually by a nurse; the menu offers a variety and quality of food far surpassing the traditional fare in hospitals; the building is almost certainly air conditioned; and so on. All of these improvements appear as added costs in the CPI measure, but the added benefits tend to disappear behind the label of "semiprivate room and board fee."

The same issues pervade almost every element in the CPI bundle of goods and services. Today's dental visit offers more highly trained dentists and staff; X-ray examinations are conducted at lower radiation levels, reducing side effects to patients; lighting is better, improving the ability of the hygienist to work carefully and with less pain to the patient.

In physician offices, nearly immediate laboratory results are available for many blood or urine tests, to complement the higher skill levels of the doctor and nurses there. Other diagnostic equipment abounds. In orthopedics offices, a wide array of rehabilitation therapy equipment supplants the old hydrotherapy unit. Only one thing remains the same in the doctor's office between the 1950s and the 1980s—the magazines. The offices of most physicians in the country still contain *Time* and *Reader's Digest*. Some of these magazines may still show dates in the 1950s.

With this lengthy preface, we can now consider the actual

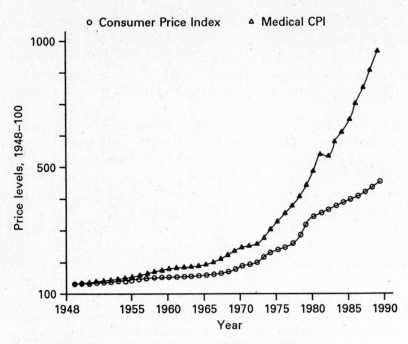

Figure 2.2 Medical prices relative to CPI, 1948 to 1990.

change in medical prices that we have observed in the United States over the past several decades. The *relative price* of medical care to the overall CPI offers the most important measure of how "costly" medical care has become. If medical care prices and all other prices and incomes had doubled over some period—say, 20 years—then the relative price of medical care would not have changed. However, medical care prices systematically have grown faster than the CPI (with a few exceptions).

One way to look at these data is to show the relative change in medical prices, compared with the CPI. Figure 2.2 plots these changes through time, setting the price level in 1948 at 100 in both cases. On average during this period, medical prices have increased at about 1.64 percent more each year than the CPI. This may sound like a small amount, but compounded over 40+ years, it leads to the nearly 100 percent relative increase in medical prices since 1948.[3]

---

[3] The arithmetic of compounded inflation works like this: The overall price level increased by 98 percent more for medical care than the CPI over 42 years. Thus where the annual rate of *relative price growth* is r, then $(1 + r)^{42} = 1.98$. The value of $r = 0.0164$ solves this equation.

This overall growth rate masks some important differences. While the overall growth of medical prices occurred 1.6 percentage points per year faster than the CPI, seven of those years showed medical prices falling relative to the CPI. Of those seven years, three occurred during the Nixon administration's price controls that constricted the health sector particularly tightly, two occurred during the Korean War, and the other two during the Carter administration, when overall inflation was at record high levels for a nonwartime economy in the United States.

The global index on medical care prices also masks some important differences between components that make up "the health care sector." Table 2.4 shows the price index (1967 = 100) for all medical care items, physicians' fees, dentists' fees, the hospital room and board for a semiprivate room, and prescription drugs. Between 1948 and 1985, prices for all medical care had increased more than eightfold, as described earlier. Over that same period, hospital room and board charges increased by a factor of 25, physician fees by a factor of 7.2, dentist fees by a factor of 5.7, and prescription drugs by a mere factor of 2.7. Recall that overall price levels increased by a factor of 4.5 in this period. Thus, physician and dental fees increased only slightly more rapidly than overall inflation; hospital room and board fees increased at a much, much higher rate; and the costs of the prescription drugs actually fell relative to overall inflation. Indeed, until more recent years, the prices for the specific prescription drugs in the CPI "bundle"

Table 2.4 PRICES THROUGH TIME

Year	Overall CPI	Medical care	Hospital room and board	Physicians' fee	Dentists' fee	Prescription drugs
1950	72	53.7	30	55	64	92.6
1955	80.2	64.8	41	65	72	103.3
1960	89.4	83.5	57.3	77.0	82.1	115.3
1965	94.5	89.5	75.9	88.3	92.2	102.0
1967	100	100.0	101.1	100.5	100.4	99.9
1970	116.3	120.6	145.4	121.4	119.3	101.3
1975	161.2	168.6	236.1	169.4	169.4	109.3
1980	246.8	265.9	418.6	269.3	240.2	154.8
1985	322.2	403.1	710.5	398.8	347.9	256.5
1990[a]	395	595	1060	565	490	375

[a] Estimated.

Source: U.S. Department of Labor, Bureau of Labor Statistics, Consumer Price Index, various issues.

actually fell considerably in nominal terms, despite increases in overall prices.

Medical Spending Patterns

Medical spending is the product (literally) of quantities purchased and the average price per unit. The patterns of spending show better than any other measure the flow of real resources in our economy. Since dollars spent represent price × quantity, it is tempting to divide spending by price to get a measure of "quantity" consumed. In part, this exercise is meaningful, but in part, misleading. The difficulty arises, as discussed before, because of technical change. The price of hospital room and board has reflected some amount of technical change, but the "quantity" of care also changes in nature. In "hospital admissions," the scope of procedures has changed greatly through time. Some types of care no longer involve hospitalization (such as cataract removal), while other types of procedures occur now that simply did not exist in 1950, including organ transplants, attachment of artificial joints, and so on. Thus even "quantity" of care has fuzzy meanings and boundaries when looking through time.

Despite these discrepancies, spending patterns can provide a useful backdrop for the study of health economics. Table 2.5 shows the personal consumption expenditures for various types of medical care through time. Table 2.6 provides the same data in constant 1950 price levels—that is, adjusted by the overall CPI. Table 2.7

Table 2.5   ANNUAL NOMINAL EXPENDITURES FOR MEDICAL SERVICES (Billions of Dollars)

Year	Total	Hospital	Physician	Drug	Dental
1950	8.8	2.1	2.6	2.2	1
1955	13.4	3.4	3.8	3	1.5
1960	20.6	5.6	5.8	4.7	2
1965	34.3	10.1	9.1	6.7	3
1970	55	19.7	14	10	4.9
1975	96.9	40.1	23.5	14.9	8.2
1980	185	82	42	23.5	14.9
1985	325.2	140.2	73.5	36	21.5
1990	541	237	124	54	32
Relative amounts, 1990/1950	61.5	112.9	47.9	20.77	32

Source: Arnett et al., 1986.

Table 2.6   ANNUAL EXPENDITURES FOR MEDICAL SERVICES IN CONSTANT
1950 PRICES
(Billions of 1950 Dollars)

Year	Total	Hospital	Physician	Drug	Dental
1950	8.8	2.1	2.6	2.2	1
1955	12.03	3.05	3.41	2.69	1.35
1960	16.59	4.51	4.67	3.79	1.61
1965	26.13	7.70	6.93	5.10	2.29
1970	34.05	12.20	8.67	6.19	3.03
1975	43.28	17.91	10.50	6.66	3.66
1980	53.97	23.92	12.25	6.86	4.35
1985	72.67	31.33	16.42	8.04	4.80
1990	98.61	43.2	22.60	9.84	5.83
Relative amounts, 1990/1950	11.2	20.6	8.7	4.5	5.8

Source: Arnett et al., 1986, adjusted by data in Table 2.4.

puts these constant-dollar amounts as if the population had
remained unchanged since 1950. Table 2.8 further adjusts total
spending from Table 2.7 as if the *relative* price of medical care
had not changed. Comparing these four tables helps convey what
has led to the overall increase (say) in total spending from $8.8
billion in 1950 to $541 billion in 1990. Adjusting for inflation alone

Table 2.7   ANNUAL "REAL" SPENDING FOR CONSTANT 1950 POPULATION
(Billions of 1950 Dollars)

Year	Total	Hospital	Physician	Drug	Dental
1950	8.8	2.10	2.6	2.20	1
1955	11.00	2.79	3.12	2.46	1.23
1960	13.98	3.80	3.95	3.19	1.36
1965	20.64	6.08	5.49	4.03	1.81
1970	25.31	9.06	6.44	4.60	2.25
1975	30.87	12.77	7.49	4.75	2.61
1980	36.99	16.40	8.41	4.70	2.98
1985	46.49	20.04	10.51	5.15	3.07
1990	60.06	26.31	13.77	6.00	3.55
Relative amounts, 1990/1950	6.80	12.53	5.30	2.31	3.55

Source: Arnett et al., 1986; data from Table 2.6, adjusted by U.S. population changes since 1950.

Table 2.8   ANNUAL "REAL/PER CAPITA SPENDING" IN CONSTANT 1950
MEDICAL PRICES
(Billions of 1950 Dollars, Also Adjusted for Relative Price Changes)

Year	Total	Hospital	Physician	Drug	Dental
1950	8.80	2.10	2.60	2.20	1.00
1955	10.16	2.26	2.94	2.46	1.22
1960	11.17	2.47	3.49	3.18	1.31
1965	16.26	3.15	4.48	4.80	1.65
1970	18.20	3.02	4.71	6.79	1.95
1975	22.01	3.63	5.44	9.00	2.21
1980	25.61	4.02	5.88	9.64	2.72
1985	27.71	3.79	6.49	8.31	2.53
1990	29.74	4.09	7.35	8.12	2.56
Relative amounts, 1990/1950	3.39	1.95	2.82	3.69	2.56

Source: Arnett et al., 1986; data from Table 2.6, adjusted by data in Table 2.4.

drops the $541 billion to $98.6 billion. Holding the population constant at 1950 levels would further reduce that 1990 total to $60.06 billion. Finally, if there had been no changes in the *relative* prices for medical care (see Table 2.4), we would have something similar to (but not identical to) a quantity measure. (It does not correspond exactly, for example, to the behavior in physician visits or hospital days.) By this measure, total spending for medical care in 1990 would be only $29.74 billion, and hospital spending would be only $4.09 billion. Nevertheless, this represents more than a threefold increase in "quantity," even after making all of these other adjustments, and a near doubling for hospital care. Note also that after holding relative prices constant to their 1950 level, the hospital sector "appears" smaller than the physician sector, and indeed even the drug sector, whereas in actual spending amounts (Table 2.5), the hospital sector is massively larger than any of the others. The difference appears in the relative price shifts (Table 2.4), which are much larger for hospitals than for the other sectors. Note also the large increase in prescription drug use over these four decades (nearly a fourfold increase). Some, perhaps a large part, of the changes may be due to unmeasured changes in quality of services provided.

Several lessons emerge from these data. First, the data show the growing importance of the hospital in our health care system. In 1950, Americans spent more money on physician care and on prescription drugs and appliances (braces, wheelchairs, etc.) than on hospital services, and under one-quarter of all medical spending took place for hospital services. By 1965—the year Medicare became law—the hospital had overtaken the physician as the leading locus of resource use, but still accounted for only 28 percent of personal spending on health care. Ten years later, the hospital accounted for over 40 percent of all medical spending, a pattern that has persisted to the present.

Second, these data mask the importance of the physician in the overall picture. While physician services directly account for about a quarter of all medical spending (varying slightly through time), physicians in fact direct a much greater portion of the resource use. In any meaningful sense, their decisions affect not only the spending on their own services, but also that for hospitals, most medicines and appliances, and at least some of the "all other" category (which mostly includes long-term care). If we count all of these categories, physicians direct about $7 out of every $8 spent on health care in this country.

The third point arises from Table 2.7, which shows medical spending in constant CPI (1950), and as if the population were constant at its 1950 size. This table shows the overall growth in medical spending in the most meaningful sense, since it adjusts for general inflation and population changes. The real level of per capita medical care spending more than quintuples over this period. In part, this is due to the overall increase in real incomes, which more than doubled over the period. This increase in "real" spending led to the doubling of the share of our nation's economic output devoted to medical care, rising from under 5 percent to nearly 11 percent of the gross national product, the most comprehensive measure of the nation's output.

Demographics and income play some role in spending patterns as well. As people age, they use increasingly more health care resources.

Table 2.9 shows the growing proportion of medical care used by persons over 65. Three things contribute to this phenomenon: (1) The elderly form a growing fraction of the total population. In 1960, only 9.2 percent of the population was over age 65, and by 1990, 12.6 percent of the population was in the same age group (see Table 2.2). In addition, "the elderly" live longer, with average life expectancy increasing by several years during this interval of three decades. (2) The health insurance coverage of the elderly expanded hugely in 1965 with the advent of Medicare, and has continued to grow through the purchase of private "supplemental" health insurance plans by many of the elderly. This insurance has provided the elderly with increased purchasing power in health care markets. (Chapter 12 gives more details.) (3) Technical

Table 2.9 AGE-SPECIFIC MEDICAL USE THROUGH TIME
(Per Capita Spending)

	1965			1970			1975			1980		
	<19	19–64	65+	<19	19–64	65+	<19	19–64	65+	<19	19–64	65+
Hospital	$23	$87	$176	$46	$153	$349	$74	$269	$628	$127	$462	$1086
Physician	$23	$49	$93	$36	$76	$150	$55	$124	$255	$91	$198	$443
Other	$37	$80	$203	$56	$108	$355	$82	$174	$593	$129	$264	$922
Total	$83	$216	$472	$138	$337	$854	$211	$567	$1467	$347	$924	$2451

Spending Relative to 19–64 Age Group's Spending

	1965			1970			1975			1980		
	<19	19–64	65+	<19	19–64	65+	<19	19–64	65+	<19	19–64	65+
Hospital	0.26	1	2.02	0.30	1	2.28	0.28	1	2.33	0.27	1	2.35
Physician	0.47	1	1.90	0.47	1	1.97	0.44	1	2.05	0.46	1	2.24
Other	0.46	1	2.54	0.52	1	3.29	0.47	1	3.41	0.49	1	3.49
Total	0.38	1	2.19	0.41	1	2.53	0.37	1	2.60	0.38	1	2.65

Source: Fisher, C. R., "Differences by Age Groups in Health Care Spending," *Health Care Financing Review.* Spring 1980. Tables A–D.

change has made a particularly large impact on the health care received by the elderly. A considerable number of surgical procedures that did not exist in 1950 (but are common now) primarily benefit elderly persons. These new procedures include hip, knee, and other joint transplants; cataract removal and lens transplants; a wide array of cardiac surgery, including coronary artery bypass grafts, replacement heart valves, pacemakers, artificial hearts, and even heart transplants; liver transplants; and many other such procedures. The proliferation of specialized intensive care units in hospitals has also heavily served the elderly.

Afterthought

In closing, we should emphasize that there is no "right" amount of spending on health care toward which we should aim. Optimally, we can evaluate spending on medical care in the same ways that we do other goods and services—namely, to ask how much our overall well-being increases or decreases as we change our use of resources in health care. The remainder of this book, and indeed much of the effort of health economics in general, provides ways to answer these questions. What forces have led to the spending patterns we now see? What benefits do we get from these activities? If we changed the patterns of resource use, how would our health and happiness change? The tools of economic analysis help us think about and sometimes answer these questions.

SUMMARY

Health care and health insurance markets are interrelated in complicated ways. Health insurance has the unique feature that it effectively alters the price in another market that consumers have to pay to buy the service (health care). No other consumer goods or services have this characteristic, making the study of health care both unique and complex.[4]

---

[4] Some financial markets have something similar—the "option" to buy or sell. For example, a standard "call" option gives the buyer of the option a right to purchase (say) 100 shares of IBM corporation stock at a specified price (say, $155) for a specific duration of time (say, until October 31). These options exist for many common stocks, and also for certain commodities such as crude oil, hog bellies, gold bullion, and even foreign currencies. Health insurance differs considerably even from these options. Most "commercial" options are written for a price near the current price of the stock or commodity. By contrast, health insurance moves prices by 75 to 80 percent or more.

Health markets also have interesting and complicated dynamic features. In addition to normal economic forces such as growth in income, health care markets change in response to the changing demographics of our society—older people use more medical care than younger people—and to changes in technology. The massive infusion of research funding by the federal government has surely created large changes in the capabilities and costs of our health care system. Understanding the patterns of spending through time will require an integration of all these factors.

PROBLEMS

1. List at least four things that will likely affect an individual's demand for medical care. Which will likely have the best ability to predict spending for *individuals*? Which for *groups* of people?

2. List at least four things that will likely affect an individual's demand for health *insurance*.

3. List at least three external forces that impinge on the production of medical care.

4. "Most of the increase in medical costs is due just to normal inflation that has affected the entire economy." Comment.

5. Since Medicare was introduced in 1965, has per capita spending on people aged 65 and over increased, decreased, or stayed about the same, compared with per capita spending for people younger than age 65? (Hint: See Table 2.9.)

6. Which sectors of the medical care system (hospital, physician, drug, dental) have had the largest increases in relative price between 1950 and 1990? (Hint: Compare Tables 2.7 and 2.8.)

7. The CPI for hospital care (as one example) measures the cost of a two-person "hospital room and board" charge as an important part of the CPI hospital index. Has the quality of that "hospital room" been constant through the past four decades? If not, what changes in quality have occurred? What does that tell us about the relevance and/or accuracy of the CPI for hospitals?

8. If you could know only one thing about an individual and you wished to predict his or her annual spending on medical care, what would you most want to know? Why?

Chapter
3

The Transformation of Medical Care to Health

We have discussed in Chapter 1 the basic idea of a production function for health, the process of transforming medical care into improved health. This chapter expands on these ideas, with several goals in mind. First, we want to understand the relationships between the amount of health care used and resultant changes in health. Second, we can explore the predictability of outcomes in this process—the uncertainty associated with the production of health. A similar issue, related in many ways, is the uncertainty in the minds of doctors and other healers about the correct ways to use various health interventions. We will study this issue as well.

THE PRODUCTIVITY OF MEDICAL CARE

Marginal and Average Productivity

For every process ever studied, the productivity of inputs varies with the total amount of inputs used. A simple but revealing example involves the productivity of labor in the harvesting of grapes. As a single worker begins to harvest a field, she must do every task for herself, and her productivity (grapes harvested per worker per hour) is low. As a second worker is added, more than twice as many grapes are harvested. The *marginal product* with two workers is greater than that with one,[1] so the amount of product *per worker* increases.[2] This can occur, for example, because the two workers can specialize in various tasks (one picking, the other carrying grapes from the field) to which they are best suited. As more and more workers are added, the effects of specialization eventually begin to diminish, and finally the workers actually begin to get into each other's way.[3] Eventually, if you were to add enough workers to a vineyard of given size, they would begin to trample the grapes due to their crowding. When this happens, the *marginal product* of more workers is negative—total product would fall as the number of workers increased. Of course, nobody would deliberately do this, unless the workers paid for the opportunity to work in the fields (rather than having the workers paid for their work).

Productivity Changes on the Extensive Margin

Just as the productivity of workers can vary with the rates of their use, the productivity of health care resources can also vary with the total amount used. Indeed, we can expect in general that the marginal productivity of health care resources will increase at low levels of use (such as might be found in a primitive or developing nation), and that the marginal productivity of health care resources will fall as more and more of such resources are used. With sufficiently large amounts of medical care used, the harm from iatrogenic illness could outweigh any gains, making the marginal product of health care resources negative.

One way to think about declining marginal productivity looks at the populations for which a particular medical treatment might

---

[1] The marginal product is technically the (partial) derivative of total output with respect to one input, holding all other inputs constant. For example, if the two inputs are labor (L) and acres of grapevines (A), and the output is pounds of grapes (G), the production function is $G = f(L, A)$ and the marginal product of labor is $\partial G/\partial L$, the rate at which G changes as L changes, holding A constant.

The language here carefully expresses an important point: It says "the marginal product of two workers" rather than "the marginal product of the second worker." The extra productivity should not be connected with the actual second worker, but rather to the presence of two.

[2] The average product per worker is G/L.

[3] In this process, the marginal product, and, eventually, the average product will begin to fall.

be used. For a particular example, consider the case of screening people for breast cancer. This disease primarily (but not exclusively) affects women, and the age profile of the disease suggests that the risks of breast cancer rise with a woman's age, at least for much of the life span. Epidemiological studies can characterize the risks that an "average" woman of a particular age will acquire breast cancer. Such studies might also sort out particular risk factors, such as dietary pattern, smoking habits, and so forth. Screening for breast cancer will have a *yield* of positive cases (say) per thousand examinations conducted that varies with both the population studied and the test's accuracy. The screening test (a mammogram, a very low-level radiation X-ray examination) might miss some actual cancer cases (false negative). The test's probability of detecting true cases—p—is called its *sensitivity*. The test might also inaccurately report breast cancer in somebody who does not have the disease, merely because something "looks like it" on the mammogram (false positive)—q. The yield of the test is composed of both true positive and false positive diagnoses. Suppose the fraction of true positives in the population being screened (as learned, for example, from epidemiology studies) is f. Then the yield rate of the test is $f \times p + (1 - f) \times q$. That is, p percent of the true positives and q percent of the true negatives will show positive on the test. Obviously, the higher the underlying probability of disease (f) in the population studied, the higher will be the yield of *true positive* cases that can lead to cures of the disease.

Now think about the populations (age groups, for example) for whom the mammogram might be used. Intelligent use of the test will begin with those most susceptible to the disease, who turn out to be women over age 50. We could expand use of the test *on the extensive margin* by adding to (extending) the population base for whom the test was used—for example, to women between 40 and 50, and then to women between 30 and 40, and so on. The number of true cases detected per 1000 tests would fall as we pushed the extensive margin, as would the number of cures achieved.

We could even extend the population receiving this screening exam to the point where its marginal product was negative. Even though the radiation used in a mammogram is very low, it creates a very slight risk that the X-rays themselves will induce breast cancer. If the population being screened has an extremely low underlying risk of breast cancer, it is possible that the *induced* (iatrogenic) cancers would actually occur more often than naturally occurring cancers would be found. This would quite likely be the case, for example, if women aged 20–30 received routine breast cancer screening with current mammogram technology.[4]

The same concepts apply to medical treatment as well as screening examinations. A common example might be the success of back surgery in eliminating the symptoms of low back pain, a common ailment of modern society. Studies show that 65 to 80 percent of Americans will have serious low back pain at some point in their lives, and at any time, over one in ten Americans has some low back pain, 10 million of whom have at least partial disability from the malady. Most do not receive surgical intervention (2 percent of American adults have had back surgery), and for those who do, the success rate reported varies considerably from location to location. One reason for this is the *case selection* methods various doctors use, another way of thinking about expanding the extensive margin of medical intervention. Some patients with low back pain are clear surgical candidates to almost any surgeon looking at them, because of the mix of symptoms and diagnostic test results.[5] For these patients, surgery has a high likelihood of success. However, some surgeons are "more aggressive" and will operate without such clear signs and symptoms. In the economist's jargon, they are expanding the extensive margin of surgery. The yield of successful cases will surely fall, as the medical literature reports with great unanimity. Eventually, as with the mammogram, further expansion along the extensive margin (operating on every patient with mild symptoms, for example) would lead to more patients with bad backs as a result of surgical complications than were actually cured.[6]

---

[4] The issue of screening for breast cancer remains controversial, not only in terms of the populations who "should" be screened, but the best frequency of examinations. David Eddy, M.D., Ph.D, has conducted numerous studies on these issues. The interested reader can find his more recent analyses in Eddy, 1980, 1988. These studies provide good examples of the use of *medical decision theory* to help guide medical decisions. Chapter 4 provides an appendix that discusses the tools of medical decision theory more completely.

[5] The symptoms for this group include intractable pain that is not relieved by lying down, pain shooting down the leg (sciatica) if the patient's leg is raised up straight while the patient is lying supine, loss of sensation, and, particularly, loss of motor function in the leg or foot, etc. Confirmatory X-ray examination of the spinal column using a myelogram (X-ray with radio-opaque dye in the spinal canal) should also show a clear bulging disk at the point in the spine associated with the particular clinical symptoms.

[6] As is common in the study of the economics of medical care, a good deal of useful information resides in the medical literature, particularly in articles that appear occasionally in the medical journals summarizing the "state of the art" for doctors. In the case of low back pain, for example, useful summaries appear in Frymoyer, 1988. These articles often provide good examples of such concepts as extensive and intensive margins in health care.

Productivity Changes on the Intensive Margin

Another way to increase the use of medical care resources is the *intensive margin*. In this case, the population being treated is held constant, and the rate at which tests or procedures are used (the "intensity" of use) is varied. As with the case of variation in the extensive margin, variations in the intensive margin can produce first increasing, then decreasing, and finally negative marginal productivity of the medical resource.

The previous example of mammograms provides an example of choices along the intensive margin. Should women receiving the test (say, those aged 50–60) have the test every ten years, five years, two years, one year, six months, monthly, or daily? Common sense tells us that daily is too often—breast cancer does not grow fast enough to show changes on a daily basis. Testing at ten-year intervals is probably too seldom as well—the tumor can begin, grow, and eventually prove fatal to the woman well within that time period. While ten-year intervals would detect some disease, it seems plausible that the marginal product would increase if the test were done (say) every five years. Increasing the interval to yearly, and then to monthly, would probably drive the test into the realm where the marginal product was falling, if not negative.

Evidence on Aggregate Productivity of Medical Care

How much health do we actually get from our current patterns of medical care use? The answer seems to be both "a lot" and "not very much" at the same time. We get "a lot" of gain on average, and considerable evidence supports this. It may also be true that we are not gaining very much on the margin, so that substantial changes in medical resource use might result in very little change in health outcomes. These ideas are not paradoxical, as the previous discussion on average and marginal productivity should suggest. Indeed, "smart" use of health care almost certainly would result in a level of health services where the marginal product was declining (rather than rising or at the highest possible level of marginal productivity).

Aggregate Data Comparisons   One type of evidence compares the health status of various nations and their use of medical care resources. These studies necessarily rely on the most simple measures of health outcomes, typically life expectancy, mortality rates, or (perhaps) age-specific mortality rates. One common indicator of health outcomes is the death rate for infants, although overall nutrition and broad public health measures (such as sanitary water supply) have a significant effect on this process, possibly more than that of personal medical care interventions.

One can also compare the life expectancy of regions of a single country, such as states, counties, or Standard Metropolitan Statistical Areas (SMSAs) in the United States and measure how variations in regional life expectancy vary with the use of medical care. These studies avoid some of the problems of cross-nation comparisons, but introduce others.

Aggregate data studies of the relationships between mortality and health care (either multicountry or within-country) invariably show that four things move in parallel: per capita income, per capita education, medical care use, and good health outcomes. All four are related to and affect each other. Higher per capita income directly creates better health through improved living conditions, including sanitary water supply, safer roads, and better nutrition. Higher per capita income also gives more buying power, which directly increases the amount of medical care used, also improving health outcomes. Higher income leads to the use of more education, which in turn leads to higher incomes in the future. Education is indeed a powerful engine for sustained economic growth. Better education also directly increases people's health, by making them more capable managers of their own lives and more adept at using those medical care resources available in the market.[7] And finally, better health increases people's ability both to learn in school and to work productively, both of which eventually create more income.

The only exception to these findings occurs in some cross-state studies within the United States, where for white males, the relationship between income and mortality is positive. One hypothesis to explain this is that the higher incomes of this group have led them to purchase many things (like fatty food and cigarettes) that reduce their life expectancy. However, in such data, complicated interactions of age, education, and migration patterns can obscure the true relationships, leaving us in a state of some ignorance about the true relationships between income and health. As the discussion in Chapter 1 suggests, it may well be that income has both positive and negative effects on health, because (in our earlier notation), X_B (bad) "goods" decrease health, while X_G (good)

---

[7] Michael Grossman's studies of the demand for health emphasize the role of education in the process of producing health. See Grossman, 1972a, 1972b.

"goods" (possibly including medical care) increase health. As income rises, one might first start purchasing X_G items, and then at still higher incomes, more X_B items. At sufficiently high incomes, the "bads" might overwhelm the "goods" in their effect on mortality.

Unraveling the pure effects of medical care on health in this kind of setting has proven to be a difficult statistical problem, because the data always show income, education, health care, and health outcomes moving relatively together through time (within a single country) or across countries (at a single time). One of the difficulties with such studies arises when trying to estimate how much medical care has actually been used. The most common approach takes reported medical spending (in the local currency) and converts that to some common currency (like dollars) using published exchange rates. However, because different countries have different ways (for example) of paying doctors, and different ways of counting the costs (for example) of hospital construction, *medical spending* may give only a fuzzy picture of the amount of *medical care* rendered.

Other studies make the same comparison (say) within a single country through time. These studies avoid the problems of finding the correct currency exchange rate, but they present a separate problem of converting spending in different times to the same "units" of measurement. Typically, the conversion of spending (say) from the 1950s to current years uses the medical component of the Consumer Price Index, just as the various versions of Table 2.7 and 2.8 did in Chapter 2. These spending rates are then compared with longevity, age-specific mortality rates, or other measures of health outcomes through time. The general outcome of such studies is similar to the cross-national studies described previously. Income, education, health care spending, and health outcomes all seem to move together through time, and separating the effects of each has proven a difficult task.

Recently, Hadley (1982) has analyzed county-level data in the United States in 1970, where the health indicator is age- and sex-specific mortality, and the measure of medical care use is *Medicare* expenditures per enrollee. He specifically assumes that total population medical care use per capita in each county is proportional to Medicare spending per Medicare enrollee. Hadley finds significant positive effects of medical care use on health outcomes; on average, he finds that a 10 percent increase in health care spending will reduce mortality rates by about 1.5 percent. (Using two other measures of medical care use, Hadley gets just the opposite finding, namely that health levels fall with higher medical use, but he dismisses those results.) He also finds insignificant effects of income on mortality rates. In terms of health habits, he also finds a strong deleterious effect of cigarettes sold per capita on mortality rates; a 10 percent increase in cigarette sales per capita increases overall mortality by between 1.2 percent (females) and 2.5 percent (black males between 45 and 64 years old).[8]

Randomized Controlled Trial Data

The cross-national and through-time data mentioned earlier all have the difficulty that many things "move together" in the data set, making it difficult to determine just what is responsible for the improved health that we see through time (within countries) and with higher income (across countries). An entirely different approach allows a completely different measure of the health-producing effects of medical care, using the results from a recently completed social science experiment using the standard techniques of the randomized controlled trial.

The study, begun in the early 1970s, was conducted by the RAND Corporation with funding from the federal government (Newhouse, 1974). The RAND Health Insurance Study (RAND HIS) had two major goals—to learn (1) the relationship between insurance coverage and health care use, and (2) the resultant effects (if any) on actual health outcomes. The study was conducted in four cities and two rural sites, and encompassed over 20,000 person-years of data. Some of the people were enrolled for three years, and others for five years.[9] The enrollees were randomly

---

[8] The statistical issues associated with Hadley's study are quite complicated. He uses multiple regression analysis to calculate the effects of each explanatory variable on the county-level mortality rate, but he counts each county equally no matter how many people live in it. Although his own analysis says that he should have made an adjustment to correct for this (see his p. 201), he does not. More troublesome is his assumption that Medicare spending per enrollee and total medical care use per capita (for the whole population) are always constant proportions of one another. As the discussion following shortly on variations in medical practice patterns suggests, there can be large differences in medical practice patterns that could easily distort this relationship. No simple conclusion can be drawn about the potential bias in Hadley's results.

[9] The different lengths of time in the study helped researchers determine whether the behavior at the beginning and the end of the experiment differed from that in the middle, a possibility given the design of the experiment. Chapter 4 provides a further discussion of this part of the HIS.

assigned to several insurance plans, one of which provided full coverage for all health care used, and others of which required some copayment by the enrollees. As we will see in Chapter 4, the group with full-coverage insurance used substantially more health care than any of the enrollees who had insurance requiring some copayment. Thus, while the study did not experimentally vary the amount of health care received by enrollees,[10] it did experimentally vary the *price* of medical care. The resultant choices by individuals created the differences in medical use that were observed, and those in turn created the opportunity to study the effects of that added health spending on health outcomes.

The RAND HIS had three different types of measurement of health outcome for the enrollees. First, there was a series of questionnaires gathering data on the ability of individuals to participate in activities of daily living (ADL), their self-perceived health status, their mental well-being, and other self-reported (subjective) health measures. These data were collected at the beginning and at the end of the experiment from all adult enrollees, and from parents for all children. There were also regularly collected data on sick-loss days (from work, from school, or from ability to work in the home). Finally, participants received a modified physical examination at the end of the experiment, designed to measure conditions that the health care system *should* be able to affect, such as weight, blood pressure, corrected vision, serum cholesterol level, hearing acuity, and the like. Some of the enrollees received the same measurements at the beginning of the experiment, allowing not only a comparison across plans (at the end of the study), but also a measure of how people's health had *changed* during the experiment.[11] Finally, all of the measurements were collected together statistically into a measure of "health-status age," a physiologic measure that reports the apparent age of people's bodies.

The measure of health-status age has a commonsense feature; it should "age" by a year for each year the enrollees participated in the study, on average.[12] The HIS had a large enough sample size so that it should have been able to detect differences as small as one

---

[10] To conduct such an experiment would probably be impossible in the United States, because of ethical and legal restrictions on medical experiments.

[11] The physical examination at enrollment was given only to part of the enrollees to detect whether the information from the exam itself changed their use of the health care system.

[12] It did, on average, almost exactly.

"year" of health status age—that is, the difference in health levels associated with a year of normal aging, if such differences occurred in the different insurance plans.

For purposes of the health-status measurement, we need know at this point only that the low-coverage group used about two-thirds of the medical care used by the group with full coverage. Given that difference, we can ask, "What health differences occurred between these groups?"

The answer, while mixed, is generally "not much, if any." For adults, virtually every measure of health status was the same for the full-coverage group and the partial-coverage group except two: the low-income full-coverage group had better corrected vision than their counterparts on the partial-coverage group, and they had a very slightly reduced blood pressure (Brook, Ware, Rogers et al., 1983). The corrected vision improvement was about 0.2 Snellen lines, equivalent to improving corrected vision from 20/22 to 20/20. On average, the blood pressure for the low-income fully insured enrollees was 3 mm of mercury lower than their partially insured counterparts. (Blood pressure is measured by the height of a column of mercury the pressure will support, usually reported at the peak and the trough of pressure as the heart goes through one cycle. A borderline high level of blood pressure will have a reading like 145/90 for the two readings. Thus a 3-mm decline in blood pressure is about a 2 percent improvement.)

This improvement in blood pressure is more important than it might seem. For persons with relatively high health risks (e.g., from obesity, smoking, high blood pressure), the risk of dying was reduced by about 10 percent in the full coverage group, but almost all of this improvement was due to the reduced blood pressure in that group. The RAND researchers conclude that targeted investment in health care activities known to produce lower blood pressure will produce a greater health yield than broad provision of free care. This conclusion underscores the importance of understanding the relationships between specific types of medical care use and specific gains in health.

The RAND HIS study has its own potential problems if we seek the definitive answer about the *marginal* effect of the medical care on health outcomes. The most obvious potential problems are (1) the short time horizon, and (2) the potential lack of power to detect true effects (since the sample included only about 5,800 persons). We are unlikely ever to know fully the importance of the time-horizon problem; to do so would require a similar experiment to be conducted for (say) 10 or 20 years, and this seems very unlikely. The failure to detect differences in health habits (like smoking, weight, and cholesterol levels) over the three- to five-year horizon of the experiment seems convincing that full-coverage insurance (and the concomitant higher use of medical care) will not alter these important life-style choices, and altering them could have made large differences in the health outcomes of the enrolled populations. In terms of the statistical power, the study was able to calculate the chances that a true effect of the medical care wasn't detected; the estimates were precise enough that the study authors concluded they could "rule out the possibility of anything beyond a minimal effect" on various health measures.

One other quasi-natural experiment reported in the literature has some bearing here. During the "medical malpractice crisis" in the 1970s (see Chapter 14), doctors in the Los Angeles area undertook a systematic "slowdown" of care to try to force changes in the state's legal system. During this slowdown, they treated only "emergency" cases, so as a result, the rate of elective surgery fell sharply. Surprisingly, so did the county's death rate, and at least one analysis attributes the decline in mortality to the slowdown, and particularly to the reduced elective surgery (Roemer and Schwartz, 1979). While this does not say that medical care has a negative marginal product (because the surgery likely has some benefits that may be worth the risks), it does cause some reason to wonder how much benefit that might be. The work described in the next section on variations in medical practice patterns further emphasizes this concern.

Summary   The evidence on the effects of medical care on health outcomes is mixed, partly because most studies have not been able to draw the distinction between an average effect and a marginal effect. Without question, medical care produces enormous positive benefit. The differences in mortality across countries make it almost impossible to conclude otherwise. In addition, a large number of studies of specific medical interventions (which we have not reviewed here) show a considerable effect of at least some of these activities. The increases in life expectancy through time also support the conclusion that medical care has some efficacy.

However, these studies still leave unanswered the important

question of "how much" medical care is appropriate. The RAND HIS results support the belief that *on the margin* the effects of substantial amounts of medical spending may be quite small. Particularly for nonpoor income groups in the HIS, health differences between fully insured people and less completely insured people were unobservable, despite large differences in total medical spending.

One way to reconcile these apparently contradictory results is to recall the previous discussion of diminishing marginal productivity. In general, studies of productive processes have found that diminishing marginal productivity invariably takes place. The use of medical care to produce health should offer no exceptions to the general rule, and indeed the observed data are fully consistent with this idea: Health care can be very productive, but we may well have moved into a rate of use of health care where the marginal productivity is quite small, at best.

CONFUSION ABOUT THE PRODUCTION FUNCTION: A POLICY DILEMMA

The previous discussions blur somewhat an important problem in the discussion of "the productivity" of medical care: Doctors themselves seem to disagree about the right ways to use medical care, based upon a growing series of studies showing different patterns of use of various medical care services. Put another way, the medical profession within this country seems to have strong internal disagreements about the marginal productivity of various medical procedures. Not only that, similar variations in patterns of use for specific services emerge in medical care systems of other countries, ranging from Norway to Canada to Great Britain (where a unified National Health Service is responsible for the provision of health care). Much of the disagreement apparently centers on the appropriate *extensive margin*—that is, how many people should receive various treatments, given that they enter the health care system with a similar set of medical problems and conditions.

Medical Practice Variations on the Extensive Margin

The studies of medical practice variations have almost universally focused on the rate at which "standard" populations have received specific medical interventions. Almost universally, the studies have used hospital admission rates for various procedures as the basis for analysis, so the studies must define some geographic area, and then measure the rates at which a specific procedure has been used for that population. Most of these studies at least control for the age and sex composition of the populations. All of the good studies also carefully measure the rates of use for the populations in question, rather than the rates at which the activities are carried out in the region. (The two could differ considerably, for example, in a city with a large university hospital, with a considerable referral practice for complicated procedures, or a rural area with no hospital or only a small hospital equipped for only routine surgery or medical treatment.)

The studies usually report the results in terms of a "coefficient of variation" in use rates across geographic regions. Box 3.1 describes the statistical basis of the COV measure. The coefficient of variation is a useful measure here because it automatically scales each medical procedure the same way. Thus, a low or high coefficient of variation (COV) has the same meaning no matter which medical activity is considered, and no matter which country the data come from. In its most simple terms, a low COV implies strong medical agreement about the way to use a specific medical procedure. A large COV implies considerable disagreement. Procedures for which large coefficients of variation repeatedly appear imply that the medical profession offers very different advice to its patients about when and how often to use those procedures. In economic language, large coefficients of variation in hospitalization rates imply substantial medical confusion or disagreement about the marginal productivity of various specific types of medical care. Since these studies look at the rates at which procedures are used in "similar" populations, the disagreement generally must arise on the extensive margin—how many people should receive the procedure.[13]

The first of such studies appeared in 1938 within the British National Health Service. That study looked at the rate at which British schoolchildren had their tonsils removed in various regions of the country. The results appear startling in retrospect: A tenfold difference existed across regions in the rate of tonsillectomies for schoolchildren. This study sat unnoticed for three decades until a growing series of studies expanded our knowl-

---

[13] A later section in this chapter offers some new evidence on disagreement on the intensive margin (hospital length of stay) as well.

Box 3.1   **STATISTICAL DISTRIBUTIONS**

Suppose a risky world contains a fixed number of events that might happen. Rolling two dice, for example, gives 36 unique outcomes (1,1; 2,1; 1,2; 2,2; 3,2; 2,3; etc.) giving 12 possible total point counts (snake-eyes through box cars). The total points on any roll of the dice occur with known probability (unless you are playing against somebody who has loaded the dice). The 11 possible outcomes occur with the following frequency:

OUTCOME	NUMBER OF POSSIBLE COMBINATIONS	PERCENT OF ALL POSSIBLE OUTCOMES
2 (snake eyes)	1	$1/36 = 0.02\overline{7}$
3	2	$2/36 = 0.0\overline{5}$
4	3	$3/36 = 0.08\overline{3}$
5	4	$4/36 = 0.\overline{1}$
6	5	$5/36 = 0.13\overline{8}$
7	6	$6/36 = 0.1\overline{6}$
8	5	$0.13\overline{8}$
9	4	$0.\overline{1}$
10	3	$0.08\overline{3}$
11	2	$0.0\overline{5}$
12 (box cars)	1	$0.02\overline{7}$

If we were to draw a frequency distribution of these outcomes, it would look like Figure A.

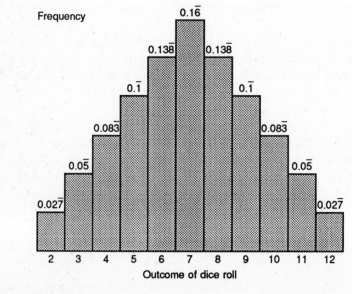

Figure A

The *average* outcome is the sum of each numerical value, weighted by the proportion of times it occurs. If each possible outcome is described as x_i and p_i is the percent of times outcome i occurs, the average for n possible outcomes is defined as

$$\mu = \sum_{i=1}^{n} p_i x_i$$

where μ (pronounced like the sound a cat makes) is the *expected value* of the random variable x. In the case of the dice, the expected value of the outcome is exactly 7.

The variance of the outcome is a measure of how spread-out the distribution is. If many events are the same (e.g., if the dice were loaded and mostly came up as 7), then the distribution would be even more peaked than that shown above. If the dice were loaded to *avoid* 5, 6, and 7, the distribution would flatten out. The more peaked the distribution (more squeezed together), the less the variance. The less peaked, the flatter it is, and the more times things occur "in the tails" of the distribution. Such distributions have a high variance.

A common measure of variance is the expected value of the square of the difference between the outcome and its average, defined

$$\sigma^2 = \sum_{i=1}^{n} p_i (x_i - \mu)^2$$

or sometimes just the square root of that, σ, known as the "standard deviation." (The Greek letter σ is pronounced *sigma*.)

Finally, it can become useful to express the variability of a distribution in terms of a *coefficient of variation,* which is simply

$$COV = \sigma/\mu$$

If the variable is continuously distributed (as with the temperature of a body of water) rather than having a specific number of possible outcomes (as with the roll of two dice), then similar measures are defined, where $\phi(\chi)$ is the *probability density function:*

$$\mu = \int_{-\infty}^{\infty} x\, \phi(x)\, \delta x \qquad \sigma^2 = \int_{-\infty}^{\infty} (x - \mu)^2\, \phi(x)\, \delta x$$

Figure B shows three distributions, each with an average value of $\mu = 10$, and variances of $\sigma^2 = 9$, 4, and 1, respectively. Thus, they have standard deviations of $\sigma = 3$, 2, and 1, and coefficients of variation of $COV = 0.3$, 0.2, and 0.1.

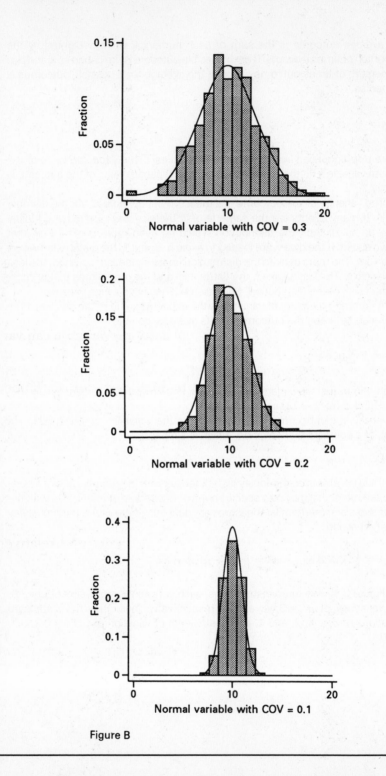

Figure B

edge of variations considerably: Kansas (Lewis, 1969), Maine (Wennberg and Gittelsohn, 1975), Iowa (Wennberg, 1990), Canada (Roos et al., 1986; McPherson et al., 1981), New England, Norway, and Wales (McPherson et al., 1982), medicare patients in the United States (Chassin et al., 1986), and New York (Phelps and Parente, 1990) all provide grist for the mill. In each of these studies, substantial variation in the rate of hospitalization occurred for specific procedures in essentially "standard" populations. In every country or region studied, the medical community shows important disagreement about the appropriate rates of hospitalization for many surgical and medical interventions.

These studies produce a surprisingly uniform picture of medical practice variations: The procedures with relatively large variations in one region or country will likely have relatively large variation in another. The *absolute* variation will differ from study to study for a variety of reasons, but the patterns of relative variation show considerable stability.

To understand how the patterns of absolute variation can vary, consider the simple effect of the size of the region chosen by the researchers for analysis. If the regions are very small (say, a city block), variations will appear quite high just due to random chance—who gets sick with which disease. If the regions are very large (say, the western versus the eastern half of the United States), the variations will be much smaller, because the "region" can actually be composed of a number of subregions that each have their own practice style. Bunching them together as a single "region" hides that variation.

The effect of aggregation on the absolute variability can be shown directly by comparing two studies of variation in surgical use within England and Wales. One study there used relatively large regions as the unit of observation, where the average population of each "region" was in the millions. Another study investigated the variations in some of the same surgical procedures in 21 health districts within West Midlands, one of the "regions" in the other study. In the health districts, the population ranged from 90,000 to 500,000, and on average was about one-tenth the size of the districts used in the other study. The coefficients of variation from these two studies appear in Table 3.1. As these data show, selecting a smaller region gives a larger coefficient of variation almost uniformly.

Table 3.2 shows the actual coefficients of variation for a variety of surgical procedures, either reported or calculated from data in

Table 3.1   EFFECT OF SCALE ON APPARENT VARIABILITY OF
             SURGICAL USE

| | Coefficient of Variation (COV) | |
Procedure	England and Wales regions	Health districts in West Midlands
Appendectomy	0.13	0.16
Cholecystectomy	0.11	0.16
Hemorrhoidectomy	0.24	0.35
Hysterectomy	0.12	0.20
Prostatectomy	0.22	0.24
Tonsillectomy	0.19	0.31

Source: McPherson, K., Strong, P. M., Epstein, A., and Jones, L. (1981).

nine different studies of medical practice variations. Among these commonly studied procedures, the greatest disagreement, almost uniformly, occurs for removal of tonsils and adenoids (T&A) and removal of hemorrhoids. The greatest agreement, in general, occurs for hernia repair and (excepting the Kansas study) gall bladder and appendix removal.

The within-region variations in surgical rates (as in Table 3.2) do not necessarily correspond well with cross-nation agreement about the "proper" rates for these procedures. For example, although each region studied has a relatively small coefficient of variations for hernia repair, the overall rates per 100,000 persons differ considerably across studies: 113 (England and Wales), 276 (New England), 186 (Norway), 235 (Canada), 309 (Kansas), 137 (West Midlands), 282 (United States).

The particular procedures shown in Table 3.2 are not necessarily those with large coefficients of variation; rather, they are the ones reported most commonly in studies. Larger coefficients of variation have appeared in several studies for some procedures. For example, in Medicare patients, Chassin, Brook, Parke et al., (1986) found high variation in such procedures as injection of hemorrhoids (COV = 0.79), hip reconstruction (COV = 0.69), removal of skin lesions (COV = 0.67), total knee replacement (0.47), and others. In a study of hospitalizations in New York State (Phelps

Table 3.2   COEFFICIENTS OF VARIATION OF SURGICAL REMOVAL PROCEDURES IN VARIOUS STUDIES

	Prostate	Tonsil	Appendix	Hernia	Hemorrhoids	Gall bladder	Uterus
Kansas	—	0.29	0.52	0.22	0.40	0.32	—
Northeast U.S.	0.30	0.36	0.26	0.11	0.30	0.18	0.22
Norway	0.33	0.48	0.16	0.20	0.47	0.18	0.31
West Midlands	0.24	0.31	0.16	0.20	0.35	0.16	0.20
Maine	0.26	0.43	0.18	0.14	0.55	0.23	0.25
England and Wales	0.22	0.19	0.13	0.16	0.24	0.11	0.12
Canada	0.33	0.23	0.15	0.14	0.35	0.14	0.18
4 U.S. regions	0.15	0.17	0.11	0.15	0.13	0.14	0.17
New York counties	0.18	0.42	0.21	0.16	0.16	0.14	0.28

Sources: Lewis (1969) for Kansas; McPherson et al. (1981) for Canada, England and Wales, and 4 U.S. regions; McPherson et al. (1982) for West Midlands, Norway, and Northeast U.S.; Wennberg and Gittelsohn (1975) for Maine; Phelps and Parente (1990) for New York counties.

and Parente, 1988), the largest coefficients of variation for hospitalization occurred for within-hospital dental extractions (COV = 0.73) and false labor (COV = 0.75).

Several studies have also provided coefficients of variation for nonsurgical procedures, and these data show that the uncertainty about hospitalization is at least as great in these areas as for surgery. The first of such studies (Wennberg, McPherson, and Caper, 1984) found very high variations (COV > 0.4) in such areas as urinary tract infections, chest pain, bronchitis, middle-ear infections and upper-respiratory infections (both adults and children), and pediatric pneumonia. In a study of Medicare patient hospitalizations, Chassin et al. (1986) found moderate to large coefficients of variation for a number of nonsurgical conditions, including diagnostic activities such as skin biopsy (COV = 0.58), coronary angiography (to detect clogging of arteries into the heart, COV = 0.32). In New York (Phelps and Parente, 1990), large variation appeared for a substantial number of pediatric hospitalizations, even after controlling for the age mix of the populations, including pneumonia (COV = 0.56), middle-ear infections and upper-respiratory infections (COV = 0.57), bronchitis and asthma (COV = 0.35), and gastroenteritis (COV = 0.42). Large variations also occurred for adult admissions in categories such as concussion (COV = 0.41), chronic obstructive lung disease (COV = 0.43), medical back problems (0.31), adult gastroenteritis (0.26), and similar diseases. Psychiatric hospital admissions were also quite variable, including depression (COV = 0.48), acute adjustment reaction (COV = 0.52), and psychosis (COV = 0.28).

In another recent study, John Wennberg and colleagues at Dartmouth Medical School reported on a comparison of the use of various medical procedures in two U.S. cities (Boston and New Haven), in both of which a large fraction of hospitalizations occur in hospitals affiliated with medical schools (87 percent in Boston, 97 percent in New Haven). One important feature of this study is that it shows substantial variations in practice patterns even within the part of the medical community—academic medicine—that should be best informed about the efficacy of various medical interventions. The cities are quite similar in terms of age profiles, the level and distribution of income, and the proportion of persons who are nonwhite. By contrast with New Haven, Boston has 55 percent more hospital beds per capita, and each hospital bed had 22 percent more hospital employees, who were paid on average 5 percent more. On average, citizens of the Boston area spent 87 percent more on hospital care than those in New Haven.[14]

The age-adjusted patterns of medical care use by citizens of the Boston area uniformly are higher than those of New Haven, with most of the variation occurring in (1) minor surgery cases, and (2) medical diagnoses where variations in admissions rate are high across the country. Small differences exist for major surgery and low-variation medical admissions, both in admission rates and lengths of stay.

Using a more finely tuned microscope, we can see diversity within this pattern of uniformity. Particularly within the major surgery category, a number of procedures appeared for which Boston had higher rates of use than New Haven, but an equal or larger number for which the use rate was larger in New Haven. For example, residents of New Haven received coronary artery bypass grafts (CABG) at twice the rate of persons in Boston, but citizens of the Boston regions received treatment using carotid endarterectomy (a procedure to ream out clogged arteries) at over twice the rate of those in New Haven.

Wennberg (1990) has documented similarly large variations in the admission rates for numerous surgical procedures in the market areas of 16 major university hospitals and large community hospitals around the country, again, in medical centers with the greatest presumed medical knowledge of any part of the health care community. The variations in admission rate correspond closely to those found in other settings. Even medicine's elite systematically disagree about the proper use of many procedures. Table 3.3 shows Wennberg's findings for 30 surgical procedures. Notice the similarity of results for procedures that also appear in Table 3.2.

Table 3.3 also reveals a useful rule of thumb: In such studies, the ratio of highest to the lowest rates of use will approximately correspond to ten times the coefficient of variation. For appendectomy, for example, the COV is 0.30, and the ratio of high to low is 2.86. This holds true for most procedures and studies in this litera-

---

[14] Interestingly, the study (Wennberg, Freeman, and Culp, 1987) was published in *Lancet*, the most prestigious British medical journal. The *New England Journal of Medicine*, published by the Massachusetts Medical Society, and often described as the most prestigious U.S. medical journal, turned down the manuscript when it was submitted for publication there.

Table 3.3   COEFFICIENTS OF VARIATION IN 16 UNIVERSITY HOSPITAL OR
LARGE COMMUNITY HOSPITAL MARKET AREAS

Procedure	Coefficient of variation	Ratio high to low
Colectomy (removal of colon)	0.12	1.47
Resection of small intestine	0.14	1.75
Inguinal hernia repair	0.15	2.01
Pneumonectomy (removal of part of lung)	0.21	2.72
Simple mastectomy (removal of breast)	0.27	2.71
Open-heart surgery	0.23	2.29
Extended or radical mastectomy	0.21	2.21
Hysterectomy (removal of uterus)	0.28	2.60
Cholecystectomy (gall bladder removal)	0.23	2.22
Embolectomy (varicose vein stripping)	0.36	4.10
Proctectomy (rectal surgery)	0.27	3.01
Pacemaker insertion	0.28	2.63
Thyroidectomy	0.34	3.35
Appendectomy	0.30	2.86
Total hip replacement	0.35	2.99
Repair of retina	0.27	3.12
Prostatectomy (prostate surgery)	0.33	3.12
Coronary bypass surgery	0.33	3.62
Mastoidectomy	0.46	4.03
Aorto-iliac-femoral bypass	0.38	3.62
Diaphragmatic hernia	0.37	3.45
Stapes mobilization	0.48	4.28
Spinal fusion with or without disc removal	0.52	5.20
Peripheral artery bypass	0.36	4.36
Cardiac catheterization	0.44	4.48
Excision of intravertebral disc	0.43	5.09
Graft replacement of aortic aneurism	0.40	6.26
Laparotomy	0.47	5.60
Total knee replacement	0.53	7.42
Carotid endarterectomy	0.83	19.39

Source: Wennberg, 1990.

ture, and provides a more intuitive way of comprehending the meaning of a coefficient of variation.

MEDICAL PRACTICE VARIATIONS ON THE INTENSIVE MARGIN

Differences in medical judgment occur in other areas than just for the choice to hospitalize patients. The decision to use diagnostic tests, including some very expensive tests such as computed tomography (CT) scans or magnetic resonance imaging (MRI), varies considerably from doctor to doctor and region to region. Similarly, within the hospital setting, the doctor continually makes choices about the use of tests and other hospital services, and for the length of stay of the patient.

Choices of medical care use within the hospital offer the opportunity to study the behavior of individual doctors, rather than "communities." For example, length of stay can be computed for a doctor's patients, and compared with the average for other doctors in the same region. One study provides such a comparison, offering new techniques to use the entire profile of a doctor's admissions (rather than looking just at one category of admissions), and to test for statistical significance of any differences that appear (Phelps, Weber, and Parente, 1990). The technique compares each doctor's length of stay on each admission with the overall average for all admissions in the same diagnostic category.[15] If a particular stay (say) was 10 days, and the average for that type of admission was (say) 7.5 days, then an index for that stay would be $(10 - 7.5)/7.5 = 0.33$. If the stay was exactly the average, the index would be zero, and if it were shorter than average, the index would be negative.

Using data from Blue Cross of Rochester, this study calculated the average index for all doctors admitting at least 15 patients a year into the hospital, using standard statistical tests to determine whether the observed differences were systematic or just due to chance. The study counted only those doctors with an index that was very unlikely to occur by chance (1 chance in 20). About one out of eight doctors in the study had lengths of stay at least 20 percent greater than the average for a "phantom" doctor with the same pattern of admissions but with an average length of stay for each patient, and about one out of six doctors had an index at least 20 percent lower than the average.

An obvious complication of studies like this is that some doctors will systematically treat the most severely ill patients, particularly those doctors in referral practices. To control for this problem, this study separately calculated the length-of-stay index for each patient after deleting all patients in the data base for whom any medical complication was indicated. This step eliminated about half of the doctors who had apparently large lengths of stay without

---

[15] The study uses the standard Diagnostic Related Group system, with 470 different categories defined.

Box 3.2 REGIONAL VARIATIONS IN LENGTH OF STAY

Hospital average length of stay (ALOS) varies substantially by region. Some of these differences undoubtedly occur because of age/sex differences in the population, but the discrepancies seem difficult to reconcile on that basis. A band of low-LOS states drops diagonally across the country from Washington (5.6) and Oregon (5.3) down through Nevada (5.9), Utah (5.1), and New Mexico (5.7). A similar band extends from Idaho and down across the mountain states into Texas (6.1), Oklahoma (6.2), Louisiana and Arkansas (6.0), and much of the "deep South." These states have neither similar weather, latitude, nor average population density.

By contrast, two cores of very long-LOS states occur, one in the north-central tier, from Montana (8-6), Minnesota (9.0), and North and South Dakota (9.0 and 8.7). The other pocket of high use occurs in New York (9.4) and Massachusetts (8.4). While these all appear to be northern states, Hawaii has a high average as well (8.3), and some northern states (Idaho, Wyoming) have much lower rates than immediately adjacent Montana and the Dakotas.

These differences do not seem closely related to the number of physicians per 100,000 persons. For example, Utah has 171 physicians per 100,000 persons, while North Dakota has 130 and Montana has 136. Age does not appear as a predominant factor either. Allegedly popular "retirement" states (Florida, Arizona, Texas, and California) all have relatively low average LOS.

These differences in LOS may well reflect "medical cultures," just as might differences in propensity to admit patients. But, as Figure A shows, the ways in which such "cultures" spread and form their territorial boundaries are somewhat mysterious.

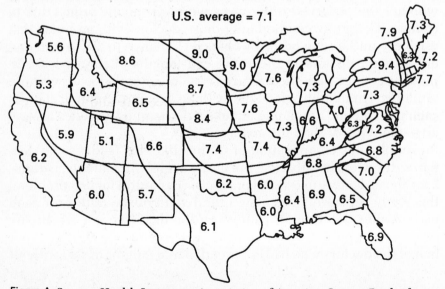

Figure A *Source:* Health Insurance Association of America, *Source Book of Health Insurance Data,* 1986–1987. Washington, D.C.: HIAA, 1987, data from Table 5.1b.

that adjustment, evidence that some doctors do treat a relatively complicated set of patients, but a group of about 6 percent of the doctors remained with high length-of-stay (LOS) indexes (at least 20 percent above the average), and over 20 percent of the doctors had an index at least 20 percent smaller than average.

The differences in hospital use between the high-LOS and the low-LOS doctors are considerable, and they represent another form of disagreement about the marginal productivity of medical care. Box 3.2 shows how LOS differs across different states. In these cases, the disagreement appears on the *intensive margin,* the intensity of resource use for patients within the health care system.

EXTENSIVE AND INTENSIVE MARGIN DIFFERENCES: ARE THEY SIMILAR?

One might wonder whether regions (or doctors) who disagree about the extensive margin of care have related disagreements about the intensive margin of use. For example, if one knew that (say) Boston was very "aggressive" on the extensive margin (admissions to the hospital), would they more than likely be high or low on choices of the intensive margin (length of stay). Unfortunately, few studies can make such a comparison. One study, looking at the use of ambulatory care by Medicare patients, concluded that the propensity to use care overall (per capita visits to physicians) was not related to the intensity of treatment rendered (Stano and Folland, 1988). This study uses a quite broad and general measure of medical care (visits), and thus may mask some important relationships. A study of Medicare patients (Chassin et al., 1986) looked at a specific surgical procedure and reached the same conclusion. The rate at which patients received coronary artery bypass grafts varied by a factor of 3 from the low-use to high-use regions studied, but the number of grafts made in each patient (the intensive margin) was unrelated to the overall use rate. Thus, from the (somewhat meager) evidence available, it appears that disagreements about the extensive and intensive margins of use may be quite different. Doctors who are "aggressive" or "conservative" on one margin may be the same or different on the other margin.

A study by Roos and colleagues in Manitoba used individual patient data from their provincewide claims system to address a similar question. They estimated the propensity of doctors to admit patients to the hospital, and also the average length of stay among those patients who were admitted. They found a slight but negative relationship between the two indexes for each doctor, suggesting that at least in this area, those who tend to admit more patients end up (on average) with less sickly patients. In other words, in their data, increases in the extensive margin (admissions) led to a slight decline in the average intensity (length of stay), as would be expected with appropriate "sorting" of patients into the hospital on the basis of illness severity (Roos, Flowerdew, Wajda, and Tate, 1986).

The Policy Question: What Should We Do About Variations?

The apparently considerable disagreement over proper use of medical interventions disturbs many analysts of the health care system. The uncertainty demonstrated by these variations highlights an important question: How do medical interventions become accepted as standard medical practice, and what causes the medical community to change its beliefs about therapeutic efficacy? More particularly, what is the proper way to create and disseminate information about the efficacy of medical interventions—their marginal productivity?

These are complicated questions, deserving careful consideration. At this point, we can merely look ahead to subsequent points of this book: Information like this is a "public good" and probably will not emerge spontaneously through private actions. Chapter 14 describes the role of the medical-legal system in affecting the quality of care in medicine. For now, we will stand on the observation already made: Most of what constitutes "modern medicine" has never been tested in a scientific fashion, and its marginal productivity in producing or augmenting health therefore remains an open question.

SUMMARY

Health can be produced by medical care, although the process is often uncertain and may not always proceed as intended. As with other productive processes, the production of health is almost certainly subject to diminishing returns to scale: The more medical care we use, the less incremental gain we get back in terms of improved health.

Available evidence shows both that the average effect of health care has been quite important in augmenting our health, and that the incremental effect of further health care might be quite small.

Further resources can be devoted to health care on both an extensive margin (more people treated) and an intensive margin (more treatment per person treated). Doctors seem to disagree considerably about the right amounts of health care to use on both of these margins of adjustment. On the extensive margin, numerous studies show substantial variations in the rate at which various medical interventions are employed. These variations *require* disagreement among doctors: They signal medical confusion. On the intensive margin, considerable disagreement exists as well about the proper length of stay (for example) for various hospitalizations. Doctors appear to have "signatures" characterizing their length-of-stay plans.

On net, the differences of opinion about proper medical practice have considerable implication for resource use. Wennberg and colleagues, for example, have estimated that if the practice patterns in Boston applied to the nation as a whole, we would spend 15 to 16 percent of our GNP on health care (rather than the current 11 percent). By contrast, if the practice patterns in New Haven applied to the country, we would spend about 8 percent of GNP on health care. At present, we have little direct evidence on the differences in production of health arising from such diverse choices in the rate of use of medical interventions. The RAND HIS results suggest that the improvement in health from such radically different spending patterns may be small, at best.

PROBLEMS

1. "Most of the variability in average medical care use across regions is due to differences in insurance coverage." Comment.

2. Thinking about a diagnostic test like breast cancer screening, describe what it means to increase use of the test (1) on the intensive margin, and (2) on the extensive margin.

3. People living in Boston are hospitalized about 1.5 times as often as those living in New Haven, yet their health outcomes appear to be identical. Does this mean that hospital care has no ability to improve health? (Hint: Think about the difference between average and marginal productivity.)

4. "Variations in medical care use probably arise from the educational level of doctors, with less-trained doctors using either too much or too little care and better-trained specialists using about the right amount of care." Comment.

APPENDIX

Marginal, Average, and Total Productivity

This appendix summarizes the ideas of average and marginal productivity, stated in the most general of production functions—that is, where output varies with a single "composite" input, for example, producing "health" with a single composite input called "medical care." Thus, $H = f(M)$ describes the type of production function we are discussing. The central idea is that the input (M) yields different *incremental* returns of output, depending on how much of the input one uses. The production function shown here displays (initially) *increasing* returns to scale, and then the more common *decreasing* returns to scale. As we will see, it seldom makes sense to operate in a realm where returns are increasing, so thinking about production functions with decreasing returns has the most common sense about it.

Figure 3A.1 shows the graph of output (H) plotted against M. The curve initially dishes up ("U-shaped"), and then tips over to be "hill-shaped." The point where it tips over is called an "inflection point," labeled M_1 on the graph. Output H increases as M increases (corresponding to the idea that more M produces more H) until the point M_3, at which point H actually begins to decline as more M is used. In medical questions, this would be called "iatrogenic illness," since health would fall as medical care use increased.

Figure 3A.2 shows the same production function, but it graphs on the vertical axis either the *marginal* productivity $(\partial H/\partial M)$ or the *average* productivity (H/M), rather than the total output H (which appears in Figure 3A.1). The marginal product curve rises initially, until it tips over at output level M_1, the inflection point in Figure 3A.1, and begins to fall. Another point of interest occurs at M_2, the point where the *average* product reaches its highest possible level in Figure 3A.2. (Note that this occurs in Figure 3A.1 at the point where a ray from the origin to the curve is at its steepest.) It also must be true that, at that point, average and marginal product are equal.[16] This is why we show the average product and marginal product curves intersecting in Figure 3A.2 at the input rate M_2.

If we say that another "unit" of H is worth P and that another unit of M costs W, then the optimal use of M says to expand its use until the *value of the incremental product* just equals the incremental cost of another unit

---

[16] The proof occurs as follows: Define average product = H/M, and find its maximum by taking the derivative and setting it equal to zero. The derivative of H/M is $(M\partial H/\partial M - H)/M^2$; setting that equal to zero gives $\partial H/\partial M = H/M$. However, since H/M = average output, this proves that when average output has reached its maximum, average and marginal product are equal to each other.

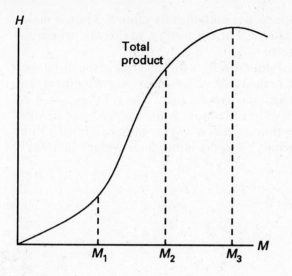

Figure 3A.1

of input—that is, $P\partial H/\partial M = W$, or in a way that we can graph in Figure 3A.2, expand use of M until $\partial H/\partial M = W/P$. Now we can see why it makes sense always to operate in a realm of diminishing marginal product (that is, where marginal product is positive, but falling in Figure 3A.2). Look in Figure 3A.2: This "optimality" condition occurs at two points, indicated by M^* (at a very low rate of use of M) and M^{**}, at a much larger rate of use, which mathematically must occur at an input rate exceeding M_2. If we "stopped" using M at the point M^*, we would "give up" a lot of output H

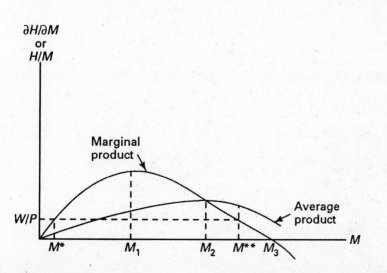

Figure 3A.2

that would cost less to produce, per unit, than its value P. Thus, it makes sense to expand use out to M^{**}, but not further, say to M_3, where the marginal product falls to zero.

Perhaps the key idea to remember from this in terms of the discussion of use of health care is that, in the realm of "sensible" use of medical care, *marginal product is less than average product*. That is to say, we'll get less than the average yield of health output when we use more medical care in the vicinity of the optimum. This is why we must continually think about *marginal* ("incremental") benefits rather than average benefits of medical care.

Chapter
4

The Demand for Medical Care: Conceptual Framework

*I*n this chapter, we derive the consumer's demand curve for medical care from the utility function described previously, and then analyze the effects of a health insurance policy on that demand curve and quantities demanded. Finally, we will study the evidence showing how demand varies with systematic features such as income, age, sex, and location. We will learn how much price and insurance coverage alter use of medical care. We will see how time acts as a cost of care, and see some evidence of how important this is in affecting people's use of care. And finally, we will see how illness events—the level of sickness people actually experience—dominate individual choices of how much health care to buy in any given year. This, in turn, sets the stage for analyzing the demand for health insurance in Chapter 10. The next few pages provide a fairly complete characterization of how the economic model goes from utility functions to demand curves. The remainder of the text will rely intensively on the demand curve, rather than the utility function, so those who are comfortable with the idea of a demand curve (or those who are willing to accept the idea on blind faith) might just skip directly to the discussion of demand curves themselves following the section on indifference curves. However, for those willing to invest the effort, the excursion through the terrain of utility theory that follows will add considerably to their final understanding of the meaning of demand curves, and how various economic and health-related events alter those demand curves.

INDIFFERENCE CURVES FOR HEALTH AND OTHER GOODS

The economist's model of consumer demand begins with the utility function, as described in Chapter 1. This model makes the consumer's own judgment of something's value the only relevant judgment—*de gustibus non disputandum est.*[1] We assume that the consumer has a stable utility function, in the sense that it doesn't change from period to period, or with new information (say) about the value of medical care. We make this assumption because without it, we lose the ability to say much about consumer behavior, but we need to remember that it is just an assumption.[2] Thus, our starting point is the utility function, Utility = $U(x, H)$.

We had developed in Chapter 1 the idea of an indifference curve, the set of all combinations of x and H that create the same level of utility. As in standard consumer demand theory, the consumer tries to reach the highest possible indifference curve possible, because utility is higher then. The budget constraint provides the limit to this process. The consumer must pay for x and for any medical care (m) used to produce H, and overall spending must be limited to the available budget. This presents the first "wrinkle" in applying standard economic theory to the demand for medical care. Somehow, we need to make the translation from health (H), which improves utility, to medical care (M), on which money is spent. The *production process* provides this translation, which we

---

[1] Literally, "There is no disputing of tastes."

[2] According to some schools of philosophy, the "realism" of assumptions matters. According to others, it doesn't matter, at least in one sense. Assumptions determine the structure of a *model* of human behavior. The *theory* corresponding to that model is that consumers act *as if* the model replicated human behavior. The *theory* can be refuted by finding human behavior that conflicts with the model, according to the philosophy of Karl Popper and others. Thus, the test of a theory is not its realism, but the predictive value it has. Of course, if a model has assumed away critical parts of the problems, then it will fail the prediction test, since it will not be able to predict behavior well. For a classic discussion of these issues in an economic setting, see Milton Friedman, *Essays in Positive Economics.*

described in Chapter 1 as the relationship $H = g(m)$. As we saw in Chapter 3, this process is almost certainly subject to diminishing returns to scale, so the relevant question is how health changes as medical use changes. To provide a shorthand way to describe this, we will define $\partial H/dm = \partial g(m)/\partial m = g'(m)$ as the *rate* at which health improves for a small change in m.[3] Thus, we can redraw the diagram in Figure 4.1a (which maps combinations of x and H) as Figure 4.1b, which maps combinations of x and m (not x and H). If $g'(m)$ were always constant for any amount of m chosen (i.e., there were no diminishing returns to producing H), then these two figures would look the same, except that the units of measurement along the vertical axis would be in medical care units, rather than health units. However, since the production of health exhibits decreasing returns to scale, the curves can also change shape.

Figure 4.1a also shows a *production possibilities curve*, labeled *PP*. This curve represents the feasible set of combinations of x and H the consumer can attain, given the available budget and the production function $H = g(m)$. It curves downward (it is "concave") because of the diminishing marginal productivity of m in producing H. Figure 4.1b shows the same situation in a map with dimensions x and m (not x and H). Instead of a production possibilities curve, Figure 4.1b has a budget line *II*, showing the budget $I = p_x x + p_m m$ as a straight line (because individual consumers can treat market prices as fixed). It is as if the original Figure 4.1a were drawn on a rubber sheet, and then stretched to create Figure 4.1b. The direction and extent of the stretching would depend on the production function for health, $H = g(m)$. In general, the map would have to be stretched in just such a way to straighten out the production possibilities curve *PP* into the straight budget line *II*.[4] The indifference curves would all get warped by the same stretching process. We need only know here that the indifference curves comparing x and m have the same *general* shape as the indifference curves in Figure 4.1a, which most closely map the utility

---

[3] In the notation of the calculus, we need the marginal productivity, $\partial H/dm$. Since $H = g(m)$, we describe this throughout the book by using the notation $g'(m)$, where $g' = \partial H/\partial m$ is the marginal productivity of m.

[4] Of course, one could begin with Figure 4.1b and then stretch it into Figure 4.1a, again using the production function (in the other direction) to describe the appropriate amount of stretching. To do this, Figure 4.1b would be stretched vertically at each point on the m-axis according to the *inverse* of the marginal productivity of m.

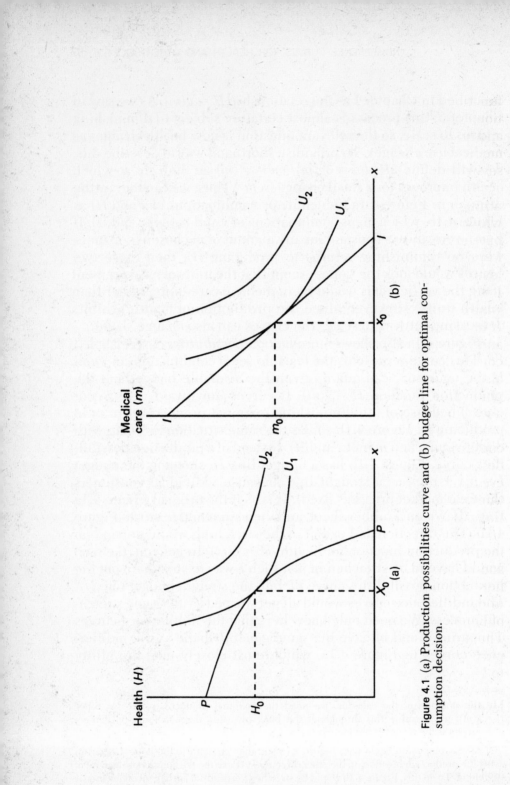

Figure 4.1 (a) Production possibilities curve and (b) budget line for optimal consumption decision.

function directly.[5] As will become clearer in a moment, the translation from Figure 4.1a (showing x and H) to Figure 4.1b (showing x and m) only has meaning for a particular illness, since the effect of medical care on health depends on the specific illness and its severity.

We will say that consumers act as if they wish to maximize utility = $U(x, H)$ within the constraints implied by their budgets, which they spend on x and m. If a consumer's income is I, then the budget constraint says that spending must not exceed income, or $I \geq p_x x + p_m m$. Since the consumer wants to reach as high a utility level as possible, the entire budget is always used, and $I = p_x x + p_m m$. This means that the consumer picks the point where a single indifference curve is just tangent to the budget line II in Figure 4.1b. This point specifies that the best possible mix of x and m is (x_0, m_0). Since m_0 produces the level of health H_0, the companion point in Figure 4.1a is the point (x_0, H_0). This point is also tangent to the production possibilities curve PP at a single point. In either figure, the consumer has spent the entire budget in a way to maximize utility.

We can now show in Figures 4.2a and 4.2b the effects of the consumer "getting sick." Begin in Figure 4.2a at the previously described optimum consumption, shown as point 1. An illness event immediately drops the level of health from H_0 to H_1. Call this health loss l. At the same time, the set of achievable combinations of x and H shifts inward *for every level of x* by the same amount l. The health loss l acts the same here as a direct loss of income—it reduces the achievable opportunities to consume x and H. However, point 2 in Figure 4.2a is not the best achievable here. After the illness, the consumer can slide along the $P'P'$ curve to point 3, giving up some consumption of x to increase the level of health from H_1 to H_2. Point 3 is the best the consumer can do. The illness initially drops the level of utility from U_1 to U_2, and the purchase of medical care (at the sacrifice of some x) raises it back to U_3. (Note that in this figure, the subscripts on the utility curves refer to the sequence in which they are reached, not comparative levels of utility.)

In Figure 4.2b, we can see the same situation in a map of x

[5] We had noted in Chapter 1 that some types of "goods" in the bundle of goods x in fact also lowered the level of health. The presence of such goods means formally that H = g(m,X_G,X_B) and we need to define g'(m) as the partial derivative of g, holding X_G and X_B constant, so g'(m) = ∂h/∂m. In diagrams like Figure 4.1a, the presence of some X_B just makes the PP curve more concave (have more of a bend in it), but the general idea remains the same.

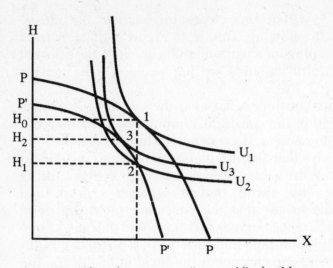

Figure 4.2a When the consumer "gets sick"—health.

and m. Several important differences occur here. First, the indifference curves *all* shift when the consumer gets sick. In particular, they change slopes, so the marginal rate of substitution between x and m shifts. While the *utility function* is stable, the *indifference map* with x and m as coordinates shifts with the illness level. In Figure 4.2b, the heavy curves show the preferences of the person before becoming sick and the light indifference curves show the

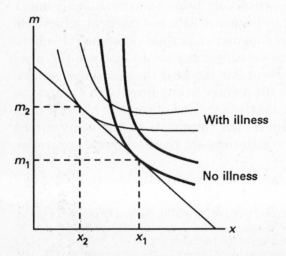

Figure 4.2b When the consumer "gets sick"—medical care.

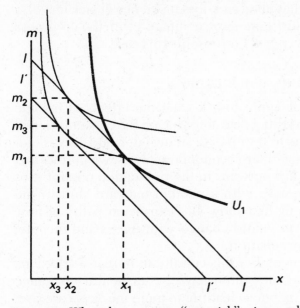

Figure 4.2c When the consumer "gets sick"—income loss.

same person's preferences after the illness event. In Figure 4.2b, the decline in consumption from x_1 to x_2, coupled with the increase in use of medical care from m_1 to m_2, shows the effect of illness on patterns of spending in the market.

In addition, *if illness harms a person's ability to earn income,* then the budget constraint in Figure 4.2b might also shift inward because of the illness. Figure 4.2c does this, with the same pattern of preferences as in Figure 4.2b, but with a decline in income as well from II to $I'I'$. The additional reduction in the consumption of both x and m reflects this loss of income. This is one of the reasons why income and good health are positively correlated. Another is that more income allows the purchase of more medical care.

In Figure 4.2c, the pre-illness choice of x and m sits at the tangency of a heavy indifference curve with the budget line II. The utility level U_1 is the highest attainable for the income I_1. When the illness strikes, indifference curves rotate and income falls to $I'I'$. The best post-illness choice is the tangency of the budget line $I'I'$ and the light indifference curve, at point (x_3, m_3). Illness has caused three potentially observable things to happen: Income has fallen, the amount of x has fallen, and the amount of m has increased. Even if illness does not cause income to decline (for example, if the person has good sick-loss insurance or sick leave), x will still fall and m will still increase, compared with the no-illness choices. The choice of (x_2, m_2) represents this case.

The Effects of an Increase in Income

We can use this sort of figure to ask what the effects of income would be, holding everything else constant, on the consumption of medical care. This is the first step in deriving "demand curves" for medical care. As in any standard economic analysis, we portray this in Figure 4.3a with a shift outward in the production possibilities (PP) line, or in Figure 4.3b with a parallel outward shift in the budget line II. These figures show an "expansion path" of how consumption of x and m would change with increasing income, *holding everything else constant.*

It is important to remember this condition: in real-world data, people with higher incomes often have better health insurance than those with lower incomes, leading to more medical care use, as we will shortly see. They may also get sick less often, leading to less medical care use. For some of them, life-style choices at higher incomes ("life in the fast lane") may cause health to fall with higher incomes. On balance, a simple plot of medical care use versus income would embed a complex set of phenomena affecting medical care use. However, if we continue to hold everything else constant, as in Figure 4.3b, a plot of the consumption of H versus income and m versus income taken from Figures 4.3a and 4.3b would yield Figures 4.4a and 4.4b, the Engel curves for H and m.[6] The pattern of indifference curves in Figure 4.3 and the Engel curves in Figure 4.4 are drawn to replicate (approximately) some empirical regularities that we will study further below—namely, that changes in individuals' incomes (all else held constant) do not seem to alter much the amount of medical care consumed.

Figures 4.4a and 4.4b show some (but not all) of the possible complications that might enter a simple comparison of income versus health or the use of medical care. For example, in Figure 4.4a, a "sanitation effect" at low incomes offers the possibility that health outcomes might rise very rapidly with initial improvements in income, as basic sanitary measures (e.g., water supply and vaccines) radically alter the conditions in which people live. This

---

[6] Engel curves, a plot of income versus amount consumed, are named after the economist Ernst Engel, to whom the discovery of the Engel curve is commonly attributed.

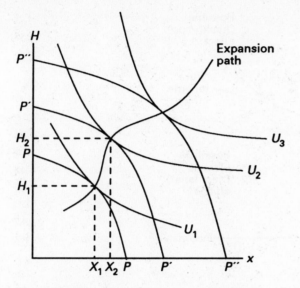

Figure 4.3a The effects of income on health.

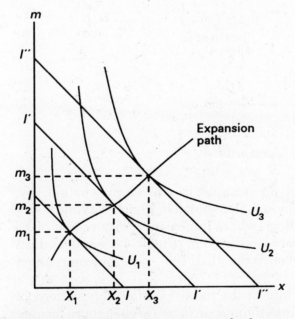

Figure 4.3b The effects of income on medical care.

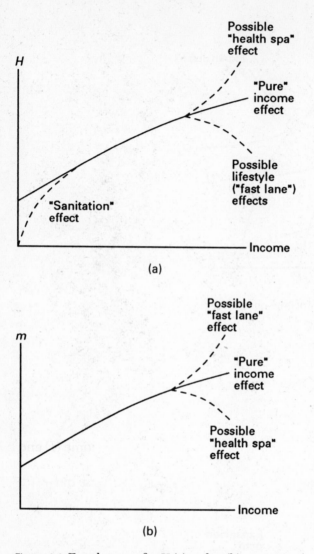

Figure 4.4 Engel curves for H (a) and m (b).

would be most pertinent, for example, in a developing nation, say, in a poor rural village. There is also shown in Figure 4.4a and 4.4b a "life in the fast lane" effect, where increasing consumption of X_B with higher income causes health to fall (Figure 4.4a) and medical expenses to increase even faster than the "pure" income effect would suggest.

These figures only hint at the many possible interactions of income and health. Some of them arise because changes in income alter the "external" production possibilities for health (e.g., public health measures). Others arise because of life-style choices, and how they change with income. Other "external" and "internal" effects may work in the opposite direction. For example, increasing income (at the level of a society) may come only with an industrial process that generates more health hazards. Similarly, nothing says that higher income "life-style" effects have to be negative. "Life in the fast lane" might just as well be "life in the health spa," so that higher income contributes more, not less, to the level of health, aside from the direct purchase of medical care.

FROM INDIFFERENCE CURVES TO DEMAND CURVES

The same sorts of diagram can help us make an important shift, from indifference curves, showing how various combinations of H and x create utility, to demand curves, showing how the desired quantity of medical care (m) changes with its own price. Indifference curves, by their nature, are unobservable, since we cannot measure utility. However, if consumers act according to the model portrayed by indifference curves, then we can infer something completely observable from this model, the standard "demand curve," a plot of how much medical care people would consume at different prices. To do this, we use a standard technique: For a given level of illness (so the indifference curves are stable), we change the price of m (p_m), holding constant money income (I) and the price of other goods (p_x). In an indifference curve diagram like Figure 4.5a, *reducing* p_m means swinging the budget line outward, holding its intercept on the x-axis constant.[7] Increasing p_m would rotate the budget line in the other direction around the x-axis intercept.

To trace out a demand curve, we simply vary p_m, holding everything else constant (in this case, income, p_x, and the illness level l), and see what amounts of m are chosen by the consumer. We begin with a particular value for p_m, say p_{m1}, which gives a

---

[7] An easy way to remember this: If the consumer spent everything on x, then the relevant part of the budget line would be where it intersected the x-axis, since m would equal zero there. At that point, the price of m would be irrelevant, and the same amount of x would be consumed. By contrast, for the same money budget, if the consumer spent everything on m, purchasing power would be increased as p_m fell. Thus, more m could be purchased. In general, a change in p_m means a rotation of the budget line around the intercept on the x-axis of the indifference map.

particular value m_1 as the best choice. In the *demand curve* diagram, we can plot that combination of p_{m1} and m_1. Now decrease the price to p_{m2}, and find the matching amount consumed, m_2. Repeating the same process for every possible price will trace out the entire demand curve D shown in Figure 4.5b.

This demand curve is completely observable in the real world. While it is derived from a fairly abstract utility theory, it offers an

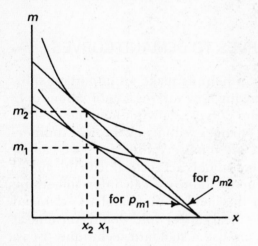

Figure 4.5a Lower price for m.

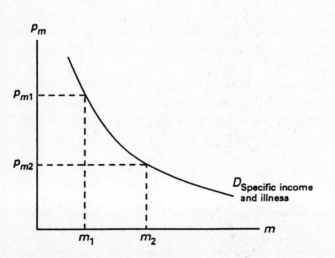

Figure 4.5b Demand curve for m.

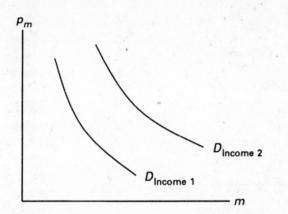

Figure 4.5c Effect of income.

observable model of consumer behavior that (in concept) is refutable. Thus, the demand curve becomes one of the most important tools of the economist. If we had begun the entire process of tracing out the demand curve at some higher level of income, we would notice that at every price p_m we looked at, more m would be chosen at the higher income than at the lower income. As we trace out the demand curves for medical care, this would create a separate demand curve for each level of income a person might have. So long as health is a "normal" good (i.e., people want more of it as their income increases), demand curves associated with higher income will always lie to the northeast of demand curves associated with a lower income. While it is common to portray such shifts as parallel, nothing requires this in general; the effects of income in shifting demand for medical care could differ considerably at different prices. Figure 4.5c shows how demand curves shift out as income rises, assuming that medical care is a "normal" good.

HOW DEMAND CURVES DEPEND ON ILLNESS EVENTS

The same theory of demand has the (commonsense) result built into it that people who are seriously sick will demand more medical care, other things being equal, than those who are less sick. The formal proof of this idea is somewhat messy, but the main idea appears in Figures 4.2a and 4.2b: The bigger the illness event (larger l), the more the health loss in Figure 4.2a, and the flatter the

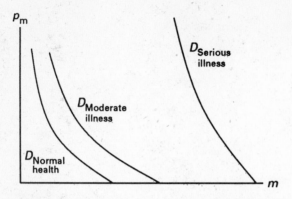

Figure 4.6 Demand curves for various illness events.

indifference curves in Figure 4.2b become. As the slopes of the indifference curves change, the tangency (optimal consumption choice) must occur at a larger level of m, given a constant slope for the budget line.

In Figure 4.6, we can see a series of demand curves for various illness events, ranging from "normal health" (when the consumer would see a doctor perhaps only once a year), to a series of headaches (several physician visits and perhaps a couple of lab tests), a moderately serious automobile accident (emergency room visit, some X-rays, stitches, and a couple of follow-up visits in the doctor's office), and a protracted bout with cancer (many diagnostic tests, surgery, therapy, etc.). Each of these demand curves also depends on the level of income. At a higher income, the demand curve associated with each event would sit somewhere to the right of those shown in Figure 4.6. Theory cannot tell us how much— empirical studies are needed to provide that information.

DEMAND CURVES FOR MANY MEDICAL SERVICES

The discussion above masks the idea that consumers have available not a single "medical care" service, but a complex variety of such services. Again, while the formal theory is somewhat messier, the ideas to deal with such a world are in fact quite similar to those described here. The only adjustments we need to make are (1) to recognize that health can be affected by more than one type of medical care, and (2) that various types of medical care can be either substitutes or complements. We discuss each of these ideas next.

Multiple Health Inputs

If multiple types of medical care affect health, then we need to write the production function for health as $H = g(m^1, m^2, \ldots, m^n)$, for the n types of medical care available. Each type of medical care would then be demanded according to its marginal productivity, *given the amount of all other medical inputs* used. Each medical input would have its own demand curve, derived in the same way we have just seen for the single good m.

Complements or Substitutes?

One important question that appears when we consider more than one type of medical care is whether the various services are complements or substitutes. Complements are goods or services that are consumed together, and "help each other out" in producing health. Gasoline and tires are complements in the production of miles driven. Substitutes are just what they sound like: Using more of one allows use of less of the other in order to achieve the same result. Cars and airplanes are substitutes in the production of passenger transportation. Formally, we define medical care of various types as complements or substitutes based on behavior in response to price changes. If the amount of service m_i rises as the price p_j (the price of service m_j) rises, then the two goods are substitutes. If m_i falls as p_j rises, the two goods are complements. Since we expect m_j to fall as p_j rises (downward-sloping demand curve), then saying that m_i and m_j are complements simply means that the use of both declines as p_j rises. In other words, their consumption patterns move together as the price of either one of them changes. Just the opposite happens if they are substitutes.

In health care, an important policy question was raised some time ago: Are hospital care and ambulatory care complements or substitutes? More generally, are preventive and acute medical care complements or substitutes? Among other things, the best design of health insurance packages depends on these relationships. If preventive and acute care are substitutes, then a "smart" health insurance program might encourage the use of preventive care, even though that care costs the insurer some money, because it would more than save that much with lower spending on acute medical care. (The same question arises with ambulatory care and hospital care.) The question first appeared in a provocative form: Was insurance that denied complete coverage for preventive/ambulatory care "penny-wise but pound-foolish"? Like other important features of the demand for medical care, theory alone cannot provide an answer. Empirical studies of the demand for care are needed to answer such a question. Chapter 5 summarizes the available evidence on these questions.

THE DEMAND CURVE BY A SOCIETY: ADDING UP INDIVIDUAL DEMANDS

All of the discussion to date really centers on a single individual, but the transition from the individual to a large group (society) is quite simple. The *aggregate* demand curve simply adds up *at each price* the quantities of each individual demand curve for every member of the society.[8] Figure 4.7 shows such an aggregation for a three-person society, but it should be clear that the process can continue for as many members of society as are relevant. The demand curve called D_1 (solid line) is for person 1, D_2 (dashed line) for person 2, and D_3 (dotted line) for person 3. The aggregate demand curve D_{total} adds up, at each possible price, the total quantities demanded. It coincides with D_3 at higher prices, since only person 3 has any positive demand for m at prices above p_2. There is a kink in D_{total} at p_2, where person 1's demands begin to add in, and again at p_1, where person 2's demands begin to add in. At a price of $p = 0$, the total quantity demanded m_{total} would equal the sum of m_1, m_2, and m_3, the quantity-axis intercepts of each person in the society.

It is important to remember that the demand curves D_1 through D_3 all depend on the particular illness levels experienced by persons 1 through 3. Thus, since their demand curves would shift if their illnesses changed, so would society's demand curve D_{total}.

---

[8] This is called "horizontal aggregation" because of the economist's usual penchant for drawing demand curves with price on the vertical axis and quantity on the horizontal axis of a demand curve diagram. Adding things "horizontally" in such a diagram produces the correct picture.

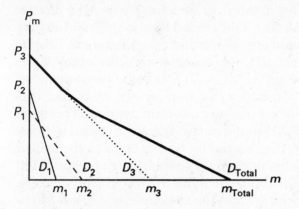

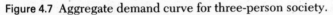

Figure 4.7 Aggregate demand curve for three-person society.

USING THE DEMAND CURVE TO MEASURE VALUE OF CARE

Marginal Value

The previous discussion speaks as if the quantity of medical care people demand is affected (for example) by price. Demand curves show this relationship, describing quantity as a function of price. They can also be "inverted" to describe the *incremental (marginal) value consumers attach to additional consumption of medical care* at any level of consumption observed. This is the *willingness to pay* interpretation of a demand curve. All that is done here is to read the demand curve in the other direction than is normally done. Rather than saying that the quantity demanded depends upon price (the interpretation given above), we can say equally well that the incremental value of consuming more m is equivalent to the consumer's willingness to pay for a bit more m. Just as the quantity demanded falls as the price increases (the first interpretation of the demand curve), we can also see that the marginal value to consumers (incremental willingness to pay) falls as the amount consumed rises. We call these curves inverse demand curves or value curves.

Inverse demand curves (willingness to pay) slope downward for two reasons: (1) the declining marginal productivity of medical care in producing health, and (2) the decreasing marginal utility of H itself in producing utility. The first issue—diminishing marginal productivity of health care—would suffice to produce downward-sloping demand curves for an individual (see the discussion in

Chapter 3 of the intensive margin), or for a society (see the discussion of the extensive margin). The second idea merely adds to the list of reasons why demand curves would slope downward. Empirically, we could separate out one concept from the other if we could accurately measure $g'(m)$ for each level of m consumed.

We can extend the same idea one step further, by noting that the *total* value to consumers of using a certain amount of medical care is the *area under the demand curve.* It may be useful to think in specific terms, such as the demand for having one's teeth cleaned and inspected. Suppose the first dental visit each year created $100 in value to the consumer (better-looking teeth, reduced concern about cavities). If the visit cost $30, the consumer would get $70 in *consumer surplus* out of that visit. A second visit per year (i.e., at six-month intervals) might create a further $75 in value, again costing $30, and creating $45 in consumer surplus. A third visit per year (every four months) might create a marginal value of $35, and a consumer surplus of only $5. A fourth visit per year would create only $20 in marginal value, and would cost $30. No intelligent consumer would do this, since it would cost more than it was worth for the fourth visit each year. We should observe three visits by such a consumer each year, unless some illness event (cavities, broken tooth) caused other visits.

Two ideas appear in this discussion. First, demand curves can tell us how to predict quantities consumed: Intelligent decision making will continue to expand the amount consumed until the marginal value received just equals the marginal cost of the ser-

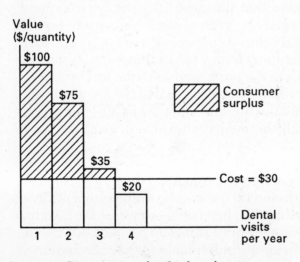

Figure 4.8 Consumer surplus for dental visits.

Table 4.1   CONSUMER SURPLUS FOR DENTAL VISITS
(Fictitious Data)

Quantity consumed per year	Incremental value of care	Net value of this visit with $30 cost	Total net value (consumer surplus)
1	$100	$70	$70
2	$75	$45	$115
3	$35	$5	$120
4	$20	−$10	$110

vice. (In the case of the dental visits, we don't quite achieve "equality" because we described the incremental value of dental visits as a lumpy step function, dropping from $100 to $75 to $35 to $20 visits, and the cost was described as $30.)

The second concept is that of *total consumer surplus* to the consumer from having a specific number of dental visits each year. As noted, intelligent planning would stop after the third visit each year. Total consumer surplus sums up all of the extra value to consumers (above the costs paid) for each unit of care consumed. Figure 4.8 shows how this would look for the dental visits described above, and Table 4.1 shows the same data numerically. It is easy to prove that consumer surplus is maximized by a simple rule: Expand the use of the service until the marginal benefit has just fallen to match the marginal cost. In this case, at the third visit per year, the marginal benefit has fallen to $35, the marginal cost is $30 per visit, and stopping at that rate of visits maximizes consumer surplus, as the last column of Table 4.1 shows.

Of course, the type of "incremental value" data shown in Figure 4.8 and Table 4.1 is just a lumpy version of a demand curve. One could easily draw a smooth line through the midpoints of the tops of each of the bars in Figure 4.8 and call it a demand curve; adding up such curves across many individuals would smooth things out even more.

HOW INSURANCE AFFECTS A DEMAND CURVE FOR MEDICAL CARE

We now come to one of the key ideas in all of health economics: The way health insurance is usually structured, it reduces the price consumers pay at the time they purchase medical care.

Health insurance lowers effective prices. If medical care obeys the normal laws of economics, providing people with a health insurance policy should increase their use of medical care. (The major function of insurance is to reduce financial risk. The mechanism of doing that—lowering the price of medical care—produces a side effect of increased medical care use. Chapter 10 discusses the demand for health insurance in detail.) Health insurance can be structured in numerous ways, but several standard features appear in many policies, and understanding how they work and how they affect demand for medical care provides an important insight into many health policy issues. The typical features are (1) copayments, (2) deductibles, and (3) upper limits on coverage. We discuss each next.

Copayments

A copayment is simply a sharing arrangement between the consumer and the insurance company, decided in advance in the insurance contract. When the consumer spends money on medical care, the insurance company pays some of it and the consumer the remainder (the copayment). Copayments come in two traditional forms, a proportional *coinsurance rate* and a flat *indemnity payment* by the insurance company.

In a *coinsurance* arrangement, the consumer pays some fraction of a medical bill (such as 20 percent or 25 percent, sometimes 50 percent in dental insurance), and the insurance company pays the remainder. If we call the consumer's copayment share C, then the insurance company pays a share $(1 - C)$. Using the methods of demand theory previously developed, we can now see precisely how coinsurance-based insurance plans alter consumer choices for medical care.

Figure 4.9 shows a consumer's demand curve for a particular illness, *without any insurance.* Suppose the price of care is p_{m1}. If the consumer has an insurance plan that pays $(1 - C)$ percent of all medical bills, then the effective price of medical care has fallen to Cp_{m1} when the consumer seeks medical care. To construct the consumer's demand curve with this insurance policy in place, find the quantity demanded at Cp_{m1} (point A in Figure 4.9). This is the quantity demanded at p_{m1} with the copayment insurance policy in place. Thus, on the insured person's demand curve, we can place the point B at the same quantity of point A, but at the price p_{m1}. Now do the same thing at some higher price p_{m2}, producing the

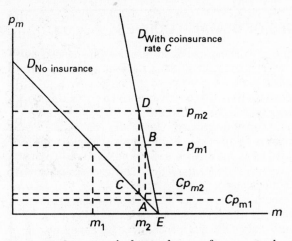

Figure 4.9 Consumer's demand curve for a particular illness.

point C on the original demand curve and the point D on the
insured demand curve. The points B and D begin to trace out the
insured demand curve. Another point—easy to find—is where
the original demand curve shows $p_m = 0$. When $p_m = 0$, Cp_m also
equals zero, so the quantities consumed are the same at a zero
market price with or without insurance (point E). Thus, at least as a
first approximation, we know that a coinsurance-type plan must go
through the same quantity-axis intercept as the uninsured demand
curve.

The effect on quantities consumed comes by comparing quantities demanded on these two demand curves (insured versus
uninsured) at a specific market price like p_{m1}. In Figure 4.9, the
uninsured person will consume m_1 at market price p_{m1}. The same
person, with the coinsurance policy in place, will consume m_2 at a
market price p_{m1}.

More generally, we can expect that the coinsurance plan will
always be found by *rotating* the insured demand curve clockwise
around the quantity-axis intercept of the uninsured demand curve.
The smaller the coinsurance rate C, the larger the rotation. At the
extreme ($C = 0$, or full coverage by the insurer), the demand curve
is a vertical line. No matter what the market price, the consumer
will always consume the same amount, because the insurance plan
pays all costs when $C = 0$. Specifically, the amount of rotation is
easy to calculate. If we were to think of the angle between the
uninsured demand curve and the vertical ($C = 0$) demand curve,
as α, then an insurance plan with some specific coinsurance rate C

Box 4.1   ELASTICITIES OF DEMAND CURVES

A demand curve shows the relationship between quantities demanded by consumers and the price, holding constant all other relevant economic variables. The *slope* of the demand curve tells the rate of change in quantity (q) as the price (p) changes. Suppose we have two observations on a demand curve, (q_1, p_1) and (q_2, p_2), where something has caused the price to change and we can observe the change in quantity demanded. Define the change in q as $\Delta q = q_2 - q_1$, and the change in p as $\Delta p = p_2 - p_1$. The rate of change in q as p changes is then $\Delta q/\Delta p$. (If we allow the change to become very small, using techniques of the calculus, we would define the rate of change at any point as dq/dp, the first derivative of the demand curve $q = f(p)$.)

The *elasticity* of a demand curve is another measure of the rate at which quantity changes as price changes. The advantage of elasticities is that they are scale-free, so you don't need to know how quantity and price were measured in order to understand the information. For example, quantity could be measured in doctor visits per 100 persons per year or doctor visits per 1000 persons per month. Changes in that measure as the price changed—the *slope* of the demand curve—would depend on how quantity was measured. To "de-scale" slope measures, we simply put everything in proportional terms. That is, instead of asking how quantity changes with price ($\Delta q/\Delta p$), we ask what the *percent* change in quantity (%dq) is in response to a *percent* change in price (%dp). That's all an elasticity is—a ratio of %dq to %dp.

Economists commonly use the Greek letter η (eta) to describe a demand elasticity, and we will follow that convention here. When we can meaningfully treat the data as having come from a demand curve of the form $q = f(p)$, with a slope dq/dp, the elasticity is defined as $\eta = (dq/q)/(dp/p) = \%dq/\%dp$. Each variable (quantity and price) is "normalized" to its own value, which makes the elasticity scale-free. (It has the dimension of a pure number, whereas a slope has the dimension of quantity/price.) Notice that the elasticity commonly will change as you move along a demand curve, so a carefully defined elasticity will also report the range of data (quantity or price, or both) to which it pertains.

In some data, only two points on the demand curve are available, in which case we usually compute an *arc-elasticity*. This measure computes the change in quantity and price relative to the average of the two observed values. Thus,

$$\text{arc-}\eta = \frac{[\Delta q]/[(q_1 + q_2)/2]}{[\Delta p]/[(p_1 + p_2)/2]}.$$

Since the 2s cancel out, the formula is commonly written as

$$\text{arc-}\eta = \frac{[q_2 - q_1]/[q_2 + q_1]}{[p_2 - p_1]/[p_2 + p_1]}.$$

This provides a way to estimate an elasticity when only two data points exist, a situation that occurs with surprising frequency in real-world data-gathering efforts.

INCOME ELASTICITY

Income elasticities measure the percentage change in quantity for a 1 percent change in income. The economics literature commonly uses the symbol E for an income elasticity. Thus, $E = \%dq/\%dI$
$$= (dq/dI) \times (I/q).$$

(say, $C = 0.2$) will rotate the demand curve by $(1 - C)$ percent of the angle α (say, 80 percent of α).

The insurance policy also makes the demand curve less elastic in general. The elasticity of a demand curve describes the percentage change in quantity arising from a 1 percent change in price. (See Box 4.1 for a summary discussion of the concept of demand elasticities.) As the appendix to this chapter proves, if the uninsured demand curve has elasticity η, then the insured demand curve has elasticity of approximately $C\eta$, where C is the coinsurance rate. When $C = 0$, the elasticity is zero—another way of saying that the consumer ignores price in making decisions about purchasing medical care. This appendix also proves that the response to changes in the coinsurance rate is just like the response to a price change. This should come as no surprise, since coinsurance operates by reducing prices. That is, if demand falls 5 percent for a 10 percent increase in price, say from $p_m = \$20$ to $p_m = \$22$ ($\eta = -0.5$), then demand will also fall 5 percent for a 10 percent increase in the coinsurance rate, say from $C = 0.4$ to $C = 0.44$. Thus, the elasticity with respect to a coinsurance rate is the same as an elasticity with respect to a price change for a person with insurance. If the price elasticity is -0.1, then the elasticity of quantity demanded with respect to the coinsurance rate is also -0.1.

One final technical addendum bears mentioning here. The demand curves just described must also be shifted inward to account for the income effects of the insurance premium. Suppose the premium costs $\$R$ per year. Paying for the insurance policy at the beginning of the year reduces the consumer's income by $\$R$ for the year. Thus, demand curves for every good, including medical care services, must be shifted inward to account for the reduced income. (See the previous discussion about income effects.) Put

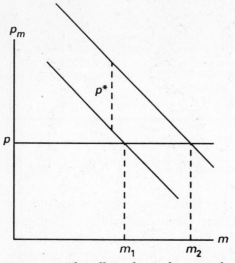

Figure 4.10 The effect of an indemnity plan on consumer demand.

differently, the demand curves must all be calculated for an income of $I - R$, instead of an income I, once the consumer has committed to buy the insurance policy.[9] Of course, if the effects of income on medical care use are small, such an adjustment would be small, and perhaps best ignored empirically.

Indemnity Insurance

Another type of health insurance, less common than coinsurance-like plans, simply pays the consumer a flat amount for each medical service consumed, where the amount is preestablished in the insurance contract. Some hospital insurance policies do this, for example, by paying (say) $125 per day in the hospital. Their effect on demand for care is easy to describe using our standard models. Suppose the insurance plan specifies that p^* will be paid each time the consumer uses a particular medical service. Then the consumer's demand curve just shifts upward by p^* for every possible quantity consumed. Figure 4.10 shows this arrangement. The quantity consumed with no insurance is m_1 and with the indemnity plan paying p^* per unit, the quantity consumed is m_2.

---

[9] Chapter 10 describes modifications to this picture when the insurance is provided through an employment-related work group, and Chapter 11 describes in detail how such insurance premiums (like R) are formed.

Deductibles

Deductibles are a common feature of many insurance plans, both in health insurance and elsewhere. A deductible is some fixed amount—say, $150—that the consumer must pay toward medical bills each year before any insurance payments are made. The deductible can be any size (ranging from $25 or $50 in some plans to over $1000 in others). The idea is to not bother insuring "small" losses, and to save insurance costs in return.

The effect of a deductible on demand is quite complicated in theory. At first blush, a deductible just puts a steplike drop in the price facing consumers, as shown in Figure 4.11. In such a world, the consumer pays the full market price if an event occurs with demand curve D_1 (for illness level l_1), and for such a minor illness, the insurance would not affect consumer demand for care at all, and the consumer will demand m_1 units of care, with or without the insurance policy in force. If a more serious illness occurs, such as l_2, then the relevant demand curve is D_2, and the consumer would pay only the predetermined coinsurance rate for all expenses above the deductible. (The figure shows 80 percent coverage above a deductible, so $C = 0.20$ for larger illnesses.) For such illnesses, the consumer will act *as if* the deductible did not exist, with the exception of a very small income effect on demand.[10] For demand curve D_2, the insured consumer will demand m_3 units of care, where without the insurance (at a market price p_m), the consumer would demand m_2 units.

Alas, life is not quite so simple if one wants really to understand the effects of an insurance policy with a deductible, because the deductible accumulates over more than one episode of illness. A simple example explains the problem more fully. Suppose the consumer has an insurance policy with an annual deductible of $200, beginning January 1. On February 15, the consumer becomes ill after a lovely Valentine's Day dinner with a date, perhaps from food poisoning. A trip to the doctor and some lab tests ($75) relieves the anxiety about the problem, which gets better. For that illness, the consumer receives no insurance payment. However, for any future illnesses during the year, the consumer has an insurance plan with only $125 remaining on the deductible. Thus, the

---

[10] Income has fallen by the amount of the deductible. However, in general, this effect would be so small as to be irrelevant empirically.

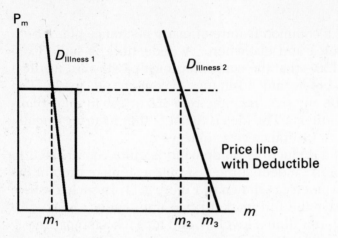

Figure 4.11 The effect of a deductible plan on consumer demand.

visit to the doctor produced not only the direct benefit of treatment, but also the added benefit of "trading in" the $200 deductible insurance plan for a $125 deductible plan. The economic value of that bonus (the "improved" insurance plan) acts as an offset to the $75 cost of the doctor visit. In a complicated fashion, the "apparent price" facing the patient falls as total spending gets nearer to the deductible amount, but the rate at which this happens varies with the time of year. (It doesn't do much good to spend your way past a deductible on December 28, if a new deductible starts up again on January 1.)

However, the clear effect of a deductible remains: Demand for care should be a lot like demand from an "uninsured" consumer for small illness events, and a lot like that from an insured consumer for large illness events. The overall effect on demand remains an empirical issue, which we discuss in the next chapter in more detail. For a more complete discussion of this problem see Keeler, Newhouse, and Phelps (1977).

Maximum Payment Limits

Some insurance plans also have a cap on the amount the insurance company will pay. Early hospital plans put this cap at 30 hospital days, after which the consumer was uninsured. Many "major medical" insurance plans also put a cap on total spending, such as $100,000 per year and $300,000 for a lifetime.

The effects of a cap on insurance are the reverse of a deductible—they make really serious medical events "uninsured." Because of this, one might think that this would be a very unpopular insurance plan, but they were quite common in the past. (See Chapter 10 for discussion). Many private insurance plans now have a "stop-loss" feature that places an upper limit on out-of-pocket spending by the consumer, even if there are copayments for some care.

Mixed-Bag Insurance

Many insurance plans pay some constant percent of all medical bills above a deductible (such as $150 per year). "Major-medical" insurance plans commonly have this feature. Some insurance plans pay with some combination of indemnity and coinsurance. An important example of this is Medicare Part B, which pays for physician services for Medicare enrollees. That plan pays 80 percent of physician bills (above an annual deductible of $70), but *only if the physician's fee is no higher than a predetermined amount*, such as $40 per visit. (The amount varies by region and type of service provided). If this occurs, Medicare Part B pays 80 percent of the maximum fee. In the example, it would pay 80 percent of $40 ($32). Thus, Part B Medicare combines a 20 percent coinsurance plan and an indemnity plan.

TIME COSTS AND TRAVEL COSTS

We should also recognize the role of time as a "cost" of acquiring medical care. As with any "service," medical care requires the presence of the patient. (Why do you think we're called "patient" while we're waiting for the doctor?) Travel to and from the doctor also creates costs, both in time and the direct travel costs.

The appropriate "value" of time to use in calculating the "time cost" of medical care can become quite complicated. For a working person with no sick leave available on an hourly wage rate, the cost of going to the doctor during the work day is obviously equal to the wage rate. If the person earns $20 an hour, and it takes 2 hours to see the doctor, he has lost $40 in wages, and the "time cost" of the doctor visit is $40. If that person has a generous sick-leave policy at work, the time cost may be much smaller, if not zero. For persons not working directly in the market (such as the proverbial "house-wife"), there is still a related time cost—the value of that person's time in the household. Since most persons working in a household (rather than market) setting have the opportunity to work in the market, we can infer that their time is more valuable to them in the home than their best market opportunity, because they have chosen the home, rather than the market, as their place of work. Thus, their time is worth *at least* as much as their market opportunity wage, and medical care visits have a time cost proportional to that amount. Of course, persons working in the household setting don't have anything like the sick leave that an employed person might have, so they bear the brunt of any time costs incurred.

Time costs act just as money costs do in affecting demand for medical care. If time costs rise (either because the actual time spent rises, or the value of time rises), then demand for medical care should fall. Sick leave from places of employment plays the same role for time costs as health insurance does for money costs. The better the sick leave policy (the better the insurance), the more medical care use we should expect.

As with other aspects of the demand for medical care, we must turn to empirical studies to determine the importance of time costs in affecting demand for care.

THE ROLE OF QUALITY IN THE DEMAND FOR CARE

Quality of care has (at least) two important facets. First, medical care "quality" assesses how well the medical care produces outcomes of improved health. Thus "quality of care" means that the medical intervention was appropriately selected and properly carried out. In a doctor office visit, for example, consumers may judge this quality by assessing the training of the doctor (good medical school? board certified? relevant subspecialty training?), or the time the doctor spends with the patient.

Consumers also place value on the "amenities" associated with medical care, since they must participate in each step along the way. In a doctor's office, this may mean that the office is neat and orderly, the magazine supply is refreshed more than occasionally, and the air conditioning works in the summer. In a hospital, the quality and diversity of the food, the quality of the TV reception,

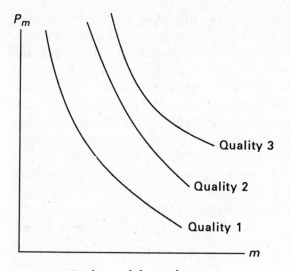

Figure 4.12 Quality and demand curves.

and the friendliness of the staff—in short, the "hotel" aspects of the hospital—are important aspects of this type of quality. Consumers should value both types of quality, although they may be able to judge the latter much better than the former. (We will return to the question of how consumers can infer the technical quality of care when we discuss licensure of physicians and other healers in Chapter 16.)

Thus, when we say that a consumer has a willingness to pay of $X for a doctor office visit, we must be certain to have in mind a specific quality of care. Perhaps the most useful idea (which will resurface later in our studies of hospital behavior more directly) is that we can actually construct an entire *family* of demand curves for medical care, each "member" of the family representing a different quality. If we think of these as willingness-to-pay curves, the idea becomes quite clear: At each quantity consumed, the rational consumer's willingness to pay will increase with quality. Thus, if we "index" demand curves on the basis of quality, the family of demand curves has a very clear structure: Demand curves sit higher up (higher willingness to pay), the higher the quality. Figure 4.12 shows a set of three demand curves for medical care (m), at three levels of quality, so that *Quality* 3 > *Quality* 2 > *Quality* 1, and so on.

We could define "quality" tautologically as those features of medical care (aside from quantity) that lead consumers to pay more.[11] Fortunately, we can avoid this approach, at least partly, in medical care, by noting that quality can denote, in part at least, the efficiency with which medical inputs produce health. Thus, for example, if more highly trained providers such as medical specialists produce health with greater precision or certainty than do lower-trained providers, then we can truly say that the more highly trained providers have higher quality. Of course, some aspects of "quality" will remain subjective, such as the "bedside manner" of a provider, but quality does have real and measurable meaning in health care, and we should expect that an informed consumer would be willing to pay more for higher quality care.

SUMMARY

The demand for medical care derives from the more fundamental demand for health itself, which produces utility. So long as medical care helps to augment health—for example, by restoring the health of a sick person—rational decision making will create systematic demand curves for medical care by individuals. These demand curves slope downward, so that less care is demanded at higher prices, other things held equal. More serious illnesses shift the demand curves for medical care-outward, so that (other things held equal) more medical care is demanded by people who are sick than by those who are well. Income (among other things) will also cause the demand curves for medical care to shift outward, so that (at constant prices) people with more income will buy more medical care. This relationship can be confounded seriously (perhaps "incredibly" would be a better word) by such things as the patterns of consumption (x_G, x_B, etc.), loss of income with illness, and other factors. At the societal rather than individual level, the relationship is further confused by such things as pollution and industrial accidents, which create more income but can also cause poor health.

Insurance coverage generally acts in some way to reduce the price of medical care, although the explicit mechanisms are very diverse. Standard economic models should prove useful in analyz-

---

[11] The old real estate joke reflects on this: "If a 'house with a view' and a 'house with no view' sell for the same price, then the 'house with a view' has no view."

ing the effects of insurance on quantities of medical care demanded.

PROBLEMS

1. For a specific consumer and for a specific illness, draw the demand curve for medical care without insurance, and then carefully draw the demand curve for the same person and illness when the person has an insurance policy that pays (1) for 50 percent of all medical costs, and (2) for 100 percent.

2. Describe the effect on demand curves for physician office visits of an insurance policy that pays a flat amount per doctor visit, for example, $25 per visit.

3. For a single consumer, show the demand curves for three illnesses, l_1, l_2, and l_3, where the degree of illness increases as the subscript increases from 1 to 3. Need they be parallel? Could they ever possibly cross?

4. For a single illness (say, sore throat and cough), show the demand curves for three different consumers with different preferences about medical care and other goods. Need they be parallel? Could they ever cross? Now aggregate those three demand curves into a demand curve for them together. If these three people constitute the entire population in a market area, then this demand curve is the "market" demand curve. Think about what would happen when you aggregated the demand curves of thousands of individuals.

5. Demands for medical care are specific to each illness, and each illness can create demand for more than one type of medical care (e.g., hospitals, doctors, drugs). Should we think about demand curves that are specific to the illness (e.g., sore throat) or the medical intervention (e.g., office visit, antibiotic injection)?

6. When the quality of a medical service rises, what happens to the demand curve for that service? What are some relevant dimensions of quality?

7. Describe the concept of consumer surplus in terms intelligible to a person untrained in economics.

8. Suppose that there's only one type of medical care (office visits) and that it costs $20 per visit. Consider now an insurance policy with a $100 deductible that pays for 80 percent of the consumer's medical costs after that. The price schedule confronting that consumer has a step in it at a quantity (five visits) such that total spending (price × quantity) equals the amount of the deductible ($100). Draw such a price schedule. Now draw a consumer's demand curve that intersects the price schedule at two visits (on the top step), at five visits (somewhere in the vertical portion of the price schedule), and at seven visits (on the lower stairstep). Question: What is the optimal quantity for this patient to consume, two visits, five visits, or seven visits?

9. What would happen to demand curves for medical care if consumer income increased? (Don't forget the effects of income on consumption of health-affecting commodities such as running shoes, cigarettes, fatty foods, seat belts, etc.)

10. What would happen to demand curves for medical care if the quality of that care increased? What if it increased in ways that were unobservable to the consumer (e.g., a reduction in the probability of adverse side effects)?
11. Suppose a demand curve has the form $q = 100 - 10p$. What is the quantity consumed at $p = 5$? What is the elasticity of the demand curve at $p = 5$?

APPENDIX

Demand Curves and Demand Elasticities

The price consumers pay for medical care with a simple insurance policy is $C \times p_m$, where C is the coinsurance rate and p_m is the market price. The insurance policy costs \$$R$ per year. Thus the budget constraint for the consumer with insurance, at the time medical care will be purchased, is $I - R = p_x x + C p_m m$. We have seen in the text how this can lead to a demand curve for the consumer of the form

$$m = f(\text{Income, Price, Illness level})$$

Suppose we take a linear approximation to such a curve, so

$$m_{ni} = \alpha_0 + \alpha_1 \text{Income} + \alpha_2 \text{Price} + \alpha_3(\text{Illness Level})$$

The derivative of m with respect to Price is α_2, and the price elasticity of the demand curve is $\alpha_2 \times \text{Price} \div m_{ni}$, where m_{ni} means the amount of m purchased with no insurance (ni), and of course this depends on income, price, and so on. If the consumer obtains an insurance policy with coinsurance C, then everywhere that "price" appears in the demand curve is replaced by $C p_m$, and income is replaced by $I - R$. Thus the demand curve with insurance (m_{wi}) is

$$m_{wi} = \alpha_0 + \alpha_1((I - R) + \alpha_2(C \times \text{Price}) + \alpha_3(\text{Illness Level})$$

The derivative of m_{wi} with respect to Price is $\alpha_2 C$, and the elasticity is $\alpha_2 C \times \text{Price} \div m_{wi}$, where $m_{wi} > m_{ni}$.

Similarly, the derivative of m_{wi} with respect to C is $\alpha_2 \times \text{Price}$, and the elasticity is $\alpha_2 \text{Price} \times C \div m_{wi}$, the same as the derivative with respect to m_{wi} with respect to price. Thus the elasticities with respect to price and coinsurance are equal.

Of course, as the text discussed, when more than one type of medical care is available, the demand for one type of care may depend on the prices (and insurance coverage) of other types of care. Suppose that two types of medical care (say, ambulatory and inpatient) could treat the same patient's illness. Then we would have *two* demand curves, here shortening Price to "P" and using subscripts of "A" for ambulatory and "I" for inpatient:

$$m_I = \alpha_0 + \alpha_1(I - R) + \alpha_2(C_I \times P_I) + \alpha_3(C_A \times P_A) + \alpha_4(\text{Illness level})$$

and

$$m_A = \beta_0 + \beta_1(I - R) + \beta_2(C_I \times P_I) + \beta_3(C_A \times P_A) + \beta_4(\text{Illness level})$$

Demand for each type of care would depend on its own price and the "cross price" of the other type of care. As discussed in the text, if the goods are substitutes, then α_3 and β_2 would have positive signs, and negative signs if the goods were complements. This idea obviously generalizes to the case of multiple types of medical care.

Notice that each type of care can have its own coinsurance rate. Hospital and ambulatory care commonly do have differing coinsurance rates, as do dental care and other specific types of care. The "benefits package" of an insurance plan determines the coinsurance rates for a given plan.

Proof that aggregate demand curve elasticity is quantity-weighted average of individual elasticities:

Define $M = \sum_{i=1}^{n} m_i$ as the aggregate spending for a society composed of n individuals, each with demand equal to m_i, rate of change in demand $dm_i/dp = \beta_i$ and demand elasticity ηi. Now, the change in M with respect to p is

$$\frac{dM}{dp} = \sum_{i=1}^{n} \beta_i = \sum \eta_i \times (m_i/p)$$

so

$$\eta = \frac{dM/M}{dp/p} = \sum_{i=1}^{n} \eta_i \times (m_i/M) = \sum_{i=1}^{n} s_i \eta_i$$

where $s_i = m_i/M$ is the ith person's share of the total quantity M. Thus, aggregate demand elasticities are a quantity-weighted mixture of individual demand elasticities.

Chapter
5

Empirical Studies of Medical Care Demand and Applications

*T*he previous chapter developed a conceptual model of the demand for medical care, showing how we can move from the basic model of utility for health (unobservable) to an observable demand curve. The key relationships that we expect to find include the following predictions about quantities of medical care demanded, in each case with all other relevant factors held constant. Quantity demanded should increase:

as price falls,

with illness severity,

with the generosity of insurance coverage (either with lower copayment or lower deductibles),

as the cost of time decreases,

as the time used to obtain care decreases, or

with age, for adults.

We also know that income increases demand in aggregate for all goods and services, but while there is no specific prediction for any good or service, including any specific type of medical care, higher income may well increase demand for care. Finally, when the prices of two services (like hospital care and office visits) move separately, or are insured differently, then how the use of one service changes as the price or insurance coverage of the other changes will depend upon whether they are complements or substitutes. With this information in hand, we can now turn to studies of how people actually have used medical care, and how these various factors actually affect demand.

STUDIES OF DEMAND CURVES

A number of economists have estimated demand curves for various types of medical care in various settings using various types of data available in the literature or (occasionally) collected for the purpose. These provide an unsettlingly wide range of information. The estimated price elasticity of demand found in these studies varies by an order of magnitude or more from the largest to the smallest. For example, this past literature finds estimates of the price elasticity for hospital days ranging from -0.67 (Martin Feldstein, 1971) to -0.47 (Davis and Russell, 1972) to 0 for some illnesses (Rosenthal, 1970). For physician care, estimates ranged from -0.14 (Phelps and Newhouse, 1972; Scitovsky and Snyder, 1972) to -1 (for hospital outpatient visits, Davis and Russell, 1972). One study reported an overall elasticity of demand for medical care as large as -1.5 (Rosett and Huang, 1973). Many of these studies had complicated statistical problems embedded in them that made their results difficult to interpret (Newhouse, Phelps, and Marquis, 1980), and the wide diversity of estimates made it difficult to know what to make of any of them individually. Perhaps the only agreement in the literature by the mid-1970s was that "price mattered."

In part because of the difficulties found in previous studies, the federal government launched in the mid-1970s a randomized, controlled trial of health insurance to learn better the role of insurance in affecting demand for care, the Health Insurance Study (HIS) conducted by the RAND Corporation. For many purposes, we can treat these results as the best available, since the statistical difficulties surrounding previous studies are essentially eliminated in the RAND HIS analysis, and the sample sizes far exceed those available in most studies. Thus, while other estimates exist, we can focus here on the RAND HIS results for most types of medical care.

The RAND Health Insurance Study (HIS)

The RAND HIS was one of several large social science experiments conducted under federal government auspices in the 1970s. This study basically followed standard "laboratory" experimental design methods, modified only as necessary for ethical and administrative purposes. A total of 5,809 enrollees was chosen from four cities (Dayton; Seattle; Charleston, S.C.; and Fitchburg, Mass.) and two rural sites (one in South Carolina, the other in Massachusetts). Enrollees were asked to participate for either three or five years,[1] and to give up the use of any insurance plan they might be holding, using only the randomly assigned insurance of the HIS for that period. The total number of person-years available for analysis was 20,190. The eligible population consisted of persons under 65 who were not institutionalized.[2] At enrollment, the persons were told which insurance plan they would use from a set including (1) full coverage ($C = 0$) for all services, (2) 25 percent copayment ($C = 0.25$) for all services, (3) 50 percent copayment for all services, (4) 50 percent copayment for dental and mental services ($C = 0.25$ for other care, $C = 0.50$ for dental and mental services), (5) an individual deductible *for ambulatory care only* of $150 per person in the family ($450 total per family),[3] and (6) essentially no coverage until a catastrophic cap had been reached (5, 10, or 15 percent of family income, subject to an overall maximum of $1000). This plan actually had a 5 percent payment rate ($C = 0.95$) in order to provide an incentive for families to file claims, from which medical use information was derived. Separately, some persons were enrolled in a health maintenance organization (HMO) in Seattle, the results of which are discussed in Chapter 11.

In every plan, the enrollees had a cap on their financial risk, just as for the "catastrophic coverage" plan, so that no family ever had to spend more than 5, 10, or 15 percent of their income (up to $1000 maximum) on medical care out of pocket. (If the 15 percent number seems large, recall that over 10 percent of the GNP is spent on medical care.) If the family's medical spending exceeded that limit, they received full coverage ($C = 0$) for all remaining expenses for the year. Of course, a family on the $C = 0.25$ plan was much less likely to reach such a cap than a family with the $C = 0.95$ plan.

---

[1] The different time horizons gave the opportunity to learn whether the relatively short time horizon of the study affected results. In general, it did not, with the exception of dental care use, which was abnormally high in the study's first year for persons on the plans with most complete coverage.

[2] The over-65 population was excluded because government rulings prohibited Medicare participants from waiving their right to Medicare coverage. Since all enrollees had to stop using their previous insurance in order to participate, this ruling eliminated the possibility of studying those over 65.

The study also eliminated persons from the very highest part of the income distribution, mostly for political appearances; the study was initially funded by the Office of Economic Opportunity, the administration's program for combating poverty.

[3] This plan was designed to test whether inpatient and outpatient care are complements or substitutes. We will see the results below.

The "cap" was put in place to provide people with an incentive to enroll, and was coupled with a "participation incentive" payment equal to the maximum risk the family faced. For example, if a family had $15,000 income and had a 5-percent-of-income catastrophic cap, they had a maximum risk of $750, and they would receive $750 just for participating in the experiment. This feature of the experiment had positive ethical properties (nobody was made worse-off financially for participating), and good experimental design properties (it successfully eliminated almost all enrollment refusals and attrition during the experiment), but it obviously complicated the analysis somewhat, compared with plans that might have been used. However, subsequent analysis based on episodes of illness (Keeler, Buchanan, Rolph, et al., 1988) rather than on simple plan comparisons allows one to project the results from the HIS for other complicated insurance plans, some with designs quite different from the small set of experimental plans.

The primary results from the HIS compare means across the plans, the simplest approach, and one allowed by the randomized-trial design. Regression analysis to control for other factors increases the precision of the estimates, and also shows the effects of other variables (such as age and income) that we will explore shortly. Table 5.1 shows the basic results for outpatient care and total medical spending. Table 5.2 converts these data into arc-elasticities (see Box 4.1 on elasticities).

More refined analysis of the HIS data has taken account of the catastrophic expenditure cap, using an analysis based on episodes of illness, rather than annual spending (Keeler, Buchanan, Rolph, et al., 1988). It breaks down the price response into two components: the number of episodes, and the spending per episode. Table 5.3 shows the resultant arc-elasticities broken down more finely by type of care received. In general, the price responsiveness of medical services is still fairly small, but larger than the simple comparison of plan means shows.[4] The patterns of demand elasticities correspond to intuition about "medical necessity." Demand for hospital care is least price-responsive (particularly at higher coinsurance rates), "well care" (preventive checkups, etc.) is most price-responsive, and acute and chronic outpatient care fall

---

[4] We should expect this to occur, because of the catastrophic cap built into the HIS plans. For serious illness events (large medical spending), all of the plans provided full coverage. Thus, there is not as much "difference" between the plans as there would be without the catastrophic cap, and differences in medical use will be somewhat compressed.

Table 5.1 SAMPLE MEANS FOR ANNUAL USE OF MEDICAL SERVICES PER CAPITA

Plan	Face-to-face visits	Outpatient expenses (1984 dollars)	Admissions	Inpatient dollars (1984 dollars)	Prob. any medical (%)	Prob. any inpatient (%)	Total expenses (1984 dollars)	Adjusted total expenses (1984 dollars)
Free	4.55	340	0.128	409	86.8	10.3	749	750
	(0.168)	(10.9)	(0.0070)	(32.0)	(0.817)	(0.45)	(39)	(39)
25%	3.33	260	0.105	373	78.8	8.4	634	617
	(0.190)	(14.70)	(0.0090)	(43.1)	(1.38)	(0.61)	(53)	(49)
50%	3.03	224	0.092	450	77.2	7.2	674	573
	(0.221)	(16.8)	(0.0116)	(139)	(2.26)	(0.77)	(144)	(100)
95%	2.73	203	0.099	315	67.7	7.9	518	540
	(0.177)	(12.0)	(0.0078)	(36.7)	(1.76)	(0.55)	(44.8)	(47)
Individual deductible	3.02	235	0.115	373	72.3	9.6	608	630
	(0.171)	(11.9)	(0.0076)	(41.5)	(1.54)	(0.55)	(46)	(56)
Chi-squared (4)	68.8	85.3	11.7	4.1	144.7	19.5	15.9	17.0
P value for chi-squared (4)	<0.0001	<0.0001	.02	n.s.	<0.0001	0.0006	0.003	0.002

Note: All standard errors (shown in parentheses) are corrected for intertemporal and intrafamily correlations. Dollars are expressed in June 1984 dollars. Visits are face-to-face contacts with MD, DO, or other health providers; excludes visits for only radiology, anesthesiology or pathology services. Visits and expenses exclude dental care and outpatient psychotherapy. n.s. = not significant.

Source: Manning, Newhouse, Duan, et al., 1987.

Table 5.2   ARC-ELASTICITIES FOR DEMAND

Range of nominal coinsurance variation	Range of average coinsurance variation	All care	Outpatient care
0–25%	0–16	0.10	0.13
25–95%	16–31	0.14	0.21

Source: Manning, Newhouse, Duan, et al., 1987.

Table 5.3   ARC PRICE ELASTICITIES OF MEDICAL SPENDING

Coinsurance range	Outpatient			Total outpatient	Hospital	Total medical	Dental
	Acute	Chronic	Well				
0–25	0.16	0.20	0.14	0.17	0.17	0.17	0.12
	(0.02)	(0.04)	(0.02)	(0.02)	(0.04)	(0.02)	(0.03)
25–95	0.32	0.23	0.43	0.31	0.14	0.22	0.39
	(0.05)	(0.07)	(0.05)	(0.04)	(0.10)	(0.06)	(0.06)

Note: Standard errors are given in parentheses. For their method of computation see Keeler, et al., 1988.

Source: Keeler, Buchanan, Rolph, et al., 1988.

in between.[5] For all medical services, the price elasticity is estimated to be -0.17 in the range of nearly complete coverage ($C = 0$ to $C = 0.25$), and -0.22 for higher coinsurance rates. The largest price response found was -0.43 for "well care" in the higher coinsurance range. As a summary measure, we can now say with considerable confidence that, while medical care use does respond to price, the rate of response is fairly small, compared with some other goods and services, with elasticities generally in the range of -0.1 to -0.2 for most medical services. The study by Keeler, Buchanan, Rolph et al, estimated the expenses for persons with a number of plans, including "no insurance," although such a plan did not actually exist in the HIS. Converting the results from their Table 5.7 to 1990 spending levels using the Medical CPI, the uninsured person is predicted to spend $830 per year, and the fully insured person $1445. Thus, their models say that a person with full coverage will spend about 75 percent more per year on medical services than a completely uninsured person. This same phenomenon occurs in very diverse settings, including even in rural China, as Box 5.1 shows.

These results stand in sharp contrast to much of the previous

---

[5] The definition of acute, chronic, and well care comes from a checkoff box on the insurance form, filled out by the physician.

Box 5.1 DEMAND CURVES IN RURAL CHINA

We should not expect to find demand curves just in the United States or other "developed" countries. The ideas of consumer demand theory ought to carry through just as well for low-income countries, even with very different ways of producing medical care. One study (Cretin, Duan, Williams, Gu, and Shi, 1988) shows how the idea of demand curves carries even into a setting of rural Chinese medicine.

In rural China, most peasants have no health insurance at all, but rather receive medical care from village or township health stations. During the last decade of reform, economic reforms included both a shift to an incentive-based agricultural system (which greatly expanded food output) and the introduction of fee-for-service medicine by village doctors. Local Cooperative Medical Funds in various regions provided the equivalent of "health insurance" to workers in their regions.

Fortuitously for our purposes, these programs have selected different levels of payment toward the fees charged by village doctors. A sample of persons within these different health plans measured their average medical care use, effectively at different prices for medical care. Some of the plans provided essentially full coverage, others paid for about half of the costs of care, some had less complete payment, and some had no insurance at all. The estimated annual outpatient expense per person appears in the table that follows:

RURAL CHINESE MEDICAL CARE USE AS A FUNCTION OF INSURANCE COVERAGE

PERCENT OF COSTS PAID BY INSURANCE	PER CAPITA OUTPATIENT EXPENDITURE (YUAN/YEAR)
0	15.36
10	17.16
20	18.96
30	21.12
40	23.52
50	26.04
60	29.52
70	33.12
80	36.96

The data on average use versus coinsurance (fraction paid by the Chinese peasant) shows the systematic effects of price on use, just as economic theory predicts. The elasticity of demand in rural China is about -0.6 for ambulatory medical care, based on these data. By way of comparision, 1 yuan is approximately $.25. Thus, annual expenses for ambulatory care of 36 yuan are approximately the same as $9 in U.S. currency.

Similarly, hospital use also varies by the proportion of hospital care paid for by insurance. The estimated probability of hospital use varies by the coinsurance rate as follows:

PERCENT OF COSTS PAID BY INSURANCE	PROBABILITY OF HOSPITAL ADMISSION
0	0.018
25	0.023
50	0.029
75	0.036
100	0.045

Again, the systematic effect of insurance is clear. The price elasticity of demand for hospital care among Chinese peasants is about −0.4 as estimated from these data.

As with U.S. data, demand for hospital care (presumably more serious illness) is less price-sensitive than for ambulatory care.

nonexperimental literature, as might well be expected, since (with a few exceptions) the previous literature confronted serious statistical problems. The few older studies that relied on "natural experiments" prove in retrospect to be remarkably close to the "gold standard" of the HIS,[6] while those relying on aggregated data now seem in retrospect to be far off target, some by as much as an order of magnitude.

Are Hospital and Outpatient Care Substitutes or Complements?

The HIS was designed to show whether hospital and ambulatory care are complements to one another (like tires and gasoline) or substitutes (like automobiles and airplanes). The $150 individual deductible provides just this test, because the deductible applied only to ambulatory care. Enrollees on this plan had full coverage

---

[6] Most notable was the data collected by Anne Scitovsky and Nelda Snyder from a natural experiment on Stanford University employees (Scitovsky and Snyder, 1972). In a multiple regression analysis of those data (Phelps and Newhouse, 1972), the estimated elasticity for outpatient care in the range of $C = 0$ to $C = 0.25$ was −0.14, remarkably close to the estimate of −0.17 in the HIS. The Scitovsky and Snyder data did not feature a catastrophic cap on spending.

for all hospital care. Thus, any differences in hospital use between that group and the overall full coverage plan ($C = 0$ for all care) can answer this important question. Table 5.4 shows hospital admissions and spending by plan. The groups with the $150 individual deductible for ambulatory care had only 10 percent fewer hospital admissions and 9 percent less total cost of hospital care than the full coverage group. While the statistical precision is not terribly strong, these results say that the two services are *complements*, not substitutes. That is, as the insurance generosity for ambulatory care fell, both ambulatory and hospital use fell.

Dental Care

The HIS data also show demand elasticities for other services, including dental care, mental illness care, and emergency room care. For dental care, previous studies provided sparse information on price responsiveness, all with potential methodological problems (Phelps and Newhouse, 1974). If anything, these data suggested that demand for dental care would be more price responsive than for other medical services. The HIS results confirm this, at least partly, as Table 5.5 reveals. For higher levels of coinsurance ($C > 0.25$), the arc elasticity of dental care demand is $-.39$, larger than all other medical care except well care. However, in the $C = 0$ to $C = .25$ range, dental demand had estimated elasticity of $-.12$, lower than most other medical services, including hospital care. Dental care was one area where behavior differed between early experimental years and later years. Intuitively, people can "store up" dental visits easier than other types of care. Those patients who went from relatively poor dental insurance onto a generous HIS plan ($C = 0$ or $C = .25$) would have a "bargain basement" price for dental care during the experiment. We should

Table 5.4 HOSPITAL USE IN THE HIS

Plan	Admissions per year	Inpatient cost (1984 dollars)
$C = 0$	0.128	$409
$C = 0.25$	0.105	$373
$C = 0.5$	0.092	$450
$C = 0.95$	0.099	$315
Individual deductible $150	0.115	$373

Source: Manning, Newhouse, Duan, et al., 1987.

Table 5.5   ANNUAL SPENDING RELATIVE TO $C = 0.95$ PLAN
($C = 0.95$ Plan = 100)

Insurance plan (nondental/dental)	Year 1		Year 2	
	Nondental	Dental	Nondental	Dental
Free/free	$200	$252	$177	$152
25%/25%	$145	$158	$128	$109
25%/50%	$144	$181	$122	$98
50%	$111	$118	$105	$112
95%	$100	$100	$100	$100
$150 per person deductible	$143	$163	$124	$94

Source: Manning, Benjamin, Bailit, and Newhouse, 1985.

expect them to take advantage of that by correcting any backlog of dental problems. Table 5.5 shows that this is exactly what happened. In Year 1, the difference in spending for any of the better insurance plans versus the 95 percent coinsurance plan was greater for dental care than for other medical care, and the differences become more pronounced, the better the insurance coverage. Yet in Year 2, the differential spending in the better-insured plans was almost always less for dental care than for "regular" medical care. Once the "backlog" of dental problems is dealt with (in Year 1), people with good insurance seem not to purchase much more dental care than those with poorer insurance, with the notable exception of the full coverage ($C = 0$) plan. These results suggest the importance of some coinsurance to control spending within dental plans, even if relatively modest in size.

Effects of a Deductible

The HIS results allow a simulation of the spending patterns under insurance policies that were not actually tested, such as a plan with 25 percent coinsurance and a $100 deductible or a $500 deductible. As briefly described earlier, the expected effects of such a plan are complicated, with the one obvious quality that higher deductibles should lead to lower total costs for care, because higher deductibles make the price of medical care higher for more types of illness events. Table 5.6 shows the simulation results from the RAND analysis (Keeler et al., 1988), relying on the estimated responses from the actual experimental plans and mathematically interpolating and extrapolating as needed. The results seem quite

Table 5.6   PREDICTED TOTAL MEDICAL CARE COSTS FOR INSURANCE
PLANS WITH DEDUCTIBLES

Deductible	Annual spending by type of medical treatment				
	Hospital	Acute	Well	Chronic	Total
$0 (free care)	$400	$226	$68	$148	$842
$50	$387	$166	$47	$112	$713
$100	$384	$152	$43	$104	$682
$200	$379	$136	$37	$92	$644
$500	$376	$121	$33	$83	$613
$1000	$291	$114	$32	$78	$515
$2000	$283	$111	$31	$76	$501
No insurance	$280	$109	$30	$73	$483

Source: Keeler, Buchanan, Rolph, et al., 1988.

striking. The initial effect of a small deductible ($50) is consider-
able, and then little happens to demand until the deductible gets
in the range above $500, particularly for hospital care.

We can compute the elasticity of demand in response to changing a deductible, just as we can for a coinsurance rate, a price, or
income. Using the data in Table 5.6, for example, shows that the
elasticity of total expenses with respect to the deductible is −0.07,
and for hospital expense, −0.04.[7] Hospital care does not respond
very much until the deductible exceeds $500. Acute care falls off
rapidly with the first small deductible, and then smoothly (and
slower) thereafter. Total ambulatory care is near the level of an
uninsured person by the time the deductible has reached $500. A
deductible of only $150 moves a person halfway between the total
medical use of a fully insured person and a person with no insurance.

To preview the results in Chapter 10 on demand for insurance,
we can simply note here that the cost-saving effects of a moderate-
deductible plan are similar to those of coinsurance plans that confront the consumer with somewhat larger financial risk. Thus,
based on these experimental results, the "best" types of insurance
plans may include deductibles of moderate size.

---

[7] These estimates come from regressions with the logarithm of total expense (or hospital
expense) regressed on the logarithm of the deductible.

Income Effects

The HIS results allow estimates of the effects of income on demand for care, including an interaction of income and insurance plan generosity. Both common sense and full use of economic theory suggest that pure income effects should be small, if not zero, with full-coverage insurance.[8] The HIS data in general show small positive effects of income on medical care use, with the exception of hospital admissions. Comparisons of use rates by the lowest, middle, and upper third of the income distribution shows a slight U-shaped pattern of care for most services, with low- and high-income families using the most care, and middle-income families the least. In part, this reflects the effects of illness on family income. In particular, persons with higher income enter the hospital less frequently than those with lower income. Using data from the RAND study, the calculated income elasticities for the number of episodes of illness appear in Table 5.7. A quick summary number would suggest that the income elasticity of demand is 0.2 or less for medical care.

When considering the effects of income on demand, we must be careful to recognize that cross-sectional studies such as the RAND HIS hold constant the available medical technology (broadly defined, including equipment, medicines, surgical techniques, etc.). As income increases, so will the demand for new approaches to healing, and technical change that emerges will alter the patterns and amounts of medical care demanded. Indeed, much of the growth in health care spending in the years since World War II has surely come from technical change.

One way to understand how important technical change is in the long run is to see how different the estimates are of income elasticities taken from time series data, which incorporate the effects of technical change, rather than taking technology as fixed (as a cross-sectional study does). Several studies have done this, and the estimated income elasticities of demand are generally much higher than those found in the HIS. Another way to derive such an

---

[8] Even with full coverage for medical care, there might be a positive income effect on demand. This can occur because better health can raise the value (marginal utility) of other goods (x), and greater income allows the purchase of more x. For example, because bicycling and golf are more enjoyable when you're healthy, higher-income people might seek more medical care so they can enjoy their sporting activities more.

Table 5.7   INCOME ELASTICITIES FOR EPISODES

Type of care	Income elasticity for number of episodes of illness
Acute	0.22
Chronic	0.23
Well care	0.12
Dental	0.15
Hospital	not significant

Source: Calculated from Keeler et al., 1988, Table 3.6 (for average numbers of episodes) and Table E.1 (for coefficients of income in regression equation). Estimated equation used SQRT (Visits + 0.375) as dependent variable and log(Income) as explanatory variable. Elasticity is calculated as β SQRT (Visits + 0.375)/Visits, where β is the coefficient in the estimated regression.

estimate is to compare the incomes and medical care use across different countries. A number of studies, discussed in detail in Chapter 17, have found an income elasticity of demand for health care of over 1.0.

EFFECTS OF AGE AND GENDER ON DEMAND

As the discussion of aging in Chapter 1 suggests, as people "wear out" faster (get older), they should demand more medical care. Almost any large data set showing people's medical care use, age, and gender will demonstrate this phenomenon. In addition, systematic differences occur between males and females, particularly for women in childbearing years (they use more care). For older adults, the situation is reversed. For example, for Medicare patients, women of any specific age use less care than men, and indeed women live substantially longer than men, on average, in our society.

One portrayal of this phenomenon is the number of hospital days used per 1000 persons per year, by age and sex. Figure 5.1a portrays the pattern of use, showing the rapid acceleration of use above age 50, and the substantial and widening gap between women and men in the over-65 population. The higher use of women between ages 20 and 50 is almost all due to hospitalizations for childbirth and related events. Similar patterns of use appear for ambulatory care in any data set. Figure 5.1b shows the relationship for physician office visits, the largest component of ambulatory care.

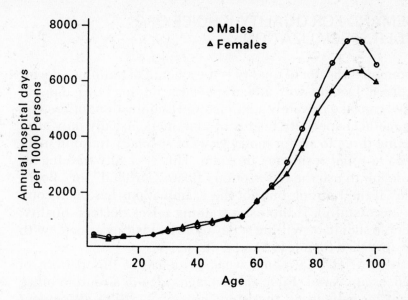

Figure 5.1a Medical care use in hospital days. *Source:* Health Care Financing Review, Spring 1986.

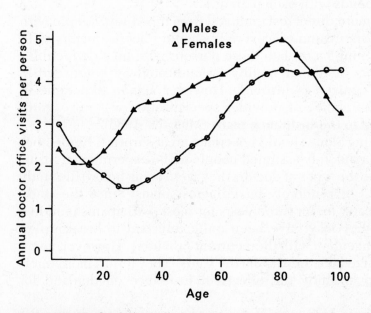

Figure 5.1b Medical care use in doctor visits. *Source:* Health Care Financing Review, Spring 1986.

THE DEMAND FOR QUALITY: CHOICE OF
PROVIDER SPECIALIZATION

One common indicator of quality is the amount of training a doctor
has received. Doctors with added specialty training beyond medical school can take privately administered examinations offered by
various medical specialty boards of examiners, usually only after
completing three to six (or more) years of specialty training in an
approved hospital residency program. This specialty training allows a doctor to use the description "Board Certified" (or "Board
Eligible" if the training but not the examinations has been completed successfully). If the extra training raises doctors' quality,
then people should be willing to pay more for visiting a doctor with
that training than for a less trained physician.

In the RAND HIS, some weak evidence of this pattern of
preferences appeared. First, the average price of a routine office
visit paid by patients in that study was about 10 to 20 percent
higher for a specialist in internal medicine (an "internist") than for
a general practice doctor, who has not received any particular
specialty training. This provides one estimate of the aggregate
willingness to pay for higher quality, although it is a weak comparison. (For example, it doesn't say anything about how much time
the doctor spends with each patient.)

Another more direct test emerged from this study. People with
more generous insurance coverage should choose doctors with
specialty training more than those persons with high deductibles
or coinsurance if higher quality (as indicated by board certification) is recognized as desirable. However, among all patients in
the study, the choice of provider specialization was not significantly related to the insurance plans (Marquis, 1985).

This finding alone should not create great surprise. Remember
that the HIS randomly assigned people to the experimental insurance plans, so the types of doctors they were using before the study
began should be randomly mixed across plans. Since the study
would only carry on for three years for most participants (and five
for others), they may have been quite reluctant to break up an
existing arrangement with their current providers. However, some
patients changed doctors during the study. They form a more interesting group to study the effects of insurance on demand for
quality.

Among those HIS enrollees who changed doctors during the
study, any effects of insurance generosity on provider choice were

still weak, at best. Only among children in the study did statistically significant effects appear; those with full-coverage insurance were 10 to 20 percent more likely to have a specialist (usually a pediatrician) as the new provider than were persons in the other insurance plans (Marquis, 1985).

On net, we know little empirically about the role of insurance in demand for quality. All of our intuition tells us that those with better insurance would more likely seek higher-quality providers, both for "medical" quality and for amenities. Unfortunately, direct studies of this question are more difficult to find.

OTHER STUDIES OF DEMAND FOR MEDICAL CARE

A wide unstudied frontier still remains to explore more about the demand for various types of medical care, and the demands in various settings. Demand will vary with the income and education of the population, the quality of care, and the institutional settings in which care is delivered. Thus, it is useful to know how to find data that produce useful information on consumer demand. The following section provides some examples.

NATURAL EXPERIMENTS I: THE DEMAND FOR PHYSICIAN CARE

Sometimes, the real world conducts a "natural experiment" that allows careful study of the demand for medical care. Perhaps the classic of such studies resulted from the work of Scitovsky and Snyder (1972) at the Palo Alto Medical Clinic. Anne Scitovsky learned that Stanford University had changed its employees' insurance coverage because the insurance was costing more than the university wished to pay. The university had the choice of either making the employees pay more for the insurance, or else changing the insurance plan's structure to make it less costly. Fortunately (from a natural experimentalist's point of view), they chose the latter. They raised the coinsurance for doctor office visits from $C = 0$ (full coverage) to $C = 0.25$ (25 percent copayment). Scitovsky and Snyder studied the medical care use (using insurance claims data) for the year before the change (1966) and the year after the change (1968). *If nothing else had changed in the picture*, any differences in use of medical care should be due to the increase in copayment.

The results (reported in detail by Scitovsky and Snyder, 1972, and Phelps and Newhouse, 1972, in a different type of statistical analysis) closely match those found over a decade later in the RAND HIS. The average number of doctor office visits fell by a quarter, and the use of diagnostic tests fell by 10 to 20 percent (depending on the type of test). The change was about the same for "preventive" annual physical examinations as for other types of care.

The potential weakness of any such "natural experiment" is that something else might have changed during the period in question, complicating the analysis. For example, if 1966 had seen an unusually large flu epidemic, then use in 1966 would have been abnormally high, and (had it been a "normal" year), the apparent effect of the coinsurance in reducing demand would have been even greater. Of course, the reverse is also true; if 1968 had an unusually large epidemic of flu, then we might have seen no effect at all, and incorrectly concluded that price "didn't matter." In the Scitovsky and Snyder data, no such unusual events seemed obvious upon study of overall health care use in the region, but it always represents something to be alert for in studying natural experiments. In this case, a repeated analysis five years later by the same two researchers found that the effects persisted almost unchanged, lending further strength to the validity of their earlier work (Scitovsky and McCall, 1977).

NATURAL EXPERIMENTS II: THE DEMAND FOR DRUGS

Mississippi Medicaid Study

Another natural experiment occurred when the Medicaid program in Mississippi began to pay for prescription drugs at a specific time.[9] A study compared the prescription drug use of Medicaid recipients for three months before and three months after the drug coverage was instituted. They obtained the data from the records of every pharmacy in town, and verified the results by looking

---

[9] Medicaid is a state-operated health insurance program for low-income citizens. Each state sets its own program scope and eligibility, within guidelines established by federal law, and the federal government helps the states by sharing in the cost of their programs. Chapter 13 discusses these programs.

directly at Medicaid records. The study showed large effects of the insurance coverage on drug use (Smith and Garner, 1974). The results appear in Table 5.8.

The number of prescriptions almost doubled, the expense per prescription increased by 25 percent, and total expense on prescription drugs more than doubled. If one computes the arc-elasticity of demand over this range of price (full price versus 0 price), the estimate is about −0.3. Interpolated down to show the effects of a 25 percent coinsurance rate (versus full coverage), the predicted effect would be about a 15 percent difference in use.

Canadian Insurance Plan Study

In a completely different setting, an association of pharmacists in Windsor, Ontario, Canada, offered a prescription drug plan providing full coverage for each prescription above a $.35 deductible. (This was in 1958; a comparable amount now would be about $1.25.) The study compared per capita use in the overall community with the use of those who purchased the insurance plan.

The results are remarkably similar to those found in the Medicaid study: About twice as many prescriptions were filled, the average cost of each prescription was higher (but only by 7 percent in this case), and total expenses were about twice as high.

One remarkable feature about this result (or its similarity to the Mississippi Medicaid result) is that we could well have expected

Table 5.8   DEMAND FOR DRUGS IN A NATURAL EXPERIMENT
(Comparative Drug Use Before and After Medicaid Coverage of Drug Expense)

Items	Three months before Medicaid[a]	Three months after Medicaid[b]
Number of prescriptions per person	5.43	9.48
Expense per prescription	$3.58	$4.49
Expense per person	$18.96	$42.54
Sample size	241	241

[a] Data obtained from records of all pharmacies in town of study.

[b] Data obtained from State of Mississippi Medicaid files.

Source: Mickey C. Smith and Dewey D. Garner, "Effects of a Medicaid Program on Prescription Drug Availability and Acquisition," Medical Care 12:7 (1974), 571–581.

an even bigger difference. The drug insurance plan would more likely be purchased by persons who planned to buy a lot of prescription drugs. (See Chapter 10, discussing the demand for insurance.) This is always a hazard with apparently natural experiments; sometimes the "subjects" of the experiment get to choose which "treatment" they receive (as was the case in this drug insurance plan), and this has the risk of contaminating the results.

TIME SERIES INFORMATION: THE DEMAND FOR DRUGS YET AGAIN

We have still another source of information available that can sometimes shed light on the nature of demand for medical care— the behavior of aggregate data over time. One example of this occurred in the British National Health Service (BNHS). Over the span of 15 years, the BNHS changed the price it charged patients for prescription drugs on a number of occasions, primarily as budget or cost-control choices. The price varied from zero to 2.5 shillings per prescription.[10] The payments by patients in the BNHS represented anywhere from zero to 23 percent of the total cost of the drugs used ($C = 0$ to $C = 0.23$). Using the techniques of multiple regression analysis, Phelps and Newhouse (1974) estimated the effect of the out-of-pocket payments on total drug use within the BNHS. (See Box 5.2 for a discussion of this statistical technique.)

The results closely match those of the Mississippi Medicaid and the Windsor, Ontario, prescription drug studies. On average, total drug use would be 15 percent higher at full coverage ($C = 0$) than with a 25 percent coinsurance ($C = 0.25$).

NATURAL EXPERIMENTS III: THE ROLE OF TIME IN DEMAND FOR CARE

One natural experiment demonstrates the importance of "time costs" in the demand for medical care. In this case, the student health service at a major university was changed from one location

---

[10] A shilling is the British currency equivalent of the American nickel, i.e., one twentieth of a pound. At that time, a pound converted to about $2.40, so a shilling was about $.12. A 2-shilling fee for a prescription would be about like a $.25 fee at the same time (in the 1950s and 1960s), or about $1 currently.

Box 5.2 REGRESSION ANALYSIS TO ESTIMATE SYSTEMATIC RELATIONSHIPS AMONG VARIABLES

Multiple regression analysis provides a tool that helps unravel systematic behavior effects in data that sometimes can contain a lot of "noise"—random information or behavior. Figure A shows this technique in a very simple data set. The 30 points shown by small circles in this diagram represent combinations of a dependent variable (one that "depends" on others in the model) and a single independent variable (one that is fixed for purposes of the analysis—i.e., set "independently"). An economic model may have a dependent variable that varies with many independent variables. The relationship shown in Figure A has only one independent variable, but the ideas are the same in more complicated models. The purpose of regression analysis is to find the straight line going through this graph that provides the "best fit" to the data points. The idea of best fit has a very precise meaning: It seeks to find the line such that the *sum* of the *squares of* all vertical distances between each point and the line through them is as small as possible. Thus, for each point, there is a distance like the one shown as d_{14} for the fourteenth data point and d_{20} for the twentieth data point. The "least squares" line has the smallest possible sum of the squares of all such distances for the points in the diagram. This technique is called "ordinary least squares" (or commonly, OLS), since there are also fancier techniques (although those are not called "fancy" least squares).

Many of the empirical studies reported in this book rely on techniques like OLS or similar methods to determine relationships among data. For example, the "dependent variable" might be doctor visits per year, and the "independent variable" might be the person's income. You should remember that "multiple regression" OLS can show the effect of a single variable (like income) while "holding constant" all other variables.

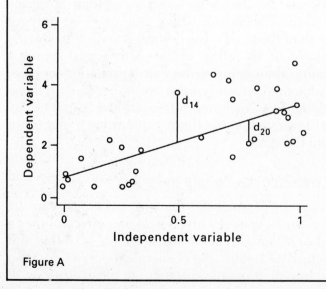

Figure A

(5–10 minutes average travel time for students) to another (20 minutes average travel time). The study reported the number of student visits (Simon and Smith, 1973). Again, using the technique of multiple regression analysis, Phelps and Newhouse (1974) estimated the systematic effect of the change in location of the health service. Visits fell by one-third from previous levels, despite the apparently improved surroundings, and (if anything) a less crowded atmosphere. The arc-elasticity of student health service visits with respect to time would be about −0.25 to −0.5, depending on whether the initial travel time is better characterized as 5 minutes or 10 minutes.

MORE ON THE ROLE OF TIME IN THE DEMAND FOR CARE

Travel Time in Palo Alto

In the studies using Scitovsky and Snyder's data, it is also possible to see systematic effects of travel time on use of medical care, since they coded the patients' home location and calculated the travel distance to Palo Alto, where the patients would have to go to receive care.

The effects of travel distance on physician visits and expense look precisely the way we would expect in theory: Those who live further away used less care, because the time and travel costs were higher. Particularly for families who lived more than 20 miles away, the effects could be considerable; they used 30 percent fewer doctor visits than those who lived very close to the clinic.

Some of this could be the effect of the patients just seeing a different doctor, rather than going to the Palo Alto Clinic for care, but the insurance plan paid only for care at the clinic. Thus, if these families did decide to go elsewhere, they would have had to pay full price (about $15 per visit at the time), rather than a quarter of that at the Palo Alto Clinic.

Time Costs for Low-Income People in New York

Another study (Acton 1975) analyzed the use of medical care by low-income families in New York City. There, the city provides a considerable number of outpatient clinics, both free-standing and in municipal hospitals. As before, the economic theory of demand predicts that those traveling greater distances to these clinics will use less care, even though it is nominally "free" to all. Indeed, some of the respondents in the survey also used (and paid for) private physician services.

The estimates by Acton showed that a 10 percent increase in the distance traveled by patients reduced annual visits by an average of 1.4 percent (elasticity of −0.14). In parallel, the same increase in distance to the hospital clinic would *increase* the use of private doctors by 0.7 percent—a small amount to be sure, and not enough to offset the decline in demand at the clinics. Those living further from the clinic would use less care in total, but make some substitution of private for "free" care.

APPLICATIONS AND EXTENSIONS OF DEMAND THEORY

The methods of demand theory allow us to expand our understanding of other events and studies in the health care area. In one important application for the tools of demand analysis, we return to the problem of variations in medical practice.

Son of Variations: Welfare Losses and Medical Practice Variations

The tools of demand theory allow us to extend further the analysis of variations in medical practice patterns that were discussed in Chapter 3. With the ability to calculate the incremental value of medical interventions using the consumer demand model, we can calculate the dollar value of the costs of variations, or at least a lower bound to such a cost.

Figure 5.2 shows the basic construction of the problem. Suppose two otherwise "identical" regions exist with different rates of use of a particular medical intervention (X), and that the *average* rate between the two regions is correct—that is, where the marginal cost (C) and the marginal value (the demand curve) are equal. (We will see below the effect of a systematic error; for now, we assume the average is correct.) Each region (by construction) has the same demand curve for X. Only mistakes in judgment lead to the differences in use rates. Region 1 uses the procedure at rate X_1, and Region 2 uses it at rate X_2. In Region 1, too little of X is used, and consumers suffer a welfare loss because they would be willing to pay more for X (at the use rate X_1) than it costs to produce. The exact loss they suffer is the area under the demand curve, but

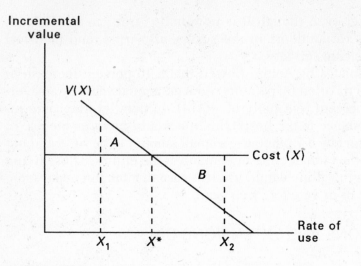

Figure 5.2 Welfare loss from wrong use of medical practices.

above the cost curve, between X_1 and X^*. In other words, they lose their consumer surplus on the consumption between X_1 and X^*, triangle A in the figure. Similarly, consumers in Region 2 lose their consumer surplus on the excess consumption between X^* and X_2, triangle B in the figure. Their welfare loss occurs because they spend more on each unit of care above X^* than it is worth to them.

The discussion of Figure 5.2 does not say what might cause people to use the wrong amount of care, but rather describes how to calculate the welfare loss once we observe a region using the wrong amount. One way to find people consuming the wrong amount is simply if they make a mistake—for example, by selecting an amount of care for which the marginal value does not equal the marginal cost. Sometimes taxes, subsidies, or supply restrictions will lead to that type of behavior. An alternative approach that we can explore here hinges on the role of information. This approach asks what would happen if people's information about the efficacy of a particular medical intervention were wrong, and then proceeds to measure the value of information that would correct those people's misunderstanding about the marginal efficacy of the intervention.

Figure 5.3 shows a society with two cities (1 and 2) that have incremental value curves (inverse demand curves) for intervention X described as $V_1(X)$ and $V_2(X)$, respectively. Their "average" value curve $V^*(X)$ is assumed for the moment to be the one to

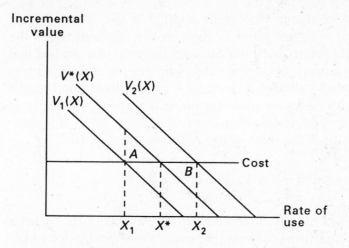

Figure 5.3 Incorrect information carries welfare losses.

which they would move if fully informed. Cities 1 and 2 hold different beliefs about the marginal value of intervention X, but given those beliefs, they behave optimally, by setting their consumption at the point where marginal value equals marginal cost. As we can see, this produces the same sort of overuse or underuse of intervention X as Figure 5.2 shows, except that we now have a particular reason identified as the cause of those mistakes—incorrect information about the efficacy of the intervention.

How much welfare loss exists from variations in practice patterns? First consider City 1 (with value curve $V_1(X)$. The triangle A has a dollar value equal to its area (since the vertical axis has dimension $\$/X$ and the horizontal axis has dimension X, their product has dimension $\$$). The area of a triangle is the product of one-half of its base times its height. Here, the base has length $(X^* - X_1)$. We can infer the height if we know how demand responds to price, or (the inverse question), how marginal value changes as the rate of use changes. In more precise terms, we want to know $dP(X)/dX$, where $P(X)$ is the marginal value at consumption rate X. If we have estimates of demand curves (where quantity varies as a function of price), we have estimates of dX/dP, the rate at which consumption changes as price changes. Here, we just want the inverse of that. The height of the triangle A is just $(dP/dX) \times (X^* - X_1)$. Thus the area of triangle A is

$$0.5(dP/dX)(X^* - X_1)^2$$

Triangle B has a similar area, except that it uses $(X^* - X_2)^2$.

Now if we consider N regions similar to the two we have discussed, each with use rates X_i, then we could characterize the average use rate for all such regions and the variance around that average. The variance is the expected value of $(X_i - X^*)^2$, adding up each observed value of X_i weighted by the likelihood that it occurs. Thus, it is easy to show (again, if the average use rate is correct) that the expected welfare loss for a large number of regions associated with varying rates of use of intervention X is

$$\text{Welfare loss} = 0.5 \frac{dP}{dX} \times \text{Variance}(X) \times \text{N}$$

where Variance(X) is the observed variance in use rates. A little algebraic manipulation converts this into a more simple formula:

$$\text{Welfare loss} = 0.5 \times \text{Price} \times X^* \times N \times \frac{\text{Variance}(X)}{(X^*)^2 \eta}$$

where η is the demand elasticity for intervention X (more precisely, the elasticity of a fully informed demand curve), and X^* is the average consumption. The *square* of the coefficient of variation (COV^2) described in Chapter 3 is equal to Variance$(X)/(X^*)^2$. Thus, the information on coefficients of variation in Chapter 3 is an important but incomplete part of the welfare loss measure. The final welfare loss measure is

$$\text{Welfare loss} = 0.5 \times \text{Total spending} \times COV^2 \div \eta$$

This formula defines a direct measure of the likely importance of information about procedure X. Good information could move all consumers to the informed demand curve, away from their current uninformed demand curves. We can think of this measure as having three components: (1) Sutton's Law,[11] (2) the Wennberg/Roos Corollary,[12] and (3) the Economist's addendum. Sutton's Law is simple: Money is important. The terms price × quantity (total spending) reflect this idea. The Wennberg/Roos corollary to Sut-

---

[11] Willie Sutton, a famous bank robber, was asked after he was arrested why he robbed all those banks. His response: "Because that's where the money is."

[12] This term receives its name from the two most prominent researchers in the variations literature, John E. Wennberg, M.D., MPH, and Noralou Roos, Ph.D.

ton's law is also simple: Confusion magnifies the importance of money. The Economist's Addendum simply says the following: For a given amount of variation, the welfare loss is greater if the incremental value changes rapidly as consumption changes. In other words, the demand curve is "inelastic," which (in its usual form) says that the quantity demanded is fairly insensitive to the price charged.

The welfare losses from variations in practice patterns have been calculated using New York State data (Phelps and Parente, 1990). The largest single welfare loss arose from hospital admissions for coronary artery bypass surgery, with an *annual* welfare loss of $4 per person. Hospitalizations for psychosis created an annual welfare loss exceeding $3 per person. Other nonsurgical hospitalizations with high welfare loss included adult gastroenteritis ($1.60), tests of heart function ($2.60 per person), congestive heart failure ($2.30), adult pneumonia ($1.35), and nonsurgical patients with back problems ($1.15).

These losses may seem small on a per capita basis, but collectively they add up to very large losses. For example, for the United States, with a population of about 250 million people, the annual welfare loss from just the variations in use of coronary bypass surgery is $.75 billion. However, since the knowledge needed to reduce variations would be valuable for many years, the value of that knowledge far exceeds $.75 billion. For example, if the knowledge persisted for 20 years before it became outmoded (at a 3 percent real discount rate), the present value of the knowledge would exceed 15 times the annual welfare loss, or about $6.5 billion. Thus, given the variations in use of coronary bypass surgery, even a study costing several billion dollars would have a positive expected payout. Fortunately, the cost of careful studies to assess the outcomes of medical and surgical interventions are only several million dollars, at most. Thus, the case for careful studies of medical practices is overwhelming: The expected returns are probably hundreds if not thousands of times greater than the costs of such studies.

All of the preceding analysis depends upon the assumption that the average rate of use is somehow correct. Of course, this is unlikely to be true, given the uncertainty about correct use. A systematic bias may have descended on medical choices. If so, then it is easy to show that the welfare loss measures indicated above are too low. The correct measure is the sum of the component described above (involving the coefficient of variation) plus the term

Welfare loss from bias = 0.5χ Spending $\times$ (%bias)$^2 \div \eta$

where (%bias) is the bias as a percent of observed use (Phelps and Parente, 1990). To see the importance this could make in welfare loss, consider a procedure like back surgery (laminectomy). The welfare loss from variations in back surgery calculated by Phelps and Parente was $1.10 per capita per year, or $.25 billion for the total country. Annually, about 190,000 such operations are performed, at an average cost of about $5,000, so total spending is about $.94 billion, or about $3.90 per capita. Using the demand elasticity for all hospitalizations in the RAND HIS, we can set $\eta = 0.15$ in absolute value. If the use rate is (say) 10 percent too high (or too low), the loss from *biased* use rates (in addition to the loss from variability) is another $.26 per year ($62 million total). The loss from bias rises with the square of the percent bias. For example, if the bias is 20 percent of the current rate, the bias-related loss is four times that arising from a 10 percent bias rate, or about $250 million per year. Again, the value of learning about such a bias is the discounted present value of all future years where the knowledge remains useful (i.e., until a new treatment for the illness emerges), and thus could be 15 to 20 times larger than the annual rate, or about another $4 billion.

Of course, we cannot learn the "proper" use rate (that at which a fully informed demand curve would indicate) by looking at variations. We can only infer that where variations exist, the likelihood is much greater that confusion exists, and therefore that a complete study of the procedure will produce considerable value. If such studies detect systematic bias in the use of a procedure, even greater value can emerge from such studies than from those just revealing and correcting the causes of inappropriate variations.

DECISION THEORY: DERIVING THE "RIGHT" DEMAND CURVE FOR MEDICAL CARE

The widespread findings of variations in the use of medical care indicate that many, perhaps most, doctors do not correctly use the medical interventions at their disposal. And if they cannot advise their patients correctly, it may be illogical to expect patients to act as "informed consumers." In part, the reason for the highly variable patterns of use may be that doctors are not taught well how to make decisions. Clinical decision making is an acquired art for most doctors, learned by example during their clinical years of medical school and in their residency training.

Fortunately, formal methods exist that could help doctors sort out the highly complicated problems of making medical decisions. Through the application of formal decision theory to medical problems, doctors are learning better how to advise patients about the use of medical interventions. In the terms of economic demand theory, medical decision theory spells out the best possible ways of using medical care to produce health, and possibly even to show the best ways of producing utility. Properly done, medical decision analysis helps find the "best" demand curve for medical care. To put it slightly differently, formal medical decision analysis derives what the demand curve for medical care "should be" for a consumer trying to maximize a utility function that includes both health and other goods—$U(x, H)$—in a consistent and organized manner.

The appendix to this chapter provides an example of how medical decision theory works, studying the problem of using a diagnostic study for a patient to help diagnose whether a disease is present. This example will show the interested student how medical decision theory and medical demand theory are intimately related.

WHY VARIATIONS IN MEDICAL PRACTICE?

The availability of formal decision theory heightens the puzzlement about the significant variations in medical practice use from region to region. Do doctors disagree about the values to patients (or do patients have different values in different regions of the country)? This hardly seems plausible, at least enough to account for the large variations in practice patterns. Do doctors disagree about the probabilities of illness? One way to answer that question is to use information on the patterns of using diagnostic tests and then the subsequent treatment. In the study of Medicare patients, Chassin et al. (1986) found that in regions where the use of tests was high, so was the use of the parallel treatment. This may just put the question back one step further: Why do doctors use diagnostic tests differently? Doctors may also hold different beliefs about important aspects of the treatment's safety and efficacy. Doctors in a specific community may have shared experiences and beliefs that begin to form a local medical "culture," identifiably different in the various regions. It seems important to learn what forms doctors' and patients' beliefs about therapeutic safety and efficacy, their beliefs about the probability of disease, and their use of diagnostic tests. As we learn more about these issues, we can better define the correct therapeutic strategies for patients, and thus give patients a better-informed judgment about medical interventions. As the welfare analysis of variations in medical practice shows, the gains from this type of study can be very large indeed.

SUMMARY

Empirical studies of demand for medical care have confirmed the role of price in people's choices of medical care of all types. Since insurance alters the net price of care to people, most of these studies have used people's responses to different insurance coverage as a proxy for different prices, since economic theory tells us that elasticities with respect to either price or coinsurance are equivalent. Early studies of demand elasticities found a wide range of elasticities, ranging from 0 (for some types of hospital care) to -1.5 or larger (for aggregate medical spending). This range of uncertainty, coupled with a strong desire by the federal government to know demand responsiveness with more certainty, led to the funding of a major randomized controlled trial, the RAND Health Insurance Study.

The RAND HIS offers the cleanest and most precise estimates of the effects of price on demand for medical care. These estimates confirm that price does matter, but probably not as much as the earlier studies have shown. Demand for hospital care is least price sensitive, and ambulatory care (acute, chronic, and well care) and dental care are more price responsive. Nevertheless, it is certain that demand is price inelastic (i.e., $|\eta| < 1.0$) for all these services. A wide variety of quasi-experimental evidence augments the data from the RAND study, adding to our knowledge of how demand responds to price, income, time and travel costs, illness, and other determinants of demand.

We can also apply demand theory to understand further the importance of medical practice variations, described previously in Chapter 3. By adopting a model of misinformation about the marginal efficacy of medical interventions as the underlying cause of variations, we can estimate the welfare losses associated with variations. If the underlying model of misinformation is correct, these estimates strongly support the value of conducting studies (and disseminating the results widely) about the correct approaches to using various medical interventions, since the value of such studies would greatly exceed their cost, even for very infrequent and relatively low-variation interventions. For common, expensive, and high-variation procedures, the yields from improving information vastly exceed the apparent costs of such studies.

PROBLEMS

1. "The RAND Health Insurance Study gave every subject a fixed amount of money that covered the maximum possible expenditure that each subject might have. Because of this, all participants could act as if they had full coverage, and the experiment was invalid." Assuming the validity of the first sentence, comment on the inference drawn in the second sentence.

2. Previous studies of demand for insurance had elasticities of demand in the neighborhood of -0.7 to -1, and sometimes -2. Compared with the results from the RAND experiment, which demand curves would lead to larger welfare losses associated with an insurance plan providing (say) a coinsurance rate of $C = 0.2$?

3. For a person who previously had no insurance and received an insurance plan paying for 80 percent of all types of medical care, what increase in use would you expect for hospital care, dental care, and physician services, on average?

4. In the RAND study, two plans had full coverage for spending within the hospital, but one had a $150 deductible for ambulatory care. The plan with the ambulatory deductible had a lower probability of hospital admission (0.115) per year than did the plan with full coverage for everything (0.128), even though both plans covered hospital care fully. (See Table 5.4.) What does this tell you about the use of hospital and ambulatory care: Are they substitutes or complements? Explain how this happens in plain English, without resorting to fancy economic jargon.

5. Calculate the arc-elasticity of demand for outpatient care in rural China medical care use in the coinsurance range of 1 (no coverage) to 0.2 (80 percent coverage). How does this compare with the arc-elasticity of demand for outpatient care in the RAND Experiment in the 95 percent to 25 percent coinsurance range? (See Box 5.1 for data.) Do you have any reason to expect these estimated elasticities to be very similar or very different?

6. Looking at Figure 5.3, explain why triangles A and B represent welfare losses, given the presumption that the demand curves V_1 and V_2 differ from each other because of differences in beliefs about the productivity of medical care between doctor-patient combinations in Cities 1 and 2.

7. Describe one or more studies that would demonstrate how time acts just as price does in its effects on medical care use, for example, higher travel or waiting time creates lower demand for medical care.

APPENDIX

An Example of Medical Decision Theory

The problem of using a diagnostic study has received considerable attention in the medical literature.[13] Even in its most simple form, the problem has some important complications, making it a good example of how formal decision theory can help clarify medical judgment. To set the problem up usefully, we need some notation. Let

f = probability that patient is sick

(1 − f) = probability that patient is healthy

This problem focuses on the use of diagnostic devices, so we need a simple way to characterize their capabilities. All diagnostic devices are imperfect, sometimes giving the wrong answer. We can follow the standard approach by measuring rates of correct and incorrect diagnosis as:

p   = probability of true positive (if sick) = SENSITIVITY

(1 − p) = probability of false negative (if sick)

q   = probability of false positive (if healthy)

(1 − q) = probability of true negative (if healthy) = SPECIFICITY

We describe benefits to patients (Utilities) as U, and costs as C. These are the patient's own utilities.[14] Our simple world contains only two treatment choices: Treat (T) or Not Treat (N). Thus utilities and costs

---

[13] The landmark study was Pauker and Kassirer, 1980.

[14] Indeed, some standard techniques help elicit these utilities from patients and use them in analyses like this one.

have a dual-state dependence, U_{ij} and C_{ij}, where i denotes the true state of health, and j denotes the treatment. Thus:

U_{ST} = Utility of sick person, treated

U_{SN} = Utility of sick person, not treated

U_{HT} = Utility of healthy person, treated

U_{HN} = Utility of healthy person, not treated

Presumably, $U_{ST} > U_{SN}$ if the treatment works. Similarly, $U_{HT} < U_{HN}$—for example, due to side effects of the therapy. Actually measuring these utilities is a complicated problem, about which a considerable amount is now known.[15]

In parallel notation, the associated costs are

C_{ST} = Cost of sick person, treated

C_{SN} = Cost of sick person, not treated

C_{HT} = Cost of healthy person, treated

C_{HN} = Cost of healthy person, not treated

We can now define some differential costs and utilities, comparing in each case the value (or cost) associated with treating people *correctly,* in accord with their true condition, compared with the incorrect action. Call these *incremental utilities* and *incremental costs*. Thus:

$\Delta U_S = U_{ST} - U_{SN}$ = utility gain by treating sick person

$\Delta U_H = U_{HN} - U_{HT}$ = utility gain by not treating healthy person

$\Delta C_S = C_{ST} - C_{SN}$ = change in cost by treating sick person

$\Delta C_H = C_{HN} - C_{HT}$ = change in cost from not treating healthy person (probably negative)

With this notation, we can evaluate the net gains from use of a diagnostic device. To do this, we need to answer several specific questions: Without additional testing, what treatment (if any) should the clinician recommend? Should diagnostic tests be used? If so, how should the clinician best interpret them? If the diagnostic test is used, how will the patient's outcomes change, and how will the benefits of any such change compare with the costs? Answering these questions provides the basis for

---

[15] The techniques use time trade-off methods, "standard gambles," and rating scale. For an excellent summary review of this work, see Torrance (1986). Torrance (1987) offers a less technically intense summary.

evaluating the benefits and costs of a diagnostic technology, and for deriving the patient's demand for the diagnostic test.

If the probability of disease is f, the expected utility from *treating* is $[f \times U_{ST} (l - f) \times U_{HT}]$ and the expected cost is $[f \times C_{ST} + (1 - f) \times C_{HT}]$. The alternative action is to *not treat*, with expected utilities and costs $[f \times U_{SN} + (1 - f) \times U_{HN}]$ and $[f \times C_{SN} + (1 - f) \times C_{HN}]$, respectively. The incremental expected utility from treating versus not treating is the difference between these,

$$[f \times U_{ST} + (1 - f) \times U_{HT}] - [f \times U_{SN} + (1 - f) \times U_{HN}]$$

$$= f \times (U_{ST} - U_{SN}) + (1 - f) \times (U_{HT} - U_{HN})$$

$$= f \times \Delta U_S - (1 - f) \times \Delta U_H \qquad (A5.1a)$$

This is really a simple idea. The expected gain from treating everybody is the gain for those correctly treated (ΔU_s) times the probability that the person *is* sick (f), minus the value of correctly not treating those who are healthy (ΔU_H) times the probability of that occurring ($1 - f$). Similarly we can define expected difference in cost from treating everybody (versus not treating everybody) as

$$f \times (C_{ST} - C_{SN}) + (1 - f) \times (C_{HT} - C_{HN})$$

$$= f \times \Delta C_S - (1 - f) \times \Delta C_H \qquad (A5.1b)$$

We can convert this to a simple decision problem *if* we know how much the consumer values in dollars a unit of utility. That is, we need something to allow us to convert "utils" to dollars or vice versa. While the process to do this can be somewhat complicated, it can be done. For example, the outcomes (the U's) can be expressed in Quality Adjusted Life Years (QALYs), and we can ask or infer how much people are willing to pay for a QALY. Another approach, valuable in some settings such as cancer therapy, uses as an outcome the increased probability of survival. We can learn from studies of people's own choices how much they are willing to pay to increase their chance of survival.[16]

We will use the label g to specify how much people are willing to pay for a unit of the health outcome. If the outcome is a QALY, for example, they may value the outcome at \$100,000. If so, then $g = 1/100,000$. Now we can convert the decision problem into a simple question: Which choice (treat or not treat) has the higher *net* expected utility, subtracting

---

[16] The most common methods look at the wages people earn in the labor force and the risks of their jobs. The higher wages people get for working in riskier jobs allows us to infer how much people value a decrease in the probability of dying. For a summary of earlier work, see Graham and Vopel (1981). A recent study puts the value of a "statistical" life saved at over \$5 million using this approach (Moore and Viscusi, 1988).

off the utility value of all associated costs? That's where the g factor comes in. Using that multiplier (if you will, an exchange rate between dollars and utils), the decision rule says to treat if the expected utility of treating exceeds that of not treating:

$$f \times (\Delta U_s - g \times \Delta C_s) > (1 - f) \times (\Delta U_H - g \times \Delta C_H) \qquad \text{(A5.2)}$$

We can reorganize the same information to tell what probability f determines the choice of treat versus no treat. To find the point of indifference between the two choices, we make equation (A5.2) an equality and solve for f, yielding the critical probability of disease:

$$f_c = \frac{(\Delta U_H - g \times \Delta C_H)}{(\Delta U_H - g \times \Delta C_H) + (\Delta U_S - g \times \Delta C_S)} \qquad \text{(A5.3)}$$

If the probability of disease f exceeds f_c, treatment is the correct choice, and conversely if f falls below f_c. We can call this choice the fallback strategy. It provides the best advice for the patient if no further information is available—that is, if there are no more diagnostic tests that might be used.

The choice of a fallback strategy sets the stage for analyzing the desirability of using diagnostic information. It turns out that the expected value of a diagnostic test to the patient depends upon the fallback strategy, as well as the accuracy of the test. Intuition gives good guidance here to the formal concepts that follow. If the doctor would treat anyway (fallback = treat), then the expected value of a test is very high if the patient's true probability of disease is low, because the test will probably reveal that the doctor would be making the wrong choice (treating a patient who isn't sick). Similarly, if the patient has a high probability of being sick, then the test isn't worth much, since it will just confirm the doctor's fallback strategy of treating.

Exactly the opposite pattern holds if the fallback is not to treat. The test is most valuable to a patient who is truly sick, and not so important to the patient who is healthy. The test should be used whenever the expected value of diagnostic information (EVDI) exceeds the cost of the test, C_t. The EVDI depends upon the fallback strategy. If the fallback is to treat, then

$$\begin{aligned} \text{EVDI}_T = {} & -(1-p) \times f \times (\Delta U_S - g \times \Delta C_S) + \\ & (1 - q) \times (1 - f) \times (\Delta U_H - g \times \Delta C_H) \end{aligned} \qquad \text{(A5.4)}$$

Similarly, if the fallback is to not treat, then the EVDI_{NT} is

$$f \times p \times (\Delta U_S - g \times \Delta C_S) - (1 - f) \times q \times (\Delta U_H - g \times \Delta C_H) \qquad \text{(A5.5)}$$

and the test should be used if $\text{EVDI} > C_T$ in either case.

These two equations, combined with the cost of the test, define the correct decision among the three alternatives of testing, treating right away, or doing nothing. Figure 5.4 shows these two EVDI curves, plotted against the probability of illness (f). The right-hand (downward-sloping) curve is relevant when the fallback is to treat, and the left-hand (upward-sloping) curve for when the fallback is not to treat. The test should be used whenever the EVDI exceeds the cost of the test (converted to utils with the multiplier g), so the relevant decision is to test if the probability of disease lies between f_1 and f_2. If the patient has a best-guess probability of disease less than f_1, then nothing should be done, and if the prior probability exceeds f_2, the doctor should treat the patient without bothering with the cost of the test. It turns out that the highest possible value of information from a diagnostic test occurs when $f = f_c$. This makes good sense, since that's the point where the patient would be just indifferent between choosing "treat" or "not treat." If you will, f_c is the point of maximum uncertainty, which must be the point where information has the greatest possible value.

The patient's individual demand curve depends on the probability of disease (and, of course, the utilities, costs of treatment, and the conversion factor g). We can see easily that the desirability of testing increases as the cost C_t falls. The range where testing is a good idea (between f_1 and f_2) gets wider as C_t gets smaller.

The aggregate demand curve for a "society" simply adds up all the demands of individuals. In any society, there will be a distribution of prior probability of disease, where person i in the society has illness risk f_i. Some of the patients (those with f_i near f_c) will have a high expected value for the test, and they create the top part of society's demand curve

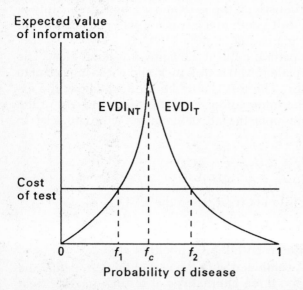

Figure 5.4 Expected value of diagnostic information curves.

for the test. Others (at very high or low values of f_i) will have only a low expected value from the test. They form the bottom part of the demand curve for the test. Filling in each person's expected value of diagnostic information gives the demand curve for the test for the entire society. This demand curve shifts upward if the willingness to pay for a QALY increases (for example, because of increased income) or if the costs of treatment fall, and of course shifts downward if the opposite happens.

This "simple" clinical problem shows how medical decision theory is linked to the demand for medical care. It takes the values of *health* as given, and then works out the "correct" use of *medical care*. If no tests are available, then the choice of fallback strategy defines the patient's demand for medical care, and an aggregate demand curve for the treatment just adds up the values of all the patients in society, each of whom has his or her own probability of disease f_i. If the test is given, then it "updates" the doctor's and patient's belief about the probability of disease, and makes a different treatment recommendation. Thus, for example, the "informed" demand curve for a medical treatment depends upon the quality of diagnostic information available.

Chapter
6

The Physician

*T*he physician stands as a central actor in production of medical care and health. Far out of proportion to the dollars actually paid to physicians, they guide and shape the allocation of resources in medical care markets, beginning with the smallest details of caring for a patient and ending with the capital allocation plans of hospitals and even the biomedical research undertaken in the National Institutes of Health, universities, and companies making drugs, equipment, and supplies for the medical care sector.

This chapter will study the physician as an economic agent. To do this well, we need to distinguish carefully among the various roles the physician can and does play in the production of medical care and health. The key we need to keep in mind is the distinction between physicians as *inputs* into a productive process, physicians as *entrepreneurs*, and the final product of *physician services*, the actual event that involves patients.

Almost every idea in this chapter applies equally well to other types of medical care providers and healers, including dentists, podiatrists, social workers, clinical psychologists, optometrists, nurse-practitioners, Christian Science healers, chiropractors, and many other specialized types of persons engaged in healing arts. Thus, while this chapter describes "physicians" and focuses on the particular work they do, nothing differs meaningfully (except the labels on some of their inputs) when other types of healers are considered.

THE "FIRM," INPUTS, OUTPUT, AND COST

Rationale for the Firm

Before beginning the actual study of the supply of physician services, a brief review of the economics of "the firm" will prove helpful. "The firm" is an economic entity designed to make decisions."The market" also makes decisions, often the same types of decisions as an individual firm makes. The boundaries of when markets versus firms make decisions are often fuzzier than one might expect. For example, a steel foundry might buy its coal from an independent supplier (in which case the market sets the price of coal), or might produce its own coal from a mine it owns. In the latter case, the production is managed internally. The extent of "vertical integration" for a firm is one of the important decisions its managers must make.

While economists are fond of saying that "the market" is the most efficient allocator of resources available, this is not always true. "The market" is really a collection of economic agents engaging in contracts with one another. Firms emerge precisely because it is more efficient to manage a particular productive process by command and control rather than through a series of external (contractual) relationships.[1] In most markets, we observe more than one firm, and economists believe that any attempt to organize the entire market through command and control would reduce the efficiency of production.[2] In many areas of resource allocation in medical care, we have drawn away from an unfettered market, for better or for worse, toward more centralized control (via regulation). In the market for physician services, this has happened much less than elsewhere in the medical care sector, with "firms" serving a much more important role in resource allocation, together with the invisible hand of the market, rather than through central command and control or regulation.

Firms hire and control the use of resources. The way they make those decisions determines the profitability and survival of the firm. The decision makers in a firm face two separate constraints in their decision making: the production process available to them, and the market demand for their final product. We can presume that these decision makers seek to maximize some objective function, subject to the relevant constraints. The normal "theory of the firm" includes only profits of the firm in the objective function. Most discussion of physician-firms has extended that objective

---

[1] The classic discussion of this problem appears in Ronald Coase's article, "The Theory of the Firm," 1937.

[2] The widespread inefficiency of the Soviet economy stands as a classic example of what happens when an economy relies too much on command and control, and not enough on markets.

function at least to include the leisure of the firm's director (and the only physician, if the firm is a "solo practice"), and sometimes also to include the health of the patient. This last issue creates a strong separation between groups of economists studying the health care sector. Some believe it is at least unnecessary, and possibly heretical. Others view it as an important addition to the careful study of the economic behavior of physicians.

These issues highlight the interaction of the physician and patient, which we consider in more detail in Chapter 7. As a brief introduction to the issues, we can note here that the question revolves around the market response to a major type of uncertainty in health care—namely, the diagnosis and recommendation of proper treatment for patients. Inherently, doctors know much more than patients about such matters, due to their professional training. With this advantageous position, they have the opportunity to deceive the patient into buying more medical care than is "optimal" in an objective sense. Defining a utility function for physicians in terms of labor and leisure alone leads to a set of conclusions about how the market might function, and (in turn) how the government might wish to intervene in the market (e.g., to protect the patient). However, if one includes the patient's health in the physician's utility function, it is easy to see that a balancing automatically occurs, whereby the physician's interests in protecting the patient's health limit any incentives for deceiving the patient that might otherwise lead to higher physician profits.

THE PHYSICIAN AS ENTREPRENEUR

The physician directing a physician-firm serves two roles, one as a labor input in the firm, and the other as entrepreneur. We can infer the value of labor by asking what other opportunities the doctor might have elsewhere, for example, on salary in a group practice of other doctors, or as an employee of some other organization. Any other returns the doctor receives as physician-entrepreneur are due to the management of the firm, and the returns (for better or worse) from the decisions that guide the firm. Those decisions include location, staffing, the product line of the firm (e.g., specialization, degree of vertical integration), pricing, and other similar choices.

THE PHYSICIAN-FIRM AND ITS PRODUCTION FUNCTION

The product of the physician-firm is "physician visits." Of course, this is as misleading about the physician-firm as saying that "hospital admissions" are the product of hospitals. Just as the hospital is a multifaceted "job shop," so is the physician-firm. The "product" of a physician-firm is really an array of treatments for patients with a mix of disease and illness, each unique in at least some ways, but having common characteristics about which the physician and other participants in the physician firm are knowledgeable.

The typical physician-firm is better known as an "office practice," because it is carried out most predominantly in that setting.[3] (The physician will also work in the hospital and may, in unusual circumstances, treat patients in their own homes.)

The physician-firm must involve at least one physician. By law, the activities of the firm must come under the direct supervision of a licensed physician. Other labor inputs include those of nurses, receptionists, bookkeepers and accountants, lawyers, laboratory technicians, and, sometimes, X-ray technicians, therapists, and other specialized personnel. Nonlabor inputs include the physical office itself, office equipment, medical equipment, computers, supplies, electric power, and insurance, including most importantly medical malpractice insurance.

This list of inputs to the physician-firm does not tell much about the economic behavior of the organization. We can ask, for example, how the combination of inputs is selected to produce a given set of outputs—that is, the mix of physician visits provided. To understand how these choices might work, we need to inquire about the production process itself.

As with all productive activities, the production of physician services allows some substitution of one type of input for another. Some types of substitution are obvious: A receptionist can replace some functions of a "general purpose" nurse. An accountant can replace some of the activities of the (physician) owner, but one

---

[3] This is sometimes called "ambulatory" care, since the patients walk in. An alternative designation is "outpatient care," as opposed to "inpatient care" in the hospital. Of course, "ambulatory care" and "outpatient care" involve not only the office practices of physicians, but also the services of hospital outpatient clinics and emergency rooms, as well as other specialized treatment settings.

hopes not those involving surgical procedures. Disposable supplies can substitute for labor and capital (the time and equipment needed to sterilize instruments). Other types of substitution are less obvious, but equally available. Careful record-keeping by the doctor may substitute for the use of legal services (by avoiding malpractice suits). Either nurses or doctors can garner routine history and physical examination information, providing the opportunity for substituting lower-cost nurse labor for higher-cost physician labor. Indeed, computer-based systems now exist that "take a history" from a patient, sometimes without the presence of either a physician or a nurse.

The extent of substitution in production is constrained both by technical and legal factors. In some cases, the technology will not permit substitution. For example, we do not have (as yet) robots that can undertake surgery successfully. Physician-firm entrepreneurs *cannot* make such substitutions. In other cases, the law will not permit it. Entrepreneurs *may not* make these types of substitution. Only licensed medical doctors may undertake surgery, for example, although the anesthetic for surgery may legally be administered either by a medical doctor or a registered nurse.

Several studies have actually estimated the opportunities for substitution (Reinhardt, 1972, 1973, 1975) using survey data from physicians. The survey gathered data on income, expenses, and the numbers of office, hospital, and home visits, as well as geographic region, medical specialization, and practice organization (solo or group). The study also asked about the numbers of aides employed and their salaries, characterized in groups including registered nurses (R.N.s), medical technicians, and office aides, all classified by full-time or part-time status. The capital owned by the doctor was approximated by the annual depreciation on furniture and equipment, plus cost of space. Thus, the data allow a reasonably clear picture of the mix of inputs in the doctor's practices.

Statistical techniques of multiple regression analysis allowed Reinhardt to determine separately the effects of the doctor's own hours, aides' hours, and capital on the total output (visits or billings) of the physician-firm. The results are somewhat startling: Doctors appear to employ far fewer aides than they should, if their goal is to maximize a utility function involving income and leisure (the same utility function economists usually use when thinking about the physician's labor supply decision). While the average number of aides hired per doctor was just under two, the "maximizing" number of aides would be nearer four, or twice the sample average. The same doctors could produce 25 percent more patient visits with these added aides.

The reasons why doctors might behave in this way are somewhat puzzling. One obvious response might be that Reinhardt's work was just wrong, but it seems carefully done and not subject to any obvious error. The other alternative is that physicians' utility function contains something beyond income and leisure. For example, if doctors really do *want* to cure patients, they might believe that their own time is more important to patients than the time of nurses or paramedics. The Hippocratic oath, taken by every medical school graduate, spells out the doctor's obligations to the patient and is widely regarded as the fundamental statement of medical ethics.[4] Extra attention to the patient (deviating from income-leisure maximizing choices) might cause this sort of discrepancy. This idea is difficult to prove or disprove, but it has attracted the attention of several prominent scholars studying the economics of health care[5] and cannot be dismissed out of hand.

The final point to make here depends on a complex economic problem confronting vertically integrated firms.[6] If such a firm owns an input that is a monopoly, then the obvious question arises as to the most economically sensible way to use the (monopolized) input. It turns out that the best way to do that is to have the monopoly-input "sold" to the parent firm at competitive prices, and then to price the final product higher than would otherwise be the case. The conclusion depends on the models of efficient use of inputs to assemble an output. Box 6.1 describes this logic in some detail. The question centers on the price at which the input should be sold to the final product firm, known as the "transfer price," versus the price at which the input would be sold to some outside

---

[4] The oath binds the doctors to honor their teachers, do their best for their patients, never give poisons, and to keep the secrets of their patients. Surely this oath is violated every day by at least some doctors (much cancer therapy involves "poisons"), but it also surely affects the behavior of some in the intended direction.

[5] Arrow's study of uncertainty and the welfare economics of medical care (1963) emphasized the nonmarket traditions of the physician service market. Arrow has won a Nobel Prize in economics.

[6] Vertically integrated firms undertake production at more than one stage. For example, if a steel company also begins to mine its own iron ore or coal, it becomes more vertically integrated. Almost all production processes have some degree of vertical integration, so the question is really one of degree.

Box 6.1 TRANSFER PRICING WITH A MONOPOLY INPUT

When a parent company owns a subsidiary that supplies an input, and the subsidiary has monopoly power, at what price should the subsidiary sell to the parent? To simplify things, call the parent company Steel, Inc. and the subsidiary Iron Ore, Inc. Suppose that Iron Ore has (for whatever reason) monopoly power, so it can set a price higher than its marginal cost. It could set the same higher price to any other company making steel, and indeed, the prices of steel in general will reflect the monopoly price of iron ore.

Since the Steel, Inc. owners own Iron Ore, Inc., they can presumably tell Iron Ore to sell them iron ore at any price they desire, including giving it away, selling it at the monopoly price, or anything in between. Remember, any iron ore sold to Steel, Inc. is not available to sell to other steel companies. What price will maximize the joint profits of Steel, Inc. and Iron Ore, Inc.? It turns out that the best plan to maximize total profits is to sell to the outside world at the monopoly price, and to transfer to Steel, Inc. at marginal costs of production.

The logic of this argument rests on the idea of substitution in production. Look at Figure A. Suppose the line AA represents the relative prices of iron and "other stuff" to Steel, Inc. if iron is priced at marginal cost. The best mix of iron and other stuff to produce (say) 80 tons of steel would occur at point *, using I_1 of iron and O_1 of other stuff. If the relative price of iron were higher, then the appropriate budget line to make 80 tons of steel would be BB, with the correct choice at **, using I_2 of iron and O_2 of other stuff. If Iron Ore, Inc. sells iron ore to Steel, Inc. at the monopoly price, then Steel, Inc. will act as if BB is the correct price line in choosing inputs, and produce 80 tons of steel at point **. However,

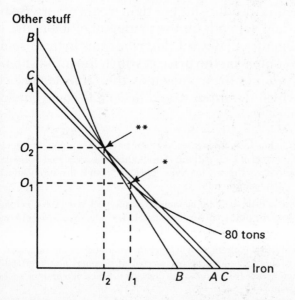

Figure A

the real production costs of "iron ore" and "other stuff" are reflected by a budget line with the same slope as AA. Thus to use the mix at **, Steel, Inc. is really on a "resource-using" budget like CC, which goes through **, but has the same slope as AA. The difference between AA and CC is a loss in profits to Steel, Inc. because they use the wrong input mix. The "right" choice transfers iron ore at the true marginal costs of production.

Steel, Inc. will then use the right amount of iron ore and other stuff, and produce at the lowest possible cost. Of course, since all steel is priced in the market reflecting the monopoly price of iron ore (which all other steel makers must pay), Steel, Inc. can sell its steel at that higher price. This lets the company recoup the profits "lost" from Iron Ore's "discount" to Steel, Inc. and also avoids the distortion in input use that ** would reflect.

The message of this story for physician markets is simple: If the market for physician services contains some monopoly rents, then transfer-pricing of physician labor at its marginal cost (lower than the monopoly price) to the physician-firm (which is, of course, owned by the physician-laborer) will permit the physician-firm to use more physician labor than if the transfer price reflected the monopoly in physician-labor. External observers of the market (such as Reinhardt, as described in the text) would "see" the market value of physician labor at the monopoly price, and thus would think that the relative prices of physician labor and "other stuff" were like the line BB above. However, intelligent physician-firm decisions would use physician labor as if the prices shown on AA were correct. Thus, to the external observer (like Reinhardt), the firm would appear to use "too much" physician labor.

firm (the monopoly price). The conclusion says that the "best" transfer price is the competitive price, because it will not distort the production process in the parent firm. One result is that, if looked at from outside, the parent firm will apparently use "too much" of the input, if the input is valued at its monopoly-level market price.

Suppose now that physicians *as labor inputs* had a restricted supply overall, perhaps due to licensure or other restrictions. This would create an artificially high "monopoly price" for that input. However, physicians also (in general) own the firms in which they work (or at least they are partners). In economic terms, they appear as vertically integrated firms with a monopolized input. Thus, we could interpret the apparent overuse of the input called physician labor in Reinhardt's studies as evidence that the physician market, at least at that time, contained monopoly rents for physicians *as labor inputs*. The data in Table 6.2 (which appears later) strongly suggest monopoly returns to physicians, since the internal rates of return greatly exceed competitive returns that typically are near

3–5 percent for low-risk investments, and the data in Table 6.3 (also later) suggest further high returns for specialties that rely on hospitals considerably—for instance, surgery and obstetrics.

Costs of a Firm

As firms assemble a final product, they generate a *demand for inputs* that will vary inversely with the price of those inputs, and directly with the amount of final product (in general).[7] By combining these inputs, using whatever substitution is legally and technically available, the firm produces its product or service. The payments it makes for those inputs are its costs, summarized in Box 6.2 from a recent survey of physician-firms. The relationship between the output of the firm and its costs forms the typical "cost curve" studied by economists.

The cost curve of a firm combines two ideas. First, it summarizes the costs of the resources actually used by the firm. These reflect the "accounting costs" that would typically appear in the financial records of the firm. The second thing that a "cost curve" reflects is the technical relationship between inputs and outputs, something that an accountant cannot describe or measure. A good engineering or operations research study might reveal these relationships, but in general what they show is that as a firm expands the scale of its activity (and hence its use of various inputs), its average costs first decline (with economies of scale) and then increase (with eventual diseconomies of scale).

If we were to draw the cost curve for a physician-firm (assuming for the moment that we knew what its output was), we would do what we would do for any economic entity with traditional production techniques, namely, to use a U-shaped curve showing how average costs fell with increasing output at first, and then increased as the scale of the firm increased. At one particular output, the firm has the lowest possible (minimum) average cost, at

---

[7] Some inputs are "inferior" in the sense that as the scale of the firm increases, demand for these inputs falls. This typically happens when the firm moves from a more general to a more highly specialized production process as its scale increases. For example, a small doctor's office might have a "general purpose" nurse who does a broad variety of tasks. A larger doctor's office might replace that "generalist" nurse with a nurse with specialized training, a receptionist-bookkeeper, and a physical therapist, rather than simply hiring three "general purpose" nurses. In this case, the demand for "general purpose" nurses would fall as the scale of the firm increased.

Box 6.2   **PHYSICIANS' COST STRUCTURE**

Physicians' cost structures differ considerably by the type of practice. The following table portrays the structure of physician costs in 1980, using data from a survey conducted by the American Medical Association.

| | | DISTRIBUTION OF NONPHYSICIAN EXPENSE[a] | | | | |
SPECIALTY	TOTAL NONDOCTOR EXPENSE[b]	NONDOCTOR LABOR	LIABILITY INSURANCE	MEDICAL EQUIPMENT	OFFICE EXPENSES	OTHER
Total	$52,900	36.6%	9.0%	8.9%	22.0%	23.5%
G.P.	$55,700	39.7%	4.9%	7.5%	21.7%	26.2%
Internist	$52,500	37.0%	5.2%	8.2%	23.2%	26.4%
Pediatrics	$51,000	44.0%	4.4%	8.6%	25.3%	17.7%
OB/Gyn	$67,000	32.7%	13.2%	7.7%	20.1%	22.3%
Surgery	$69,300	34.3%	11.8%	9.3%	21.9%	22.6%
Psychiatry	$23,100	25.0%	5.4%	5.8%	27.6%	36.2%
Anesthesia	$26,300	30.7%	37.0%	2.4%	13.2%	16.7%

[a] As percent of all nonphysician expenses.

[b] Excludes physician net earnings. For calendar year 1979, average physicians' net incomes were $78,400, ranging from $62,000 for G.P.s to $96,000 for surgery.
Source: Profile of Medical Practice, 1981. Chicago: American Medical Association, 1981, Table 38.

The structure of "primary care" doctors' costs (general practice, internists, pediatricians) all look quite similiar. Surgeons and obstetricians/gynecologists have similarly appearing "office" costs, but higher malpractice insurance. Psychiatrists and anesthesiologists both have low total costs, but a very different structure of costs. The typical psychiatrist's practice would have mostly space rental with little specialized medical equipment and probably only a part-time receptionist/bookkeeper. The anesthesiologist spends no time in "an office" with patients, but rather has all patient contacts in the hospital, thus incurring office rental space only for clerical-secretarial assistance, and so on. However, anesthesiologists' professional liability insurance is the highest of any speciality listed, reflecting the risks associated with participating in surgical activities. Chapter 14 discusses medical malpractice issues in more detail. A more detailed study of medical costs appears in Becker, Dunn, and Hsiao, 1988, pp. 2397–2402.

---

the bottom of the U. This is the output rate sought by purely competitive firms, but as we shall see below, it seems unlikely that physician-firms operate at this point.

We could also draw the marginal cost of the firm, the rate that total costs change as output changes. Marginal cost curves *always* rise and fall more steeply than average cost curves, and *always* go through the average cost curve at its precise minimum.[8]

This cost curve would shift up and down as the cost of any input shifted up and down. For example, if nurses' wages increase, then the average cost *curve* for a doctor's office will shift up. At every possible output, it will cost more to produce "doctor visits" than at a lower wage for nurses. This relationship between input prices and output costs should also translate directly into a higher product price, although not necessarily on a one-for-one basis. An appendix derives the effects of an input price increase on product price in several settings. The results appear as follows: In a long-run equilibrium in a competitive market, each \$1 increase in the marginal cost of production of a good or service leads to a \$1 increase in product price.

In any "short" run, the price increase will be less than the costs. In particular, the percent passthrough (P) in any market will be $P = \epsilon/(\epsilon - \eta)$, where ϵ is the elasticity of supply of the final product and η is the demand elasticity at the market level. This provides a nice characterization of the "long run" as well. When all inputs are free to vary, the elasticity of supply in a competitive industry should become very large. As ϵ approaches infinity, the passthrough rate P approaches 1 for any finite demand elasticity. Thus, the previous statement about dollar-for-dollar passthrough in the long run is just a special case of this more general model. By contrast, in a pure monopoly, an increase in the marginal cost of production of \$1 will lead to an increase in price that exceeds \$1. An appendix proves both of these assertions about cost passthrough.

Steinwald and Sloan (1974) have studied fees charged by physicians throughout the United States, including the effects of wages of office workers on fees. They used a composite measure of the wage rates for five "typical" office workers in physician offices to measure wages. Their results showed that fees always increased proportionately with wage costs, in a way quite close to that expected in a competitive (or monopolistically competitive) industry. For example, for fees of G.P.s, a 1 percent increase in wage costs would lead to a 0.84 percent increase in fees. For general surgeon office visit fees, a 1 percent increase in wage costs would lead to a 0.93 percent increase, and for pediatricians, a 1 percent

---

[8] Problem 1 at the end of the chapter asks you to prove this statement. This result holds so long as the production function eventually exhibits decreasing returns to scale.

increase in wage costs would lead to a 0.48 percent increase. In a purely competitive market, the proportionate increase would always be less than 1.0 percent unless the firm was experiencing pronounced diseconomies of scale.[9] The elasticities for several specific in-hospital procedures are also available from the Steinwald and Sloan study. For appendectomies, the elasticity is 0.85; for a normal delivery by obstetricians, the elasticity is 0.75.

NONPHYSICIAN PRIMARY CARE PROVIDERS

Many of the overall physician visits each day involve relatively simple medical conditions, and the practice of medicine that deals with these events is commonly described as "primary care." The medical practices of general practitioners, specialists in family practice, internal medicine, pediatrics, and obstetrics/gynecology contain much of this sort of activity. For these "primary care" visits, even more extreme substitution in the production of health may be possible than Reinhardt suggested, by breaking out of the model requiring a physician as the central actor in face-to-face patient visits. While state licensure generally requires the presence of physicians in such a setting, some evidence suggests that other ways to produce primary care visits may be at least as good, and at lower cost.

One organization that *may* use different models of production of primary care is the military health care system of the United States. Each of the major uniformed services (Army, Navy, Air Force) operates its own health care system, responsible not only for the care of active duty personnel, but also for the dependents of such personnel and any retirees from the military. Because the federal government operates these programs, state licensure laws do not apply to them. They can set aside legal *may-not* concerns, and concentrate on finding the boundaries of *can-not* in the production function.

During the 1970s, the U.S. Air Force instituted an experimen-

---

[9] To a first-order approximation, the elasticity of price with respect to the cost of an input "L" (for "labor") would be $(\epsilon/(\epsilon - \eta)) \times$ (cost share) $\times (\%dL/\%dQ)$, where $\%dL = dL/L$ and $\%dQ = dQ/Q$. The term involving elasticities of supply and demand must be less than 1.0, as must the cost share. Thus, the elasticity of fee with respect to wages must be less than 1.0 in a competitive industry unless $\%dL/\%dQ$ is larger than 1 by enough to offset the other two factors that are less than 1. This would occur only if the firm experienced significant diseconomies of scale in labor.

tal model of primary patients visits in a number of their clinics where the patient saw only a trained paramedic. While a physician was "in the vicinity" for consultation, the paramedic determined when to call the physician for assistance.

Studies of this alternative way of producing patient visits revealed three clear features (Buchanan and Hosek, 1983). First, the cost was obviously considerably less; the paramedics earned about a quarter of what the physicians earned. Second, the quality of care, as rated by panels of physicians reviewing the medical records, was at least as good by the paramedics as by the physicians. Third, patients were at least as happy with the paramedic visits as with the doctor visits (Goldberg, Maxwell-Jolly, Hosek, and Chu, 1981).

Several states now allow paramedics to practice under the relatively loose supervision of physicians, but none allows freestanding practice by paramedics, for example, that includes the prescription of drugs. Another new model of production of health visits uses the nurse-practitioner as the entrepreneur. These approaches are so new at this point that no complete evaluation of their costs and efficacy exists, but as they become more widespread, such studies will surely emerge.

THE SIZE OF THE FIRM—GROUP PRACTICE OF MEDICINE

Physician-firms can obviously contain more than one physician. Indeed, most doctors now practice in multidoctor groups. These groups can in turn include only doctors of a single specialty (such as pediatrics) or multiple specialties. The extent to which such groupings are desirable depends on the economies of scale and the economies of scope of medical practice.

Economies of scale in physician-firms, like those for hospitals (and any other productive firm), occur when the average costs fall as the scale of the firm rises. It is possible for both average *variable* costs and average *total* costs (including fixed costs) to fall, and finally to rise, with the firm's scale. The idea that costs can fall with size occurs most commonly to people—for example, by observing the possibilities for grouping doctors together to share "overhead" items (such as telephone and secretarial staff) that might not be fully utilized in a single-doctor office. The idea that costs might rise as the size of the group increases is less intuitively obvious.

One way in which groups can eventually confront increasing costs (decreasing returns to scale, or "diseconomies of scale") arises through problems of coordination, control of costs, and monitoring of work effort—which in turn depend on how the doctor is paid within the group.

Suppose at one extreme that each doctor in a group is paid on a flat salary. This is a common arrangement in large multispecialty groups, and is often true for "junior" members of smaller single-specialty groups. The "product" of each doctor is somewhat difficult to observe directly (patients' health), so any attempt to monitor the work level of a doctor within the group usually falls back on some intermediate measure, like patient visits per hour or per day. Some groups control this by tightly scheduling the patient visits for each doctor.[10] Of course, the doctor's total workload depends in part on the complexity of patients seen, so one possible response is to try to get the patients to return more often for an easier "repeat" checkup.[11]

Salaried doctors also have little incentive to control the use of costly resources within the group, because their earnings are independent of the group's other costs. If they can make their own work easier—for example, by using more R.N. or secretarial time—they can be expected to do so.

The difficulty in monitoring such behavior by the group stems from the job-shop nature of medical practice. Each patient is unique, at least partly, so it becomes more difficult to monitor excessive use of resources or other costly activities with great precision. Sometimes, it will be difficult to tell whether a particular activity is patient-related or merely "goofing off," as Figure 6.1 suggests. Obviously, a serious "shirker" can be fired from the group, but within some bounds, the problem is difficult to detect and control.

One way to help reduce these problems is to pay all doctors

---

[10] The military health care systems use this technique to the extreme. See Phelps, Hosek, Buchanan, et al., 1984.

[11] In the military health care system, where doctors are on salary, this "quota-meeting" behavior produces some very unusual patterns of physician visits. For example, where a normal private-practice doctor might see a diabetic patient three to four times per year for routine visits, the same patient might return monthly in the military setting. Doctors can enforce this by giving prescriptions that last only one month. The patients, relatively captive in this system, are thus forced to cooperate with the doctor's activities designed to minimize total daily work effort, subject to the constraints imposed by the system.

Figure 6.1 The problem of detecting "shirking" in large organizations.

within the group on a profit-sharing basis, so that their use of costly resources cuts (at least partly) into their own pockets. Newhouse (1973) showed how this works. If each of N doctors in the group has $1/N$ of the patients, then each of the doctors will worry about $1/N$ of the costs. In a totally noncooperative situation, a large group looks more and more like a group paying doctors on salary, with each having little concern about the costs of the group. Of course, many management methods exist to help control such behavior (i.e., shirking and resource waste), but these are also costly to undertake, in terms of both the direct management activity and the corresponding added work of the doctors in the group (filling out forms, attending meetings, etc.).

One piece of evidence on the role of incentives in work effort of physicians emerges by comparing the rate of patient visits produced by physicians in private (fee for service) practice and in a fee-for-service payment plan ("independent practice association," or IPA) with the rate for those on straight salary in a "staff model" HMO setting. (Chapter 11 discusses the IPA and HMO concepts more fully. For now, we need only understand that the staff model HMO doctors receive an annual salary, not directly dependent on the number of patient visits they produce, whereas IPA and fee-for-service doctors receive more money for each patient visit they produce.) In a 1980 survey of physicians, salaried HMO physicians produced office and hospital visits at only about 80 percent of the rate of their IPA colleagues (see Table 6.1), while working for 93 percent as many hours per week. Most of the difference arises because HMO physicians apparently spend a longer amount of time with each patient. However, the "time per patient" was not directly measured here, but rather arrived at by dividing time spent in office-visit by the number of visits. Thus, the larger time per visit calculation could also reflect, for example, more trips to the water cooler, chatting with colleagues, or other forms of "shirking" difficult to monitor in a professional setting.

The eventual diseconomies of large groups of physicians are yet another manifestation of the pervasive role of uncertainty in medical care. The problem extends not only to cost control, but also to the monitoring of a doctor's quality of care. Because the final output of each doctor is difficult to observe, even a poor doctor can exist for a long time within a group before it becomes apparent. A series of bad outcomes might easily appear as a run of bad luck. As with quality, similar difficulties occur with detecting doctors using resources excessively. A doctor might successfully hide ex-

Table 6.1   PHYSICIAN PRODUCTIVITY IN ALTERNATIVE PRACTICE SETTINGS

	Hours per week	Patient visits		Minutes/visit	
		Office	Hospital	Office	Hospital
Salaried HMO M.D.s	46.4	67.2	24.5	23	40
IPA M.D.s	50.6	83.0	30.0	20	33
All doctors	49.6	78.7	33.1	20	30

Sources: HMO and IPA data from "Organizational Structure and Medical Practice in Health Maintenance Organizations," by F. D. Wolinsky and B. A. Corry, *Profile of Medical Practice 1981.* Chicago: American Medical Association, 1981. Data for "all doctors" from same publication, "Part III—Selected Tabulations," Tables 12, 13, 19, and 20.

cessive use of a group's resources for some time, because of the difficulty in knowing what "should" be used for a given set of patients.

The problems of shirking exist within any organization, but they diminish rapidly in small firms because detection is easier. They are more difficult to control and monitor within an organization that produces services, rather than physical products, because the tasks are more difficult to measure. The job-shop nature of the physician-firm compounds the problem. These issues eventually lead to increasing costs as the size of the physician-firm increases, and ultimately limit the optimal scale of the firm.

Another important limitation on the scale of the physician-firm is a cost that does not appear on the firm's books: patient travel time. As with any service requiring the participation of the client-patient, travel time will ultimately limit the sphere in which the physician-firm's market can meaningfully operate, providing yet another diseconomy of scale. This particular diseconomy of scale is not directly apparent from physicians' cost accounting, since patients provide their own travel to the doctor office. However, an equally feasible arrangement technically would have the doctor provide transportation to each patient, in which case the costs of expanding the scale of the doctor's office would soon become apparent.

Referrals Among Doctors—"Subcontracting" in Medicine

Often, a patient will come to a doctor who then determines that the patient's illness will be best treated by a physician of another specialty or subspecialty. Typically, the primary care doctor will then suggest a referral to another more specialized doctor, who treats the patient for that particular illness only. Thus, one of the particular assets of the primary care physician-firm is the set of specialists to whom it refers.

The networks of referral doctors maintained by a primary care doctor arise through a variety of ways. In a sense, the primary care doctor takes the place of the patient in searching for another doctor. This aspect of a doctor's activity is considered further in Chapter 7, where we more fully discuss search in physician markets.

One feature of the referral market particularly draws the attention of economists, the prospect of "fee splitting." Fee splitting simply means that the doctor making the referral receives part of the fee received by the specialist as a reward for sending the patient to the specialist. The practice of fee splitting is strongly discouraged in medical ethics. One reason why medical ethics views this practice as a problem—again, further discussed in Chapter 7—is that it could distort the physician's judgment about the "best" referral to make for a particular patient. If the doctor is supposed to act on the patient's behalf in selecting the specialist for consultation, then any fee-splitting arrangement can only distort that choice. As such (for example) it could violate the Hippocratic oath's requirement to do the best possible for the patient.

Fee splitting has a potential good side as well. For example, some procedures can be carried out either by a first-contact "generalist" or a referral specialist. If the specialist offers higher quality (e.g., lower chance of side effects or iatrogenic illness), then something that increases the rate of referral may be desirable for the patient. (Notice that we can't say "will be desirable," since the patient may prefer to pay less for lower quality.)

The actual extent of fee splitting is almost impossible to measure, because no doctor will readily admit to it.

Multispecialty Firms

Some doctors work in multispecialty groups, usually with a large number of physicians. Several of these groups have become quite famous nationally,[12] despite their relatively rare nature. An advan-

---

[12] The Mayo Clinic in the other Rochester (Minnesota) is probably the most famous. Few people know the correct origin of the name, incorrectly attributing it to the name of the founding doctors, two of whom were brothers named Mayo. Actually, the group's fame arose from a large picnic in the town one summer, at which the mayonnaise in a chicken salad went bad, creating widespread stomach illness. The doctors' success in treating that event eventually led to the formation of the "Mayo" Clinic.

tage of these multispecialty groups from the patient's perspective is "one-stop shopping" for a wide variety of medical diagnosis and treatment alternatives, with (presumably) closer coordination of care. The physical proximity of all doctors within a multispecialty group also offers an advantage to patients, reducing travel time and costs, compared with other arrangements. The location of many physicians in a single office building offers the same advantage, and may help explain why doctors "bunch up" so much in the same area. Of course, doctors' office buildings also are usually near hospitals, so the doctor can go to see patients within the hospital easily.

From the group's perspective, the guaranteed "referrals" within the group may increase profitability by reducing reliance on outside referrals. However, the overall scale of the group usually must be quite large to make this work well, particularly in rarely used subspecialities. The problem that can arise is that the multispecialty group hires (say) a neurosurgeon, but then the number of illnesses requiring that type of care may be too small to carry the doctor's salary.

Most likely because of the large volume of patients required to maintain a proper balancing act, most multispecialty groups have arisen in combination with an insurance plan that inherently attracts many patients. These combinations—health maintenance organizations and similar arrangements—are discussed in Chapter 11, after we have discussed the insurance market more completely.

THE PHYSICIAN AS LABOR

The previous discussion illuminated issues decided by physicians acting as entrepreneurs. They must also act as workers in a labor force. Thus, we can now turn to the questions of how the physician decides to supply labor into the market. Undoubtedly, many people decide to enter medical school for reasons not wholly related to the financial rewards. Indeed, as we will see later, some of the decisions about specialty training appear to show important factors in addition to "dollars" that affect some specialty choices. However, there are also systematic financial questions associated with the practice of medicine that deserve study.

The two major decisions we can study are (1) the initial decision to become a physician, and (2) the decision about how much to work (hours per year) once the physician has completed training. Both of these have systematic economic components.

The Decision to Become a Physician

Entering medical school is an important investment. The physician-to-be invests the tuition payments and other costs of schooling (over $20,000 per year currently) for four years. More important, the physician-to-be invests four years in medical school, plus three or more years in specialty training, that could be used in some alternative work. Thus, the *opportunity cost* of entering medical school is the forgone earnings from some other career that would be relevant for a person with a college degree.

Strictly as an investment, one can evaluate the desirability of medical school by comparing it with alternative investments the physician-to-be might make. The "medical school" investment has heavy front-end costs, and higher earnings in later years. The "college" investment has higher initial earnings, and lower average earnings in later years. The desirability of one or the other investment will hinge on the *time preferences* of the physician-to-be. Those with relatively strong preferences for current consumption will be less likely to become physicians. Those with relatively similar preferences for current and future consumption will gladly give up some current consumption now for substantially larger consumption opportunities (earnings) in the future.

The most compact way to summarize the desirability of an investment is to use the "internal rate of return" (IRR). In its most simple terms, the IRR asks, "If I put the same amount of money into a bank account as in this investment, what rate of interest would the bank have to pay to make me indifferent between the bank account and the alternative investment?" When studying the economic returns to postcollege education, studies commonly use the "typical" college student's economic opportunities as the basis of comparison. A recent study (Burstein and Cromwell, 1985) has made just such a calculation for physicians, dentists, and lawyers, in each case compared with the alternative of going to work directly after college.

This study adjusted for the average costs of tuition, the wages received during post-medical-school ("residency") training, and the average hours worked by physicians in their posttraining years. Physicians work many more hours per week than the average college graduate, the reasons for which we will discuss below. If one ignored this larger work effort, the apparent returns to schooling would be overstated.

Table 6.2 shows these economic returns. As is commonly per-

Table 6.2 RATES OF RETURN TO PROFESSIONAL TRAINING

	All physicians, hours adjusted		Dentists, hours adjusted		Lawyers, hours adjusted
	Yes	No	Yes	No	No
1980	12.1	14.0	—	—	7.2
1975	11.6	14.2	12.3	16.7	7.1
1970	11.8	14.7	12.1	16.8	7.0
1965	—	24.1	—	—	—
1955	—	29.1	—	—	—

Sources: Burstein and Cromwell (1985) for 1970–1980; Sloan (1970) for 1955–1965.

ceived, the financial returns to becoming a physician are considerable. An IRR of 12 percent means that you would have to find a bank paying 12 percent *real* rate of interest (i.e., 12 percent above the inflation rate) in order to make it comparable. By contrast, the real rate of return on low risk investment our economy has seen over this century has fluctuated nearer 3 percent or so.

Some of the higher incomes received by physicians reflect their long hours of work. When that is not adjusted for, the apparent rate of return for physicians exceeds 14 percent in most of the years studied by Burstein and Cromwell. On average, adjusting for work effort reduces the rate of return calculated by 2 to 3 percent. Nevertheless, the return to the investment in education for physicians is considerable, and exceeds that available in most alternative investments.

The Decision to Specialize

Most physicians in the United States continue their training beyond that required by law (in order to obtain a license to practice medicine). Licensure requirements vary from state to state, but they invariably include (1) graduation from an approved medical school, and (2) one year of training in an approved "internship" or "residency" program.[13] Most doctors continue for further training, in order to become eligible to take voluntary certification examina-

---

[13] Until recently, all of these first "postgraduate" programs were called "internships." They then became specialized as "medical internships" or "surgical internships" or "pediatric internships." Now such programs almost all are part of a multiple-year program, and are just called "first year residencies."

tions offered by "specialty boards," that is, organizations that administer advanced tests in special areas of medical competence, and then allow those doctors passing "the boards" to indicate that they are members of the appropriate specialty board. These specialty boards exist for virtually every area of medicine, and indeed, comparably for many nonphysician providers of health care as well. For physicians, specialty boards exist for internal medicine, surgery, obstetrics/gynecology, family medicine, pediatrics, psychiatry, radiology, pathology, and many "subspecialty" boards.[14]

"Board certification" adds many years to a doctor's period of training. Almost all specialty boards require at least three years of training beyond the four years of medical school. Some complicated surgical specialties (such as neurosurgery) have requirements extending for seven or more years beyond medical school. Some doctors seek still further training in "fellowships" to learn specific techniques, or to become more adept in an area of research.

This specialty training takes place in "residency programs" in hospitals throughout the United States. Almost all of these residency programs have some affiliation with a medical school, although in many programs, the affiliation is quite weak, and the residency training is provided "on a voluntary basis" by the doctors on the hospital's medical staff, rather than by full-time medical school faculty. Each specialty board approves the corresponding residency programs, and successful completion of such an "approved residency" is prerequisite to taking the final specialty board qualifying examination.[15]

The economic returns to specialty training are often quite high. Just as we could analyze the decision to enter medical school as an investment, so can we analyze the decision to specialize. Marder

---

[14] For example, in pediatrics, subspecialty certification is available in such areas as pediatric allergy, cardiology, surgery, and newborn intensive care. In surgery, specialty boards exist for orthopedic surgery, cardiac surgery, thoracic surgery, plastic surgery, neurosurgery, and many other portions of the anatomy. For a good picture of the distribution of such surgeons, open up the Yellow Pages of any major city in the United States to the heading of "Physicians and Surgeons—MD" and you will get a good sampling of the number of various specialty areas and the number of doctors in such a community with specialty training.

[15] A doctor who has completed the requisite training program is usually called "board-eligible." Upon successfully taking the examination, the doctor can use the title "board certified."

and Willke (1991), Burstein and Cromwell (1985), and Sloan (1970) have estimates of these economic incentives that are remarkably similar, despite the studies' reliance on data from quite different periods (pre- and post-Medicare). The returns to specialization question can be summarized as follows: essentially "If a doctor goes immediately into private practice instead of specializing, a certain expected path of income is achievable. If the doctor specializes, some added years of training take place, at reduced incomes (during the period of residency), and then the doctor can earn higher incomes. What discount rate makes these two choices equivalent?" As before, the internal rate of return (IRR) summarizes that choice. See Table 6.3.

The internal rates of return for specialization substantially exceed competitive returns, particularly for the surgical specialities. Note that all three studies uniformly find low rates of return—sometimes negative—for pediatrics, and that all uniformly find relatively high rates of return for hospital-based specialities.

Hours of Labor Supplied

Any study of physicians shows that they work more hours per week and per year than most persons. What accounts for this high level of effort? One explanation lies in the simple economics of labor supply.

To think about how many hours a person might work in some period of time (a week or a year), consider what else they might do with that time: attending baseball games or the opera, reading,

Table 6.3 RATES OF RETURN TO SPECIALTY TRAINING
(Hours Adjusted)

	Internal medicine	General surgery	Obstetrics and gynecology	Pediatrics
1987	12.7%	22.1%	25.9%	1.5%
1980	9.8%	13.6%	14.8%	—
1975	12.5%	11.6%	12.1%	—
1970	9.3%	11.2%	11.8%	2.4%
1967	8.3%	7.4%	7.5%	1.6%
1965	1.5%	5.2%	4.8%	<0
1955	<0	5.7%	6.7%	<0

Sources: Marder and Willke (1991) for 1987 estimates; Burstein and Cromwell (1985) for 1967–1980, Sloan (1970) for 1955–1965.

cooking, going on vacations, hunting and fishing, golfing (on Wednesday afternoon), repairing the roof, fixing the plugged drain, or playing with the children. In sum, they can engage in "leisure" activities—those not associated with earning more income.

Of course, any income they earn helps pay for those activities, plus food, shelter, transportation, and the other things included in "consumption." Hence the problem of deciding about the "right" amount of labor supply: Every hour of work the physician gives up provides more leisure but less income: What is the correct balance?

If we think of a person as having a utility function $U = U(X, L)$ where X = consumption goods and L = leisure activity, then we can make some sense out of this.[16] We can draw the tastes of this person in an indifference map, just as we did in Chapter 1, when talking about X and H. The price of X is just its cost in the market. The price of an hour of leisure is the amount that the person could earn at work, net of taxes. Thus, we can draw a budget line for the person that shows (for a given wage rate) how much X he would buy if he worked "all available hours" (the intersection on the X-axis) with no leisure. Its slope is the (negative) wage rate of the person.[17] In a society with nonconstant marginal tax rates, of course, the relevant cost of leisure changes with income, because the marginal tax rate changes with income. Thus, where t is the marginal tax rate and w is the implicit or explicit hourly wage rate of the wage earner, the marginal cost of an hour of leisure is $(1 - t) \times w$.

If a person gets a higher wage rate—for example, due to investment in medical education—then that budget line does two things: It shifts outward, and it also becomes *steeper*, since an hour of leisure costs more. The two effects push in opposite directions. The higher cost of leisure (higher wage rate) pushes hours of work upward. The higher income can push in the other direction, since leisure (like other goods) is probably something people want more of as their incomes rise.

---

[16] In the spirit of earlier work, we could easily say that utility was a function of $U = U(X, H, L)$. Then, assuming for the while that H was held constant, we can study the labor-leisure choice by temporarily ignoring H and focusing on X and L, which means we can legitimately talk about a utility function where only X and L change.

[17] For convenience, we can set the price of $X = 1$, since we haven't picked any particular dimension in which to measure X.

Combining these phenomena, one can draw the labor supply curve of an individual as in Figure 6.2. It will initially slope up (higher wages leads to more work hours), but eventually, the higher income *may* lead to a diminution of hours supplied. If this point is reached, we say that the labor supply curve "bends backward." If this happens, higher wages will lead to *fewer*, not more, hours of work.

Such labor supply curves are different for each individual in a market. The labor supply curve for an entire market is the sum of all the curves for individual participants.[18] Notice that the labor supply curve *for the market* need not bend backward even if the curves of many individual participants do. At higher wage rates, some individual's curves may bend backward, others may still retain their upward slope, and indeed some people may newly enter the market at higher wages, adding new "suppliers" to the total.

There is almost no serious evidence available on whether the supply curve of "real" physicians has a relevant backward-bending portion or not. The theory of labor supply says that it might. Empirically, the problems associated with finding that supply curve are considerable. The particular problem is this: Few doctors are actually hired "by the hour," such that an hourly wage can readily be determined. Many are entrepreneurs in their own firms, and others work for physician groups on an annual salary, for which the hours worked are "as needed." One can calculate the "implicit" hourly wage by comparing the net receipts a physician-firm receives for carrying out typical "procedures" (such as an office visit, a hernia repair surgery, etc.) with the number of hours worked, or else by calculating the average dollars of physician earnings per hour of work supplied. Each of these approaches provides some assistance in understanding hourly returns to physician labor, but creates technical difficulties when such measures are used to try to estimate physician labor supply curves. The concept of a backward-bending supply curve uses an hourly wage rate as the focus, and we just don't observe this concept very often.

One setting allows a direct study of the effect of higher wages on labor supply, and it suggests that the labor supply curve is

---

[18] The labor supply curve adds up just as the demand curves for medical care were added up in Chapter 4—by adding "horizontally." At any wage rate, the aggregate supply adds up the hours offered by each individual.

Hourly
wage

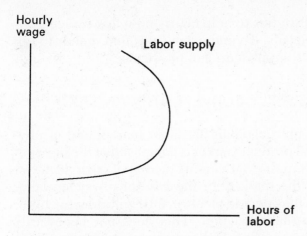

Figure 6.2 Labor supply as a function of wage.

upward sloping (not backward bending) for this group of doctors. The opportunity to study this phenomenon occurs with "moon-lighting" doctors, usually residents in advanced training, who work by the hour (on a specified hourly wage) in emergency rooms or similar settings. Several studies of the moonlighting effort by residents have been conducted. The most recent of these studies (Culler and Bazzoli, 1985) shows a large positive supply response by residents as the hourly wage increases. This study finds no backward bending supply curve. This stands in some contrast to the results of Sloan (1975) from an earlier study, which found only very small "price" effects and more-than-offsetting income effects, thus producing the classic backward-bending labor supply curve.

The empirical results we have available apply only to a select subset of doctors, whose incomes we know are lower than at other parts of their careers. For doctors in a more normal career, almost nothing is known. Thus, we should probably leave this as one puzzle not yet solved empirically.

Policy warning! Some people, assuming that because labor supply curves have a backward-bending portion for individuals they also necessarily do at the aggregate level, draw policy conclusions that are unsupported empirically. This can result in consider-able mischief. For example, one could argue that raising taxes to physicians, freezing the wages they receive by law (as was done in the 1970s) or lowering the fees they receive from Medicare (as was also done in the 1970s) would *increase* their labor supply. This *might* happen, but their response might be just the opposite. We do not yet have good studies to help clarify how this choice actually is made by physicians. To understand this, one needs to understand the aggregate supply curve of physicians.

THE AGGREGATE SUPPLY CURVE: ENTRY AND EXIT

Aggregate supply of physicians in the labor market and of physician-firms in the final product market is composed of the horizontal summation of all supplies (at various prices) of each doctor or firm participating in the market. In the case of physician labor supply, the physician may decide to retire, thus withdrawing his or her services from the labor market. This does not necessarily change the final product market. If the physician is employed in a group, the group may well simply hire another doctor. If the physician has been an entrepreneur directing a physician-firm, the firm may well be sold to another physician. An active market exists for the latter type of exchange, including a number of brokerage firms that advertise weekly in widely read medical journals such as the *Journal of the American Medical Association,* the *New England Journal of Medicine,* and numerous specialty journals.

THE OPEN ECONOMY: UNITED STATES- AND FOREIGN-TRAINED PHYSICIANS

The supply of physicians' services at any time comes from all of those who have entered practice but not yet retired. To change the *stock* of physicians takes time, since the annual *flow* of new physicians moves slowly, relative to the stock. On average, medical schools in this country graduate only a number matching (at most) several percent of the aggregate stock.

In the U.S. economy, where overall demand for medical care is growing (and has been for the last several decades), two sources can supply new doctors to the U.S. market: U.S. medical schools, and foreign medical schools. In an open economy (involving foreign trade) a large increase in demand (as accompanied the introduction of Medicare and Medicaid in the United States in 1965) should generate a rapid increase in supply from all available sources. The most "elastic" supply is the rest of the world, since we can draw physicians not only from other countries' medical school output, but also from the stock of practicing physicians in other countries.

Figure 6.3 Foreign medical graduates as a percentage of new doctors, 1945–1985.

At least some sources allege that U.S.-trained physicians are of higher quality than foreign trained physicians. Rates at which foreign medical graduates (FMGs) pass licensure examinations are below those for U.S. medical graduates, for example.

Figure 6.3 shows the proportion of newly licensed physicians in the United States since World War II attributable to foreign competition. The large demand stimulus associated with Medicare clearly created a massive influx of foreign-trained physicians.[19] At the peak in 1972, nearly half of all newly licensed physicians came from foreign medical schools.

This increase in immigration of FMGs could not have taken place without changes in U.S. immigration rules. Partly in response to a perceived "doctor shortage," these rules were greatly relaxed in 1968, particularly allowing for almost unrestricted immigration of physicians from the Eastern Hemisphere, and allowed any physician into the United States on a training visa (e.g., for a residency program) to apply immediately for permanent citi-

---

[19] One rumor during the early 1970s held that the entire graduating class of a medical school in Thailand had chartered a Boeing 747 to fly en masse to the United States after their graduation from medical school.

zenship. Increasing restrictions in immigration laws, particularly as part of the Health Professional Educational Assistance Act of 1976, reversed many of these liberalizations in immigration policy, precipitating the decline in new FMGs shown in Figure 6.3 after 1976.[20]

SUMMARY

Physicians control much of the flow of resources in the U.S. health care system. We can think about physicians in separate roles, first as the entrepreneurs of *physician-firms* and second as labor input into the production function of physician-firms. In the former sense, physicians make the same sorts of decisions as any entrepreneur—for instance, which input to use, how much to produce, how to price one's product, and so on. In the latter sense, physicians behave much as other skilled labor behaves. We can think about the economic returns to specialization, for example, as a rational investment in education, much like any other worker's decision. Physicians as workers also have decisions to make about labor supply to the market, and (unless protected by legal barriers) will likely face competition from foreign sources of supply.

PROBLEMS

1. Prove that average cost and marginal cost are equal only at the point of minimum average cost. Hint: Define $TC(Q) =$ total cost, $AC(Q) = TC(Q)/Q$, and $MC(Q) = dTC(Q)/dQ =$ marginal cost. To find the point where average costs are at a minimum, take the derivative of AC with respect to Q and set it equal to zero. When you do this, you will find that $AC = MC$.

2. (The result from this problem becomes useful in the next chapter.) Suppose that the cost curve of a physician firm has the form $TC = a_0 + a_1Q + a_2Q^2 + a_3Q^3$.
    (a) What is the average variable cost curve?
    (b) What is the marginal cost curve?
    (c) Find the level of output where the variable AC reach their minimum. (Hint: Take the derivative of the variable AC curve, set it to zero, and solve.)
    (d) Find the level of output where MC reaches its minimum. (Hint: Use the same strategy, only using the MC curve.)

---

[20] For an extended analysis of the roles of immigration and U.S. government support of medical schools as another source of competition for existing doctors, see Noether (1986).

(e) Express the output where marginal cost reaches its minimum, relative to Q^*, the point where AC reaches a minimum.

3. What features distinguish the physician-firm from physicians themselves?

4. "Physicians earn above-competitive wages, and the best evidence of this is the very high annual incomes that some doctors earn." Comment.

5. "The increasing costs of medical care can all be directly attributed to increased medical malpractice insurance costs." Comment.

6. When one thinks casually about economies of scale, one might conclude that physician-firms should be very large. What economic forces might cause physician-firms to limit themselves to relatively small sizes? (Hint: What would happen to firm size if doctors went to patients' homes rather than patients coming to doctors' offices?)

7. If the demand for physician services increases dramatically (as happened in the 1960s), what are the potential sources of increased supply of physician labor?

APPENDIX

Cost Passthrough

I. *Competitive Markets.* The behavior of prices in a competitive market follows a simple rule: Cost increases are passed through to consumers as higher prices by a ratio that depends upon both the supply and demand elasticity in the market. The percent change in price ($\%dP$) for a 1 percent change in marginal cost of production ($\%dMC$) is:

$$\%dP/\%dMC = \epsilon/(\epsilon - \eta)$$

The proof follows, using linear demand and supply curves, although the proposition holds generally:

Let the inverse supply curve (supply price curve) be of the form

$$P_s = \alpha + \beta Q$$

and the inverse demand curve be

$$P_d = \gamma + \delta Q$$

The intercepts of the supply and demand curves are α and γ ($\gamma > \alpha$), and the slopes of each curve are β and δ, respectively ($\beta > 0$ and $\delta < 0$).

In competition, $P_s = P_d$, so we can set these two equations equal to one another and solve for the equilibrium quantity Q^*:

$$Q^* = (\gamma - \alpha) / (\beta - \delta)$$

Note that Q^* must be positive, because $(\gamma - \alpha) > 0$ and δ is negative, since it is the slope of the inverse demand curve.

Once we know Q^*, we can find the equilibrium price P^* using either the supply or demand relationship. Insert Q^* into (say) the supply curve, giving

$$P^* = (\beta\gamma - \alpha\delta) / (\beta - \delta)$$

Now we can find the effect of changing marginal cost. The simplest way to change marginal cost is to shift that whole cost curve upward—that is, change α. To see the effect on P^*, take the derivative of P^* with respect to α, giving

$$dP^*/d\alpha = (-\delta) / (\beta - \delta)$$

Now if we multiply both the numerator and denominator of this expression by Q/P, we get

$$\frac{dP^*}{d\alpha} = \frac{(-1/\eta)}{(1/\epsilon) - (1/\eta)}$$

where $\epsilon = \%dQ_s/\%dP$ = the elasticity of supply and $\eta = \%dQ_d/\%dP$ = elasticity of demand. Again, after a bit of algebraic manipulation to complete the proof, this becomes

$$\frac{dP^*}{d\alpha} = \frac{\epsilon}{(\epsilon - \eta)}$$

II. *Monopoly Markets.* In a monopoly market, the monopolist sets marginal revenue equal to marginal cost, so $P(1 + 1/\eta) = MC$. Slightly reformulated,

$$P = \frac{\eta MC}{(1 + \eta)}$$

No monopolist will willingly operate where demand is inelastic, because she could always raise profits by reducing output (thereby reducing costs and increasing revenue), so $(1 + \eta)$ is always negative, and the ratio $\eta/(1 + \eta)$ always exceeds 1. It is then simple to see by taking the derivative of P with respect to MC that the change in price for a \$1 change in MC always exceeds 1 for a profit-maximizing monopolist.

Chapter
7

Physicians in the Marketplace

*C*hapter 6 describes physician-firms, without considering much how these firms interact with one another and with patients. This chapter analyzes the market for physician services further, emphasizing the interactions between producing firms, and the interaction between firms and consumers.

We begin this chapter with the most general "long-run" question confronting doctors: "Where should we locate our practices?" The answer to this question contains important economic information, but also represents a long-standing issue in public policy. For decades, public policy discussions have concerned themselves with getting more doctors into "under-served" rural and low-income urban areas. Any mechanism to achieve this in the United States (and any other moderately decentralized health care system) must operate within a market structure where doctors can freely migrate and select their locations of medical practice.

The second central focus of this chapter deals with the "intermediate-run" process by which individual patients and doctors get matched with one another. This process of search and "matching" of doctors and patients plays a prominent role in how well "the market" will function in matters of medical care and health. As we shall see, it also speaks importantly to the third question we shall study in this chapter—the episode-by-episode relationships between doctors and patients.

This third question centers on the interaction of physicians and patients and the role of information in that interaction. Physicians have more knowledge—arguably, massively greater knowledge—than patients about the interchange between them. A broad line of inquiry in health economics seeks to know how doctors use this informational advantage. Some believe that doctors use this advantage to create more demand for their services, possibly stopping only when they have reached a "target income." Others believe that, while such demand creation might happen, it is not an important part of the interchange between doctors and patients, and is surely self-limited by market forces. Still others deny the existence of demand generation. We will study the logic and evidence behind the phenomenon of induced demand in some detail. The phenomenon or concept of induced demand represents yet another of the important areas in health care where uncertainty and information centrally affect the way the market functions and the ways in which we should think about the medical care market.

These three questions form the basis for our discussion of physician markets. We begin with the long-run decision of geographic location, and then continue to the shorter-run questions of matching patients and doctors, and finally to the question of demand inducement.

PHYSICIAN LOCATION DECISIONS

A general phenomenon of markets is the spreading of suppliers across demanders. The process was first described by Hotelling (1929). A simple model to think of is ice cream vendors on a beach with people (potential ice cream customers) uniformly distributed along it. (This simplifies the problem because the "world" is linear, rather than having two dimensions.) If only one vendor exists, the optimal location for that vendor is in the center of the beach, with equal numbers of customers on each side. This minimizes the average travel cost of the customers, and hence maximizes the demand confronting the seller. If more than one vendor exists, the new ones will try to compete with location by situating themselves between the existing vendor(s) and the largest possible set of customers, since (by presumption) customers will shop at the closest possible vendor.

The key idea with this type of spacial competition is that every seller will confront the same expected number of customers. If new sellers find it profitable to enter, then the long-run allocation of sellers to locations must adjust to make the number of customers equal for each. Of course, if some of the customers have more income than others, they "count" more in this process, since they will buy more of the sellers' goods. Thus, more precisely, each seller should face the same effective demand when a process of spacial competition occurs.

Table 7.1   HYPOTHETICAL DISTRIBUTIONS OF DOCTORS ACROSS TOWNS

City	Population	Number of doctors (1 per 10,000)	Number of doctors (1 per 5,000)
A	100,000	10	20
B	20,000	2	4
C	5,000	0	1

In markets for physicians, we can understand this process unfolding by thinking of a number of cities with various populations. Table 7.1 shows a small society with three cities and 125,000 total population. If the total doctor/population ratio is just under 1/10,000 (12 doctors total in this small society) then cities A and B will have the same ratio of doctors, but City C will have none, because with a population of 5,000, it cannot compete with either City A or B to attract a new doctor. Even if the total number of doctors increased to 23, each would find it more profitable to enter Cities A or B than C, because the effective demand per doctor would be higher there.[1] Only when the total number of doctors exceeded 24 would City C finally attract a doctor. (The twenty-fourth doctor will just drop the equilibrium doctor/patient ratio to 1/5,000 in both City A and City B.)

We can observe this process in action because of the natural experiment conducted by the U.S. medical schools during the 1970s. Because of the doubling of medical school enrollment plus the influx of foreign-trained physicians, the number of active doctors per 100,000 persons actually increased by almost 50 percent (from 146 to 214 per 100,000). At present, the physician pool is still growing at about 2.5 percent each year, over triple the rate of growth of the overall population. Thus, we can actually examine where doctors have chosen to practice medicine, and how that has changed in response to this large outpouring of new graduates.

Under one extreme model, the new physicians would select the most desirable locations for practice, establish themselves, and "induce" enough demand to keep themselves content. (This requires almost unlimited ability to induce demand in the extreme

---

[1] This simple example obviously ignores the possibility for monopoly pricing, which would change the picture somewhat, but not the general idea. Thus, we can proceed with the simplified belief that every doctor will charge the same price, no matter which city is selected.

case.) Under the other extreme model (pure spatial competition), doctors would resettle themselves in a way corresponding to that shown in Table 7.1 between the last two columns. This sort of change would manifest itself in the emergence of doctors into smaller cities, where none had previously been in practice.

In most cities and even most small towns, at least one doctor is present, but the same ideas of "spacial competition" should appear as well in terms of location of doctors within individual specialties. Once having selected a specialty, doctors must also make a choice of location. How much does economics play a role in this decision?

The results correspond closely to that expected by spacial competition models (Schwartz, Newhouse, Bennett, and Williams, 1980; Newhouse, Williams, Bennett, and Schwartz, 1982a, 1982b). As Table 7.2 shows, drawn from their work, the diffusion of specialists into small towns occurred only when the total doctor/population ratios became sufficiently high to make the small towns effective competitors for a new physician. Moreover, the aggregate ratio of doctor/population of each of these specialists corresponds closely to the size of the town that can effectively compete for a doctor within that specialty. For example, in 1979, there were about 9 pediatricians per 100,000 population in aggregate, or about 11,000 people per doctor. The spacial competition model says that towns of under that size should not be able commonly to attract a pediatrician. Indeed, as Table 7.2 shows, only a small fraction of towns with 5,000 to 10,000 people contained a pediatrician, even in 1979. Almost half of the towns of 10,000 to 20,000 citizens had attracted a pediatrician by then, and nearly all towns of larger size had one.

Similarly, there were about 65,000 patients per neurosurgeon in that year overall in the United States (1.5 per 100,000). As Table 7.2 shows, only cities in the 50,000–200,000 population size regularly contain a neurosurgeon. Indeed, the same phenomenon appears in all the specialties studied. The overall population/doctor ratio for any given specialty in the country as a whole provides a good prediction of the size of city needed to attract a doctor in that specialty.

These studies clearly demonstrate how powerful economic forces direct the location of doctors. Doctors respond to effective demand by locating in regions with the highest population/doctor ratio yet available. As the outpouring of doctors from medical school and immigration increased the overall doctor/population ratio (and similarly in each specialty), they diffuse across the coun-

Table 7.2 PERCENTAGE OF COMMUNITIES WITH NONFEDERAL PHYSICIAN SPECIALTY SERVICES IN 1970 AND 1979[a]

Specialty	Number of full-time equivalent physicians in 23 sample states[b]	Population in thousands								
		2.5–5	5–10	10–20	20–30	30–50	50–200	200+
Group 1								
General and family practice								
1970	11,514	89	96	99	100	100	100	100
1979	11,869	86	96	99	100	100	100	100
Group 2								
Internal medicine								
1970	5,242	17	40	69	96	100	100	100
1979	9,467	23	52	84	97	100	100	100
General surgery								
1970	5,214	42	79	97	100	100	100	100
1979	6,071	44	77	96	100	100	100	100
Obstetrics-gynecology								
1970	2,928	13	32	74	96	100	100	100
1979	3,978	15	35	77	97	100	100	100
Psychiatry								
1970	1,990	3	12	28	46	91	100	100
1979	3,203	9	17	40	59	96	100	100
Pediatrics								
1970	2,263	6	17	57	92	100	100	100
1979	3,429	12	25	68	92	100	100	100
Radiology								
1970	1,823	5	22	60	88	100	100	100
1979	3,042	9	30	73	97	100	100	100

(Continued)

Table 7.2 (Continued)

Specialty	Number of full-time equivalent physicians in 23 sample states[b]	Population in thousands						
		2.5–5	5–10	10–20	20–30	30–50	50–200	200+
Group 3								
Anesthesiology								
1970	1,527	11	19	34	65	90	97	100
1979	2,303	11	19	40	83	100	100	100
Orthopedic surgery								
1970	1,380	2	6	29	67	91	100	100
1979	2,409	7	17	47	88	100	100	100
Ophthalmology								
1970	1,539	4	15	54	87	100	100	100
1979	2,147	4	14	62	89	100	100	100
Pathology								
1970	1,073	1	8	36	71	95	100	100
1979	1,840	4	15	50	85	95	100	100
Urology								
1970	950	1	7	29	62	98	100	100
1979	1,340	2	10	47	89	100	100	100
Otolaryngology								
1970	902	2	9	38	85	95	100	100
1979	1,127	2	6	29	79	98	98	100
Dermatology								
1970	528	1	3	10	31	79	100	100
1979	795	1	3	15	59	96	98	100

Group 4

Neurology								
1970	365	1	4	6	25	48	73	100
1979	724	0	4	13	24	70	98	100
Neurosurgery								
1970	349	0	1	2	8	28	78	100
1979	523	0	1	2	18	56	88	100
Plastic surgery								
1970	210	0	1	1	2	16	51	97
1979	430	1	1	8	20	46	83	100
Any physician								
1970	41,325	92	97	99	100	100	100	100
1979	58,911	92	98	100	100	100	100	100
No. of towns in each population range								
1970	—	615	352	182	52	58	37	33
1979	—	644	379	206	66	57	40	34

[a] Population of towns is specific to the relevant year. Data are from the following 23 states: Alabama, Arkansas, Colorado, Georgia, Idaho, Iowa, Kansas, Louisiana, Maine, Minnesota, Mississippi, Missouri, Montana, Nebraska, New Hampshire, North Dakota, Oklahoma, South Dakota, Tennessee, Utah, Vermont, Wisconsin, Wyoming.

[b] These values include physicians in towns with populations of fewer than 2,500.

Source: Newhouse et al., 1982a.

try into increasingly small towns, just as spacial competition requires. By the end of the period studied by the group at RAND, the ratio of patients to doctor had fallen considerably, so smaller towns became more likely to contain a specialist that fit into their size category.

A study of the migration of physicians and dentists (Benham, Maurizi, and Reder, 1968) provides a different but supporting portrait of the same market forces. That study looked across states (not towns), and found that doctors and dentists tended to move *from* states with relatively low economic returns *to* those with relatively high economic returns. Thus, direct study of the movement of physicians across states also supports the importance of market forces in determining the location of physicians and other healing professionals.

The Health Service Corps

The same forces that create this diffusion of doctors into small towns create a general futility for a recent public program designed to induce doctors to establishing practices in small towns. The program, the National Health Service Corps, helped doctors pay for the costs of medical school in exchange for their accepting an assignment in an "underserved" area for a fixed period of time after their graduation. Such programs can succeed only temporarily in locating a doctor in a town where market forces will not otherwise direct one. The idea of the program was presumably to get doctors to "settle in" to an area, and then open a private practice after their tour of duty was completed. While some doctors did stay in the towns where they had been assigned, a study of the presence of doctors in previously "underserved" areas showed that these areas were equally likely to contain a doctor whether or not they had a Health Service Corps doctor assigned to them (Held, 1976). Moreover, since we know the general effect of market forces in directing more doctors to small towns now, we cannot say for sure that towns that have a Health Service Corps doctor would *not* have had a doctor purely from market forces. Of course, if the HSC had assigned a doctor to a small Town X, that would "fill up" the spot that might have been taken by a freely migrating doctor.

Similarity to Decision to Specialize

The decision about location of a medical practice has considerable similarity to the decision to select a specialty. In both cases, economic forces direct the aggregate patterns, while individual preferences may well influence individual choices. Whether we think about a choice between "neurosurgery" and "radiology" or a choice among "Spokane, Washington," "Washington, D.C.," "Washington, Iowa," and "Phelps, New York," the same economic forces affect the choices of doctors. In the case of specialty choice, as we have seen previously, one must consider the present value (at the time of decision) of the future income stream, since the decision to specialize represents a time-consuming investment. A geographic locational choice has lower costs and is easier to adjust. Despite these differences, the general proposition remains the same: Doctors respond to economic forces in systematic and predictable ways.

CONSUMER SEARCH AND MARKET EQUILIBRIUM

Having studied how doctors select a location, we can turn to the next question in our analysis of physician-service markets: How do doctors and patients "match up," and how is the price determined?

In a classic "textbook" competitive market, several clearly distinguishable features would appear. First, we could unambiguously determine the market·supply and demand curves by adding up the supply and demand curves from all participants in the market. The quantity and price that we observe in the market would be determined by the intersection of these market supply and demand curves. We could also readily calculate the effect of changes in cost on the equilibrium quantity and price (see Appendix to Chapter 6). Each consumer and provider would of necessity take the market-determined price as given and would respond accordingly: Suppliers would select the amount to produce, and consumers would select the amount to consume, based on this market price. Their collective choices would lead to the "right" quantity being produced and consumed. Finally, if we could cleanly and clearly measure the quality of care, only one price would prevail in the market for each level of quality observed. This would be "the" price of a physician office visit in the market.

The contrast with what we observe in actual data seems remarkable as Box 7.1 shows. For the most part, many prices prevail in a given market area. The dispersion of prices seems far too large to correspond to quality differences. Indeed, several studies have shown substantial price dispersion for *identical* goods in the same geographic area (see, for example, Pratt, Wise, and Zeckhauser, 1979).

Box 7.1   **PRICE DISPERSION IN MEDICAL MARKETS**

In medical markets, this dispersion of prices is very common. While some of the differences in price are surely related to quality, it seems hard to accept that *all* of these differences are quality-related. For example, in the city of Dayton, Ohio, in 1975, the average difference in price for a "standard office visit" between general practice doctors and internal medicine specialists was only $1.50, while the spread in prices was much larger. The specialists on average received only 10 percent more for an office visit than the GPs (Marquis, 1984).

Figure A shows distributions with the same averages and standard deviations as the distributions of prices in Dayton. The averages in this figure are general practice (GP), $15.30; internists, $16.90; other, $15.80. The range of prices for GPs extends from under $9 to over $21. For internists, the range extends from under $10 to nearly $23. Thus, while the *average* difference between specialists and GPs was only 10 percent of the price, the dispersion of prices within each specialty group was such that the high price was over twice the low price.

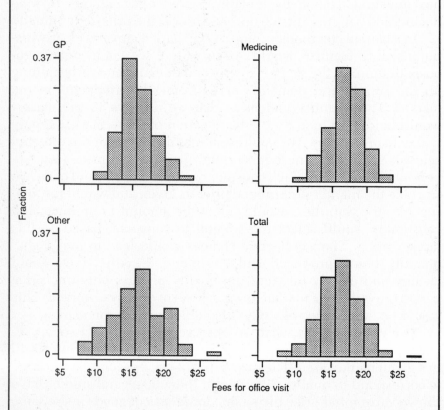

Figure A

> The same phenomenon appears pervasively in other markets and for other medical services. In Seattle, for example, the same study showed a slightly bigger difference between specialists and GPs, but an even larger within-specialty dispersion. In dental markets, insurance data commonly show the dispersion in prices for "routine" services like cleaning, extraction, and fillings have similar characteristics.*
>
> Surely, some of these differences are due to quality differences perceived by patients, but not measured by the researcher. Tautologically, one could say that *all* of these differences in price are due to such unobserved quality differences, and there would be no way to refute that. However, given the small difference between the average price of GPs and specialists, it seems difficult to believe that the entire dispersion is quality-related.
>
> ---
>
> * The same thing occurs in other markets as well. Auto insurance companies commonly require that you get three quotes on any auto repair job, and then pay for the lowest of those three. A rule of thumb says that when you get three quotes, the highest price will be twice as high as the lowest price for body work.

Second, doctors seem to face a downward-sloping demand curve for their services. There is a widespread belief (albeit poorly documented) that doctors price-discriminate—charge different people different amounts for the same service.[2] Such discrimination is impossible in a purely competitive market. (Box 7.2 describes monopoly pricing and price discrimination in more detail.)

The model of monopolistic competition describes the market for the physician service market quite usefully. The basic idea of monopolistic competition is that each producer faces a downward-sloping demand curve that shifts inward (or outward) as the number of other producers in the market increases (or decreases), or as the market demand curve shifts inward (or outward). The number of other sellers increases or decreases as opportunities exist for profits. The "monopolistic" part of the name comes from the downward-sloping demand curve. The "competition" part of the name comes from the idea of free entry into the market by other competitors. In monopolistic competition, entry occurs until everybody is just making a competitive rate of return (zero monopoly profits).

---

[2] The classic article on this topic (Kessel, 1958) describes the pervasiveness of the phenomenon. However, the logic used by Kessel blurs the distinction between a monopoly in an input market (physician labor) and the way the output market functions (physician services).

Box 7.2   MONOPOLY PRICING AND PRICE DISCRIMINATION

A monopolist is the only seller in a market, and thus the demand curve facing the seller is identical to the market demand curve. If the monopolist wishes to maximize profits, the best solution is to reduce output below the competitive level, and raise the price in parallel.

For each intended increase in the volume of sales, the monopolist must lower the price on *all* units sold. Thus, the addition to total revenue ("marginal revenue") coming from one more unit of sales is the revenue on that sale *minus* the reduction in price on all other units sold. Since the monopolist confronts the demand curve of the market, the amount the price must fall can be determined from the market demand curve.

The marginal revenue curve in Figure A shows the additions to total revenue associated with the market demand curve. (For straight-line demand curves, the *MR* curve just bisects the angle between the demand curve and the axis of the price-quantity diagram.) Intuitively, what the monopolist wants to do is expand production until marginal revenue just equals marginal cost of production. At that point, any further additions to output would create less incremental revenue than the incremental costs of production. Profits reach a maximum when $MR = MC$. A monopolist picks this quantity of output, and then sets the price *by moving up to the demand curve* at that level of output. Thus, in the monopoly market, output is lower and price is higher than in a competitively priced market with the same demand curve and marginal cost curve.

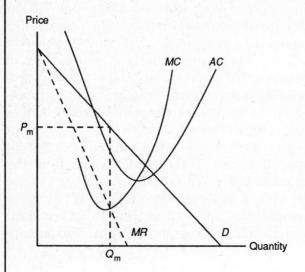

Figure A

The marginal revenue function can be described formally in terms of the demand elasticity. $MR = P(1 + 1/\eta)$, where η is the demand elasticity. Note that

MR is negative if the demand curve is *inelastic* $(-1 < \eta \leq 0)$. Thus, no monopolist will willingly operate in the realm of inelastic demand. It always would pay to reduce output and raise price. The general pricing rule for a monopolist is to set *MR* = *MC*, so $P(1 + 1/\eta) = MC$. Another reformulation of this says that $P = \eta MC/(1 + \eta)$. Since $(1 + \eta)$ is negative when demand is *elastic*, $P > MC$ at all points where a monopolist will operate. For example, if $\eta = -1.5$, $P = 3 \times MC$. If $\eta = -2$, $P = 2 \times MC$. If $\eta = -20$, $P \approx 1.05 \times$ MC. Thus, while it is strictly correct to describe a firm as a monopolist if $\eta = -20$, it's not very interesting, since we can't measure *MC* of firms within a 5 percent accuracy. The general rule is that the larger the demand elasticity confronting the monopolist, the closer the price will be to marginal cost.

Discriminating Monopoly. In a price-discriminating monopoly, the monopolist can identify different segments of the market with different demand elasticities, and sets a price to each of them according to the price rule just described. Those with more elastic demands get lower prices, and conversely.

To maintain a successful discrimination, the monopolist must be able to prevent resale of the product from those getting the low price to those getting the high price. Medical care is ideally suited for this: It's hard to buy a few extra appendectomies and resell them to your friends.

Figure 7.1 shows how the typical firm appears in the monopolistic competition equilibrium. The firm faces a downward-sloping demand curve and has an average cost curve of typical U-shape. In equilibrium, *entry by other firms will have occurred until the demand curve for each firm is just tangent to its average cost curve.* At this point, each firm just covers its cost, it faces a downward-sloping demand curve, and no further entry into the market is economically attractive. Price equals average cost ($P = AC$), but production does not take place at a minimum average cost (i.e., not at P_{min}). As Chamberlin (1962, the originator of the idea) described it, there will be persistent excess capacity in the market, in the sense that every producer could expand output and obtain lower costs.

The physician's decision about geographic location described in the previous section precisely describes this type of entry-exit decision. When a market has above-average economic potential, physicians will enter. Otherwise, they will disperse themselves uniformly around the effective demand in all other markets.

Search in Monopolistic Competition

The most powerful analyses of market behavior have employed the concept of consumer search in tandem with the monopolistic

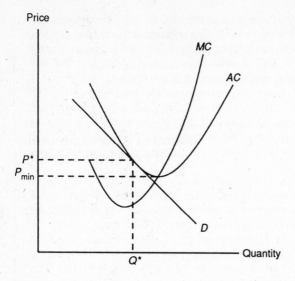

Figure 7.1 Typical firm in monopolistic competition equilibrium.

competition model. To understand the role of search, we can begin with an extreme case: Suppose *no* consumers in a market ever engaged in search for a "better" doctor (improved price, quality, or both). Patients and doctors would be matched at random or on some other basis that did not correspond to price or quality. Each firm's demand curve would be a scaled-down replica of the market demand curve. If the market demand curve had an elasticity of −1.5, so would each firm's demand curve. Each would price according to the pure monopoly model, since each would know that the consumers who arrived at the firm were "theirs."

Now consider a consumer who engages in search. Suppose, for example, that the consumer decides (by some process) to search three firms, and take the firm with the lowest price (or most desirable price-quality combination). Now the firm faces a dilemma when setting price: At higher prices, it gets more profits from the sales it makes, but it increases the risk that a comparison shopper will find a lower-priced firm, and thus the higher-priced firm will lose business. Lower prices obviously do just the reverse. Selecting the firm's best price requires a balancing act between these two forces.

It should seem clear (and can be proven) that as more and more of the consumers in a market engage in search, the closer the market comes to a purely competitive market. The fewer consumers who search, the higher the price that can be expected, with monopoly prices as the extreme outcome when nobody engages in search.[3]

The other useful feature of this model is that it allows different prices for different producers, depending on their cost structure. For example, differences in fixed costs of each firm will lead to different optimal prices for each firm. Thus, at least when a "small" proportion of the consumers in a market engages in price searching, the market should have a distribution of prices (rather than everybody charging the same price) even if everybody's quality is the same. Those with higher fixed costs have higher prices, and conversely. Of course, when a "large" fraction of consumers shops around, the distribution of prices collapses to the competitive price.

Health insurance plays two roles in this story. First, as a technical matter in the mechanics of the monopolistic competition model, as the demand curve becomes less elastic (less price responsive), the price dispersion will increase and average prices will become higher. Intuitively, we can see how this happens by noting that demand curves become more vertical (less elastic) with insurance (see Chapter 3). Hence, the point at which they become tangent to any firm's AC curve *must* be higher up to the left, rather than lower down near the bottom of the U-shape of the curve. This is one effect we can expect from health insurance on the equilibrium dispersion of prices on physician markets.

Second, at least with some forms, insurance may also change the incentives for people to shop. The obvious effect (say) of full-coverage insurance is that it reduces the incentives to shop for price. Some forms of insurance preserve at least some of the incentives to search for price, but at the extreme, a person with full coverage has no incentive to search for price. As one study has pointed out, however, insurance can also *increase* the amount of comparison shopping going on. By eliminating much of the financial cost of "trying out" a different doctor, insurance might actually increase search (Dionne, 1984). Particularly if some consumers wish to sample the "quality" of a different provider, insurance can

---

[3] The most comprehensive study of this is by Sadanand and Wilde, *Review of Economic Studies*, 1982. This paper is highly technical, and not recommended for the novice economist. A more accessible version of this work appears in Schwartz and Wilde, 1979, written for lawyers in the *Pennsylvania Law Review*, or Schwartz and Wilde (1982) in the *American Economic Review*.

promote search. Given the relatively wide dispersions in prices, it would be difficult to *presume* that quality and price were strongly correlated in this market, even though they are clearly connected.

There is growing evidence that doctors create a "style" of practice that best matches their own preferences, and then attract patients who also prefer that style (Boardman, Dowd, Eisenberg, and Williams, 1983). One "style," for example, might be the "busy office practice" with "standard" amenities. Another might be "Medicaid patients," which would probably decrease the amenities and increase the rate at which patients were seen in the practice.[4] A third style might be called "Beverly Hills Doc," with a low patient volume and extended time (and high prices) for each patient visit.

Several studies have characterized the search process patients might undertake in a specific way (see, for example, Pauly and Satterthwaite, 1981, and Satterthwaite, 1979, 1985) and concluded that there will be *less* search in markets with a high density of doctors. Their logic is that it will be difficult for any individual person to find anybody else who uses a particular doctor when there are large numbers of doctors in the city. The style of search in this analysis is exemplified by a patient expressing an interest in Dr. Johnson, and then starting to ask friends, "What do any of you know about Dr. Johnson?" In a small city with only a few doctors, somebody to whom the patient speaks will surely know Dr. Johnson and the search will be successful. In a larger city with many doctors, the answer will probably be "nothing," in the sense that no single patient is likely to know much about a randomly named doctor.

Two problems appear in this search model, however. First, if the search concentrates on prices, then it seems no more costly to acquire price information from a sample of five doctors, no matter whether there are five or five hundred doctors in the city. Each requires only a phone call to ask (say) "How much do you charge for a physical exam?" Thus, the Pauly-Satterthwaite type of search would be most relevant for gathering information on quality.

The second issue is how patients actually conduct their

---

[4] The pejorative term *Medicaid mills* has been applied to some doctors or clinics. Clearly, not everybody who adopts a Medicaid-intensive style necessarily practices medicine deserving of scorn.

searches. In an alternative search process to the one embedded in the Pauly-Satterthwaite model, a person would begin by asking several friends who their doctor is, how well they like the doctor, and something about the doctor's "style." By selecting friends with similar preferences, the person could gather at least some useful information about some doctors in town quite readily. The costs of gathering such information seem about the same, as with the price information search, no matter how many doctors are in the city. If so, then search costs should be about the same in any city. Box 7.3 extends the discussion of patients' searches.

The main motivation for the analysis by Pauly and Satterthwaite is to explain a common empirical phenomenon—namely, that prices for doctors are higher in big cities than in small cities, even after adjusting for costs of inputs. They explain this with the logic that search is more costly in the bigger city. The debate among economists (like the debate on the evidence for "induced demand" that we will review shortly) hinges on numerous technical issues and has not been resolved clearly.

We should be careful not to embrace any economic model just because the data are "consistent" with it. Other equally plausible models also may match a particular empirical finding, but those models might disagree on other important dimensions. Key tests among models occur when they make different predictions about the world, and then critical data can be found or produced that focus precisely on the differences in predictions, rather than on their similarities.

In the case of Pauly and Satterthwaite's model, for example, another very simple economic model predicts precisely the same phenomenon (higher prices when higher physician density occurs). This study, by Devany, House, and Saving (1981), emphasizes the trade-off between price charged and the rate at which patients can be served by doctors. In their model, a community where patients have a relatively high value of time will lead physician-firms to establish a practice style with relatively low waiting time for patients, but at higher costs (and hence higher price), and vice versa in communities with a low value of time. They conclude that "we must observe a positive association between money price and firm density." Thus, the empirical finding that prices are higher when physician density is higher can appear without any of the differences in search cost that Pauly and Satterthwaite depend upon.

Box 7.3   CONSUMER SEARCH AND PRICES

The technical debate between economists about the role of consumer search and prices can get quite complicated, but some key ideas appear throughout this work. The following contains some selected sections from a discussion by Mark Satterthwaite, telling why he believes that consumer search is more difficult in a market with many doctors, and a response by Jeffrey Harris, offering an alternative explanation for the same phenomenon. The discussion revolves around two pervasive phenomenon, as characterized by Satterthwaite (1985):

1. At any time, physicians in communities that are well supplied in terms of physicians per thousand population are likely to charge higher fees than physicians in communities that are less well supplied.

2. Increasing the aggregate supply of physicians per thousand has had no obvious downward effect on physicians' prices.

The debate takes the following form:

Consumers when they are seeking a new physician . . . generally rely on the recommendations of trusted relatives, friends and associates. . . . If the cost of eliciting detailed and useful recommendations from trusted relatives and friends is low, then the problem of finding a new physician is easier than if the subjective cost is high. . . . The ease with which a consumer searches for a new physician . . . depends on two factors: the number of people in the community from whom he or she feels comfortable seeking recommendations, and the level of relevant knowledge possessed by those whose recommendations are sought. Both of these levels are likely to be low in communities that are experiencing rapid growth, that have a great deal of population flux, and that do not have a high level of social cohesion. . . . In addition to these straightforward social and demographic determinants of the cost of search, the number of physicians in the community, N, may by itself affect the ease or difficulty of consumer search. . . . Because so many physicians practice in a large city and their practices overlap geographically to a greater degree, every urbanite hears stories about a large number of different doctors. Except for those blessed with unusual memories, the stories [heard from friends about doctors] become a jumble; people do not remember which story is about which physician. Therefore, when a person is asked for a recommendation, oftentimes they can only report on their own personal physician, because they cannot remember any specific information about any other physician. This makes search more difficult and means, in conclusion, that [other things being equal] the more physicians serving a community, the poorer consumer information is likely to be.

In a discussion of this article (which was presented at a conference in 1985), Jeffrey Harris, who is both an economist (at MIT) and a physician (at the Massachusetts General Hospital) offered the following ideas:

In the Chamberlinian story of monopolistic competition, an expansion in the size of a market is ordinarily accompanied by an increase in the extent of product differentiation. At the start, microcomputers offered a few basic packages. As the market grew, specialized home computers, scientific computers, and computerized games appeared. At the start, Camels, Chesterfields, and Lucky Strikes were the three main brands of cigarette. As the market grew, menthol cigarettes, filter cigarettes, 120mm cigarettes, and low-tar cigarettes emerged.

The same phenomena are taking place in the physicians' services market. Starting in the 1960s, the U.S. government markedly increased its subsidy of both the cost of medical care and the cost of medical education. The resulting growth of physician supply has been accompanied by a proliferation of "physician brands." This product differentiation has not been confined to the emergence of new subspecialties based on new medical techniques. The emergence of new residency programs in primary care and family medicine were also manifestations. Physicians moved increasingly into . . . group practices. New practice styles, with enhanced use of paramedical personnel, have evolved. . . . We are seeing more physician-managed diagnostic centers, emergency care centers, and ambulatory surgical centers. . . . We have no unequivocal economic basis for proclaiming that the proliferation of new styles is beneficial or deleterious. . . . My guess, in any case, is that the welfare consequences of such increased product differentiation overshadow the consumer search effects that Professor Satterthwaite has identified. To me, the issue is whether surgicenters offer better medical care, not whether the proliferation of such centers makes it more difficult to pick a surgeon.

In a separate study of physician practice styles and prices, Boardman, Dowd, Eisenberg, and Williams (1983) discuss the positive association of physician fees and the number of physicians, noting that Satterthwaite explains this in terms of search costs. On the basis of their study, however, they offer another idea—namely, that the association rests on the link between the scale of the market and the degree of specialization that one would expect to find:

There are other possible explanations. In towns with few physicians there would be tremendous pressure for . . . physicians to "compete at the median" and adopt [a busy, successful ambulatory practice style]. In larger cities with many patients with different tastes and many physicians willing to meet these diverse needs there would be greater opportunity to segment the market. In particular, we would expect some physicians to see fewer patients, but charge higher fees.

Their empirical studies support these ideas. The propensity for a doctor to adopt a few-patient, high-fee ("Beverly Hills Doc") practice was strongly increased with the physicians per square mile, while the propensity to adopt the "standard" practice fell. Of necessity, this will increase the average price in a community as physician density increases. This corresponds closely to the model of DeVany, House, and Saving (1982).

ACTUAL PATIENT SEARCH

One study shows how often (and for what reasons) patients actually do search for a new doctor (Olsen, Kane, and Kastler, 1976). In a survey of 632 households in the Salt Lake City region in 1974, residents were asked about their use of doctors, including whether they had ever changed doctors. About 60 percent of the respondents had changed doctors at some time, at about the same rate for both low and high socioeconomic-status (SES) groups, and about another 10 percent wanted to change doctors, but were deterred in some way (e.g., fear of offending current doctor). The reasons given for changing doctors appear in Table 7.3. Note that the most prominent reasons all are the "right" kind from the point of view of economic search models. Patients switched doctors because the delays to appointment were too large, they didn't like the quality of care, they sought verification of a diagnosis, or the price was too high. Note also that about 10 percent of those who switched used the single recommendation of a friend (see discussion above about search in the Pauly-Satterthwaite model).

Table 7.3   REASONS FOR DOCTOR SHOPPING BY SOCIOECONOMIC STATUS

Reasons for shopping	Percentage of shoppers	
	High SES (n = 234)	Low SES (n = 153)
Couldn't get appointment within week	53%	51%
First doctor not helping patient	49%	44%
To verify a diagnosis from first doctor	38%	34%
Friend recommended another doctor	26%	26%
Sought nontraditional healer because doctor wasn't helping	19%	16%
Care too expensive or could do with less expensive care	12%	22%
Personal friends with doctor	16%	5%
Belong to same club or church as doctor	6%	5%
Preferred doctor of different sex than first doctor	3%	3%

Source: Olsen, Kane, and Kastler, 1976.

ADVERTISING AND THE COSTS OF INFORMATION

One way to increase the amount of search in a market is to reduce the cost of information. Having consumers search is intrinsically desirable because it moves the market closer to the amount of medical care delivered (and hence received) that is "correct" from an economic standpoint. (The "right" level occurs when the marginal value to patients of X amount of care just equals the marginal cost of the same amount. This will maximize total social well-being, which is the economists's usual definition of "right.")

Advertising for professional services has been prohibited by many states for years. These prohibitions have been struck down by many states recently as anticompetitive, but there still remains strong pressure from most professional groups against advertising. Advertising, like "the force" in a recent science fiction movie series, has both good and bad sides to it. Certainly it has the capability of distorting patients' choices, and it is this "dark side" of advertising that professional groups emphasize in their opposition to advertising. The word *quack* commonly appears in such discussions, and the prohibition against advertising allegedly protects consumers from quacks.[5] Some people even fear that advertising, by increasing the demand for a product, might cause higher prices to appear. (Of course, when advertising is common in a market, the costs of that advertising form part of the cost structure of the firm, and hence are part of its pricing strategy.)

The other thing advertising might do, however, is to reduce the costs of providing information to consumers. In the monopolistic competition-search model, more consumer search leads to lower prices and a smaller dispersion in prices. In a classic study of the effects of advertising, Benham (1972) used survey data to measure the price paid for eyeglasses by consumers throughout the country, repeated with more sophistication three years later on another data base (Benham and Benham, 1975). Some of the states prohibited advertising, while others permitted either limited or unlimited ads. Their study found that the price paid for a pair of eyeglasses in states that permitted advertising was about 25 percent lower than

---

[5] Despite the connotation of the sound of a wild fowl, the word *quack* comes from elsewhere (although not the Greeks, this time). *Quack* is a short form of the German for *quacksalver*, a person who attempts "quickly" to heal with salves, but without any scientific basis for the cure.

in states that prohibited advertising (a difference of about $8.50 in 1970, corresponding to about $25 currently), even after controlling for other factors that might affect price. In states with advertising, the average size of firms selling eyeglasses considerably exceeded that in the nonadvertising states. Apparently, one effect of advertising was to allow some firms to attract enough customers that they could take advantage of existing economies of scale.

Subsequent studies have extended the ideas contained in Benham's original work. Cady (1976) has shown the same phenomenon for prescription drugs, but with a smaller proportional effect (about 4 percent higher prices in restrictive states). Benham and Benham (1975) showed that the higher prices from advertising restrictions also reduced the quantity of the product consumed. Feldman and Begun (1978) held quality of the product constant and compared prices across states, finding prices on average to be 16 percent higher in those states without advertising for eyeglasses. Kwoka (1984) measured the quality of service provided, and found that one way advertising reduced the cost of eyeglasses was a parallel reduction in quality of service. Apparently, when firms could not advertise with price, they resorted to quality-based competition that led to quality that was "too high" from a market perspective (in the sense that when advertising was allowed, the market reverted to a lower-price and lower-quality combination).

Some interesting puzzles remain in the question of professional advertising. For example, advertising by dentists appears much more common than advertising by physicians. Inspection of the weekly television guide in the newspaper of many cities will reveal ads for dentists commonly, but almost never ads by physicians. The Yellow Pages telephone directory shows the same phenomenon, although advertisements by physicians are now much more common than they were (say) 10 years ago. Mostly, when physicians do advertise, they either do so because they are in direct competition with nonphysicians firms for a similar service (quit-smoking or weight-loss clinics), or because they have a new service to offer of which few people are aware. (Radial keratotomy, a surgical procedure to correct poor vision without contact lenses or eyeglasses, offers a prime example. This procedure was developed in the former Soviet Union.)

THE ROLE OF LICENSURE

Licensure by the state for professionals has a long if occasionally checkered history. Our governments (at various levels) license physicians, registered nurses, dentists, and many other healing professions. Governments also require licenses for barbers and beauticians, airplane pilots, automobile drivers, and civil engineers, but not college professors, hockey players, financial advisors, rock musicians, or economists. It is not readily apparent that in the absence of licensure the licensed groups (e.g., beauticians) would be intrinsically more dangerous to society than the unlicensed group (e.g., rock musicians) but licensure is commonly supported on the belief that it protects the safety of the public from incompetence by practitioners of a profession. Interestingly, licensure commonly occurs for *inputs* into a productive process, but seldom for the firm responsible for the *output*. In medical care, of course, doctors and nurses receive licenses, but not physician-groups. While hospitals receive safety inspections, these relate more to fire hazard and food safety than to the quality of the actual product. In air travel, pilots and airplanes are both rigorously certified, but not airline corporations.

Like advertising, licensure has both potential good and bad features. The obvious "good" is the maintenance of quality and the prevention of harm to patients, which loom particularly large if patients have difficulty in assessing the quality of a provider of care. Licensure can provide a "floor" on quality of care upon which consumers can rely without any investigation of any particular provider. To the extent that it succeeds in this, licensure can also increase search, by making price information seem more useful (in the sense that a low price would not imply too-low quality).

Licensure also can stand as a barrier to entry, decreasing competition and creating monopoly rents to those who obtain a license. Taxicab licensing in some cities has clearly done this, for example, as evidenced both by the presence of unlicensed taxicabs (despite the risks of conviction) and more important, by the resale price of a taxicab "medallion," the license attached to the vehicle (Kitch, Isaac, and Kaspar, 1971).

In a classic study of physician markets, Kessel (1958) argued that licensure restrictions on entry into the medical profession created entry restrictions that led not only to monopoly pricing by physicians but also to price discrimination. (See Box 7.1 describing monopoly pricing.) While Kessel's evidence on price discrimination may be accurate, the imputation of cause to medical licensure cannot be correct. Licensure might create a barrier to entry of an *input* in the production of physician services, but it says nothing about the production or market organization of the *output* market. If licensure does create an economically important entry restriction in the input market, it will raise the cost curve of every firm producing physician services. Licensure cannot, however, create the opportunity for such firms to price monopolistically, because (1) physicians can and do migrate (see Benham, Maurizi, and Reder, 1968), and (2) physician-firms can (and do) substitute other inputs for those of physician labor (see Reinhardt, 1972). Other economic phenomena, primarily including the extent of search by patients, will determine whether physician-firms act as monopolists or competitors.

On net, licensure can have both economically desirable and undesirable features. The capability for limiting entry obviously has some economic liabilities associated with it. However, the quality control capabilities of licensure are probably positive. On net, one cannot say that licensure is necessarily a benefit or a harm to consumers.[6] The potential gains in quality information may outweigh any costs from monopolization. Indeed, it may even be that the market operates more competitively because the "minimum quality" guarantee that licensure produces may increase consumer willingness to search for lower prices.

Medical specialty boards have similar aspects to licensure, except that they are voluntary rather than mandatory. Thus, while every doctor is supposed to have a license, specialty certification is completely optional. Of course, since specialty boards cannot limit entry into the medical profession, they primarily serve as indicators of quality. It would be difficult to construe them as barriers to entry, since one can practice a specialty without having the board certification (but a doctor cannot say that he or she is certified by the board unless that is true). In addition, there are occasionally multiple (competing) organizations providing certification for the same area of medical practice, although in general, only a single specialty board exists in any given area of medical practice.[7]

---

[6] The earliest proponents of the monopoly-restriction hypothesis were Kessel (1958) and M. Friedman (1962). The importance of licensure for quality control was emphasized by Arrow (1963), and more recently by Leffler (1978).

[7] In some areas of the economy, multiple certification is common. For example, in scuba diving instruction, three common certification bodies include the National Association of Underwater Instructors (NAUI), the Professional Association of Diving Instructors (PADI), and the YMCA. In the area of medical care, certification for people prescribing eyeglasses has arisen at least in three areas; for M.D.s, the Board of Ophthalmologists; separate certification exists for optometrists (who have different training in separate schools), and most recently, for "prescribing opticians," i.e., a group of people who formerly specialized in making lenses, but now also prescribe them.

ESTIMATES OF THE DEMAND CURVE FACING PHYSICIAN-FIRMS

The various concepts discussed previously all have some effect on the scope of physician practices and where they locate. Given all of these factors, we know that doctors will disperse themselves into various regions. What market conditions do they face then? One recent study has estimated the demand curve confronting primary care physician-firms using survey data from the American Medical Association (McCarthy, 1985) for firms located in large metropolitan areas (population > 1 million). The results support much of the previous discussion and add empirical specificity to the concepts previously discussed. Perhaps of most importance, the price elasticity of the demand curve facing a typical large-city primary care physician is quite large. In all of the various forms estimated, this study found the *firm's* demand curve to have an elasticity of −3 (or larger, in absolute value). We should be careful not to confuse this with the elasticity of the *market* demand curve, which (according to the RAND HIS results) is about −0.2 to −0.3. The difference, of course, is that the individual firm loses customers to other firms by raising prices, where "the market" as a whole loses customers only when they drop out totally due to higher prices. An elasticity of −3 certainly confirms the validity of using something other than a "perfectly competitive" model to explore physician markets, but it also shows that the departures from purely competitive markets cannot be huge.[8]

The estimates in this study also showed the effects of waiting time on prices just as one might expect. Longer waiting times in the office reduced demand for physician visits, holding price constant. The estimated elasticity confronting physician firms ranged from −0.4 to −1.1 in various versions of the model estimated.

In the context of the monopolistic competition and search models, another interesting result emerged: Contrary to previous empirical findings (which show higher fees with higher density) this study, which used data on individual firms rather than data at the county or SMSA level, found that higher physician density in a region *reduced* the demand confronting a physician. This finding

---

[8] The optimal price for a monopolist is inversely related to the demand elasticity confronting the firm. In general, where MC = the marginal cost of production, the optimal price is found by setting $P = MC/(1 + 1/\eta)$, where η is the demand elasticity facing the firm. If $\eta = -3$, then $P = MC/(1 - 1/3) = (3/2)\ MC$. Put differently, there would be a 50 percent "markup" of price above marginal cost.

corresponds precisely to the standard monopolistic competition model, where more firms in a community imply fewer customers per firm. This, of course, shifts inward the demand curve facing any firm, thus reducing the price it can charge. This finding helps strengthen the belief that something other than "difficult search" makes prices higher in markets with more doctors.

Excess Capacity

A final aspect of the monopolistic competition model is the predicted presence of "excess capacity" in the market. This sort of excess capacity is most likely to arise in areas where the demand curve facing a single physician-firm is least elastic (see Figure 7.1, and imagine the output rate where the tangency exists with the U-shaped AC curve, compared with the output producing minimum AC: The steeper the demand curve, the further the actual output will be below the point of minimum AC). Markets where search is least likely to take place and where insurance coverage is greatest offer likely targets to find considerable excess capacity. In one study, Hughes, Fuchs, Jacoby, and Lewitt (1972) actually examined the working hours and practices of surgeons. They found persistent excess capacity among surgeons, often with as much as 40 percent "slack time" in surgeons' practices.

INDUCED DEMAND

> Lather. Rinse. Repeat. [Common attempt to induce demand by manufacturers of consumer product]

It is probably fair to say that the question of induced demand separates "academic" study of health economics from "policy-related" questions more than any other topic. Academics find the topic fascinating, and spend nearly endless resources devoted to its study.[9] By contrast, probably few members of Congress could even define the topic, although perhaps they should learn what it is. The three key political issues (and the government studies evolving from those interests) are cost control, cost control, and cost control (to paraphrase the standard statement about real estate).[10] However, if the proponents of the induced demand con-

---

[9] The editor of the *Journal of Health Economics*, Joe Newhouse, once said that he had considered renaming the journal as the *Journal of Induced Demand* because he had so many articles submitted for publication on this topic.

[10] Location, location, and location.

cept are correct, perhaps the political issues should at least also contain the concept of induced demand.[11]

The idea of induced demand is not limited to the provision of medical care. Almost any person who has ever owned an automobile has probably felt at some time that he had been sold a repair service that was not needed. From the consumer's point of view, however, it's sometimes just not worth the trouble to check up on a mechanic (Darby and Karni, 1973). In medical care, the large differences in knowledge between the doctor and the patient certainly suggest the possibility that doctors could use their position of superior knowledge for their own financial benefit.

The idea of "demand inducement" by doctors was given its most prominent boost by studies of Robert Evans (1974) and Victor Fuchs (1978). Since that earlier work, literally dozens of studies that have appeared in professional economics journals discuss aspects of the question.[12] The central idea stems from the observation that (with hospitals) areas that had greater hospital bed supply had greater hospital utilization (Roemer, 1961). Economists were quick to dismiss the relevance of such simple data with the observation that a competitive market would produce such results, with supply *following* demand into areas of high demand.

Fuchs's study offers a good example of the type of evidence available on the extent of inducement. He estimated the demand for surgical procedures in a number of metropolitan areas (SMSAs) holding constant price, income, and other relevant variables, and using the physician supply *predicted* to be in a region from economic forces that could be measured.[13] He found that as that predicted supply increased by 10 percent, the number of surgical

---

[11] Recent proposed changes in Medicare payments to physicians actually did take account *in advance* of a presumed increase in quantity of services in response to a reduction in physician fees. Thus, demand inducement has, in some ways, entered health policy.

[12] For the truly fascinated student of the topic, the list contains (in alphabetic order, but not comprehensively) Auster and Oaxaca (1981); Boardman et al. (1983); Cromwell and Mitchell (1986); DeVany, House, and Saving (1983); Dranove (1988); Evans (1974); Evans, Parish, and Sully (1973); Farley (1986); Feldman (1979); Fuchs (1978); Fuchs and Kramer (1972); Green (1978); Hay and Leahy (1982); Held and Manheim (1980); McCarthy (1985); McCombs (1984); Newhouse et al. (1982a, 1982b); Pauly (1980); Pauly and Satterthwaite (1981); Reinhardt (1985); Rossiter and Wilensky (1983); Satterthwaite (1979); Schwartz et al. (1980); Stano (1985, 1987); and many others. The price of paper argues against providing a complete listing of relevant citations.

[13] This technique involves the simultaneous estimation of supply and demand curves for a product. In econometric terms, one uses Two Stage Least Squares (TSLS) or some nonlinear equivalent.

Box 7.4   ECONOMISTS' PERSPECTIVES ON DEMAND INDUCEMENT

The idea of demand inducement creates strong feelings among economists, some of whom accept it as a simple matter of course, and others of whom adamantly deny its very existence. The reasons for worrying about it are not trivial: Demand inducement complicates the use of inverse demand curves as measures of consumer's private valuation of medical care. In turn, this makes welfare-economic applications in health care somewhat suspect. Demand inducement also means that "markets aren't working," and this intrinsically bothers many economists as well. Thus, the idea of demand inducement and research surrounding it seems to evoke strong passions among health economists. Two commentaries from the literature on this issue follow. The first, from a study by DeVany, House, and Saving (1983), discusses their own research on this issue:

Proponents of the "supplier induced demand model" argue that physicians inflate demand whenever the supply of medical manpower is expanded. This interpretation has gained attention because it appears to offer an explanation for the observed cross-sectional positive relationship between fees and the density of health care providers. . . . We explain the positive association of fees and density using an expanded version of standard competitive market theory and offer some interesting empirical support collected from dental markets. . . . The cross-sectional pattern of fees and density reflects the willingness of patients to pay higher fees in exchange for lower time cost. This positive relation between high density and low time cost gives rise to the observed cross-sectional relation between fees and provider density. . . . We believe that our results clearly indicated that the observed cross-section positive relation between fees and provider density is not the result of demand creation. Rather, this seemingly perverse relation is simply a reflection of the preferences of patients for times supplied to the dental practice.

Victor Fuchs (1986), commenting on a particular study of demand inducement (which estimated that an important amount of demand inducement existed), offers the following:

Whether the shortcomings [of this study] lead to an over-estimate or an under-estimate of the "inducement effect" is not certain. What is certain, however, is that many economists will react to the study . . . as they have in the past on this issue, with the fervent hope that maybe there is no inducement.

This reaction has always reminded me of the story of the Frenchman who suspected that his wife was unfaithful. When he told his friend that the uncertainty was ruining his life, the friend suggested hiring a private detective to resolve the matter once and for all. He did so, and a few days later the detective came and gave his report: "One evening when you were out of town I saw your wife get dressed in a slinky black dress, put on perfume, and go down to the local bar. She had several drinks with the piano player and when the bar was closed they came back to your house. Then they sat in the living room, had a few more drinks, danced and kissed." The Frenchman listened intently as the detective went on: "Then they went upstairs to the bedroom, they playfully undressed one another, and got into bed. Then they put out the light and I could see no more."

The Frenchman sighed, "Always that doubt, always that doubt."

procedures increased by 3 or 4 percent. This type of result was more disquieting to neoclassical economists, because it dealt with the earlier refutation, and still produced the result that supply could apparently create its own demand, at least to some extent. Box 7.4 provides examples of the existing degree of disagreement on the subject.

The debate in the health economics literature has become arcane and complicated, with various alternative explanations offered by economists showing how the positive association between supply and quantity demanded could sustain itself in a competitive structure: Perhaps more doctors (through their increased density) lower the time cost to patients, thereby increasing the quantity demanded. Perhaps there are hidden flaws in the statistical analysis.[14] One model showed how increasing physician supply could lead to both more office visits *and* fewer hospital admissions, through a profit-maximizing substitution of outpatient for inpatient care (McCombs, 1984), but without any inducement. This model of the market has the distinction of actively predicting a phenomenon (reduced hospitalization) that has been observed in parallel with the increasing physician supply, rather than merely being another "noninducement" explanation of the phenomenon originally ascribed to inducement (the positive association of physician supply and M.D. visits). Not uncommonly, the studies focused on prices, which (in the pure classical sense) should decline as supply shifts outward, but are commonly found to be higher in areas with greater per capita supplies of physicians. Several studies (including DeVany, House, and Saving, 1983, and Boardman et al., 1983) offered explanations for the positive association of price with supply in a "classical" framework, without resorting to in-

---

[14] One recent empirical foray in this literature (Cromwell and Mitchell, 1986) dealt with most of these issues. Even then, complicated statistical problems remained (Phelps, 1986). Some natural experiments provide further evidence (Rice, 1987).

ducement. However, none of these studies could show that inducement *was not* occurring, but only that an alternative explanation existed for the phenomena said to arise from inducement.

Several conceptual ideas emerged with nearly unanimous support. First, if inducement could occur, some limit to inducement had to take place, or doctors would own the entire world. Some analysts posited a "target income" for doctors (Newhouse, 1970; Evans, 1974), while others posited a doctor who felt more and more guilty with greater inducement (Sloan and Feldman, 1978). Pauly (1980) and others noted that under any of these models, any doctors who would induce at all would induce "to the max," or as much as any constraint present would allow, and this idea has received widespread support (although some people believe that the constraints are totally binding, so that there is no inducement). Dranove (1988) developed a model in which patient wariness about inducement would create natural limits to inducement, even for a profit-maximizing doctor.

The Physician as "Agent" for the Patient

A conceptual approach that has proven quite fruitful in these discussions relies on the game-theory concept of "agency," in which a principal (the patient) delegates authority to an agent (the doctor) to make crucial decisions. The problems between principals and agents arise when the principal cannot fully monitor the behavior of the agent, and yet the agent may sometimes confront situations when the goals of the principal conflict with those of the agent. Various arrangements (contracts, agreements, and rules) emerge to try to "make" the agent do what the principal wants, but these are difficult to enforce in many settings. Dranove and White (1987) provide a good discussion of the role of agency in health care delivery.

The Doctor/Agent Making Referrals

In the previous chapter, we discussed how doctors make referrals for some treatments, and how fee splitting might alter the decisions of the doctor about such arrangements. Within the context of a principal-agent model, the potential importance of fee splitting is easy to understand. When Dr. C (a cardiac surgeon) agrees to split any surgical fee with Dr. A (a referring cardiologist) or Dr. B (a referring internist) for any surgery undertaken, then the advice from Drs. A and B will be distorted. Particularly if the patients of Drs. A and B do not know about the arrangement, they will be too willing to consult with Dr. C, and therefore too willing to undergo surgery at the recommendation of Dr. C.

By the same token, an "honest" principal-agent relationship between patient and doctor can go a long way toward solving the inherent problem of asymmetric information. If the agent (the doctor) really does not provide unbiased advice about medical treatments, particularly those regarding referrals, then it really wouldn't matter how uninformed the patient was, because the agent (the doctor) would always provide good advice. Alas, we have not yet discovered a contractual agreement that turns all doctors into perfect agents. One important role of "medical ethics" may be to encourage such honest agency performance.

Empirical Findings Regarding Induced Demand

Over the past decade or so, a fairly broad set of empirical findings has shaped the portrayal of induced demand better. The work on physician location decisions (described earlier) has shown that market forces play a strong role in directing the location of physicians. This evidence alone supports the belief that any inducement present has been fully exploited, since physicians who could induce further demand would probably choose practice locations as their predecessors did, in larger cities rather than in small towns. Other studies provide at least the suggestion that inducement is difficult to achieve for primary care doctors (McCarthy, 1985), but plausibly relevant for surgery (Fuchs, 1978; Mitchell and Cromwell, 1986; Rice, 1987).

Several studies also attempted to measure the role of consumer information, by comparing the rate at which *doctors and their families* received medical treatment, compared with a more general population. The general idea of these studies is that doctors could not "fool" other doctors into accepting "unnecessary" care. Bunker and Brown (1974) made the first of such studies, using other professionals (teachers, ministers, etc.) as "controls," and found that doctors and their families received *more* care than the controls. Unfortunately, for a variety of reasons, the comparison is flawed.[15] A more recent study (Hay and Leahy, 1982) used national

---

[15] Most important, doctors had higher incomes and probably had better insurance coverage than the comparison groups. Doctors also receive "professional courtesy" from other doctors, thereby reducing the out-of-pocket price to zero. This would normally make the doctors' use higher than that of others if the other groups faced any positive price. If a study found the same rate of use (which Bunker and Brown did), this might imply that doctors had resisted some inducement.

survey data to make a similar comparison, but controlled for the extent of insurance coverage and other relevant economic factors, using the techniques of multiple regression analysis. They found the same thing that Bunker and Brown did—namely, that if anything, doctors and their families received more care than others. This tends to cast doubt on the extent to which consumer ignorance plays a strong role in inducement, although the studies still cannot directly correct for the role of "professional courtesy," which should create larger demands by physicians and their families, other things being equal, than by otherwise comparable persons.

Finally, a randomized, controlled trial of Hickson, Altmeier, and Perrin (1987) described in the next section clearly shows that providers not only try but succeed in altering the rate of patient visits as the financial incentives to the doctor change. This seems to be the cleanest available evidence of the *presence* of demand inducement, although we can surely expect that the equilibrium magnitude of inducement will differ from setting to setting.

THE ROLE OF PAYMENT SCHEMES

The idea of providers' altering their treatment recommendations on the basis of financial reward sits at the heart of the "induced demand" concept. One analysis develops specifically the ways in which payment methods would alter the treatment recommendations of "ethical" physicians who knew what the "right" standard of care was, and who had a utility function that included leisure *and* the provision of appropriate care to their patients (Woodward and Warren-Boulton, 1984). They consider three types of payments to physicians: an annual salary, a time-based wage payment, and a payment based on the number of procedures performed (like the usual fee-for-service system). This model unambiguously predicts that physicians paid by the first two methods will deliver *less* than the "right" amount of care, and physicians paid on a fee-for-procedure system will deliver *more* than the "right" amount of care.

Remarkably, a randomized controlled trial exists that tests precisely this idea (Hickson, Altmeier, and Perrin, 1987).[16] Using the

---

[16] This study is not well known in the economics community because it was reported in *Pediatrics*, a journal to which few economists subscribe, and which is not cross-indexed in economics literature data bases. The only solutions available for an economist who is interested in health care (1) to read medical journals, or (2) to marry a doctor who reads medical journals.

residents in a continuity care clinic at a university hospital, this study randomly selected half of the doctors to receive a fee-for-service payment, and half to be paid by flat salary. Patients coming into the clinic were also randomized as to the type of doctor they would see. Once assigned to a given doctor, the patients continued with that doctor for all their care, unless the doctor missed an appointment, in which case another doctor would see the patient. The payments were set so that on average, every doctor would receive about the same income from this activity, and also to correspond to the "profit" per patient achieved by physicians in private practice in the community (about $2 per patient).

This study directly supported the model of Woodward and Warren-Boulton. Physicians paid by fee-for-service scheduled more visits for their patients (4.9 visits per year versus 3.8 visits per year), and saw their patients more often (3.6 visits versus 2.9 visits).[17] Almost all of the difference in behavior was due to well-care visits (1.9 visits vs. 1.3 visits).

The American Academy of Pediatrics has a schedule of recommended treatment for children (well-care visits for routine examination, vaccinations, etc.), corresponding to the idea of a standard of "correct" care. Doctors on the fee-for-service payment system missed scheduling any of these recommended visits only 4 percent of the time, whereas doctors on the salaried group missed scheduling recommended visits for their patients 9 percent of the time. (The difference is strongly statistically significant.) Thus, the prediction about salary-paid doctors providing "too little" care is supported.

In addition, the fee-for-service doctors scheduled excess well-care visits (beyond those recommended) for 22 percent of their patients, while the salary doctors did this for only 4 percent of their patients. (This difference is also very significant statistically.) Thus, the prediction about "too much" care for fee-for-service doctors is also supported.

SUMMARY

Economic forces direct many of the important decisions doctors make and how they interact with patients. In particular, we have seen (in the previous chapter) how economic forces guide the

---

[17] Doctors "missing" a patient visit and turning them over to another doctor on duty account for the difference between scheduled visits and actual visits.

choice of physician specialty, and even the decision to enter medical school. In this chapter, we expanded the discussion to show how economic forces affect the locational decisions of doctors, in a way actually similar to the choice of specialty at a conceptual level.

We next investigated the way in which prices are set in markets like those for physician services. A model of monopolistic competition with incomplete search seems to fit this market well: Rather than a single price, a dispersion of prices exists, often quite wide. While some patients change doctors, many do not, and some would not consider it, not wishing to offend their doctor. Thus, search is probably incomplete, and each doctor has some price-setting power.

The price-setting power also interacts with the possible ability of doctors to "induce demand" by their patients—that is, to shift the demand curves outward, thereby increasing economic opportunities. Numerous studies of demand inducement exist, many with statistical difficulties or flaws. Nevertheless, a controlled trial in a pediatric clinic conclusively demonstrates that the mechanism of payment alters the number of visits both recommended to patients and the number of such visits actually undertaken by the patient. Market forces apparently limit actual demand inducement, however, so that many observable phenomena (such as physician location) correspond closely to those that would occur without inducement.

PROBLEMS

1. In a pure profit-maximizing monopoly, no monopolist will willingly operate where the demand curve is inelastic, as Box 7.2 describes. Since careful studies of the demand for physician services all show that the elasticity of the *market* demand curve is less than 1 in absolute value (e.g., $\eta = -0.4$), what does that tell us about the forces of competition (if any) operating in the market for physicians' services?

2. "Physicians will never go to work in rural areas, because they can always generate more demand for their services in big cities, where they prefer to live." Comment.

3. The American Medical Association (AMA) has had a long-standing effort to suppress advertising by physicians, asserting that it was "unethical." What effects might you expect advertising to have on consumer search? What effects does consumer search have on the elasticity of demand confronting individual firms? In a monopolistic market, what happens to the optimal price (and returns to the monopoly) as the demand elasticity becomes large in absolute value? Do these ideas help explain the AMA position on advertising?

4. Suppose a society consists of three cities (A, B, and C), with populations (respectively) of 99,000; 51,000; and 6,000 people. To begin, suppose also that the society has 15 doctors total.
   (a) What cities will have how many doctors?
   (b) If the number of doctors doubles, how many will live in each city?

5. What experimental and nonexperimental evidence can you cite about the extent of induced demand? Is there any? Is it unlimited in extent? Does the experimental evidence also reveal anything about the tendencies for "shirking" in a physician-firm that pays doctors a flat salary, rather than on a piece-work basis? On net, can we say for sure whether a salary or fee for service arrangement is better for patients?

6. Why do discussions about induced demand always focus on doctors, rather than (say) nurses?

Chapter
8

The Hospital as a Supplier of Medical Care

*T*he hospital stands as the center of modern medicine, for better or for worse. Almost all people who become seriously ill will find themselves in the hospital, and $1 out of every $2 spent on health care in the United States is spent on hospital care. Remarkably, most of the decisions made about the delivery of medical care in hospitals—whom to admit, which procedures to use, which drugs to give to the patient, how long the patient should stay in the hospital, and where the patient should go upon discharge—are made by persons who are not employees of the hospital nor under their direct control or supervision. In this chapter, we examine the organization of the hospital, study the various forms of ownership of the hospital (not-for-profit, for-profit, government) and the effects of such ownership on the behavior of the hospital.

THE HOSPITAL ORGANIZATION

It may seem odd to begin the discussion of the economics of the hospital with an organization chart (Figure 8.1), but indeed we must, in order to understand the hospital meaningfully. As we shall see, it may be more appropriate to draw two organization charts for the same hospital. Drawing appropriate links between them has proven almost beyond the capabilities of most artists or chartists.

This discussion will focus on the typical not-for-profit hospital that dominates the market in the United States (and some other countries). Important distinctions between this type of hospital and others (such as for-profit hospitals) will appear throughout the discussion. To begin with, we need to understand just what a not-for-profit hospital is and is not, and what it may and may not do. Not-for-profit hospitals may, can, and do earn profits. They may

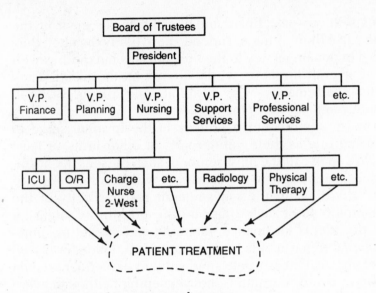

Figure 8.1 Hospital organizational structure.

not, cannot, and do not distribute such profits to shareholders (as would commonly occur in for-profit organizations of all types), because their form of organization does not allow distribution of profits to shareholders. In a typical for-profit organization, the shareholders are the "residual claimant," receiving any revenues of the organization after *all* costs have been paid, including labor, materials, supplies, interest on bonds, taxes, and so on—the "profits" of the organization. In the not-for-profit organization, there are no shareholders, and hence no legally designated residual claimant.

Not-for-profit hospitals differ importantly from standard corporations. In the absence of a residual claimant, their profits must be distributed to somebody else. How and to whom they do this affects the product mix, the costs, the input mix, and possibly the size of the hospital.

Numerous models of the not-for-profit hospital have sprung up, depicting various ways this might happen, which we will review momentarily, and then synthesize. A good place to begin this discussion, it turns out, is the hospital's organization chart.

Sitting at the top of the organization chart is the board of trustees, empowered by the hospital's legal charter to direct all that goes on within the hospital. The board is self-replicating (members elect their own successors, including themselves), and typically serves without pay. Board members own no stock in the hospital, because there is none. Indeed, more likely than not, they are expected to donate money to the hospital at some time, as well as providing overall direction while serving on the board. They choose who shall manage the hospital, and provide overall strategic policy and advice to those managers.

The primary administrative officers of the hospital serve in much the same role as their counterparts in other firms, at least nominally. We shall call the most senior of such persons the hospital's president.[1] Reporting to the president will be a series of vice presidents,[2] dividing up the management responsibility of the hospital. A prototype of such division of responsibility might include vice presidents for finance, planning and marketing, nursing, professional departments (emergency room, laboratories, social service, physical therapy, etc.) and support departments (such as food service, laundry, supplies, housekeeping), although every hospital's organization chart is unique. To each of these, in turn, report middle managers in their various areas.

Much of the hospital's activity focuses around subunits serving particular types of patients, commonly described by the physical location of the unit or its function: 2-West (2nd floor, west wing); OB (where obstetric patients stay); Delivery (where babies are born); Newborn Intensive Care (where babies go if they are very ill or premature); the Emergency Room (ER). A physically based designation (such as 2-West) usually implies that this unit serves "basic" adult medical and surgical patients. These "units" commonly have 20 to 40 beds, and operate under the immediate supervision of a head nurse on the unit, around whom all of the other activity revolves. These "charge nurses" (the ones in charge during their shift) direct all of the nursing care, and coordinate almost all other patient care given on their units.

If the patient receives medications, the pharmacy delivers them to the floor, where a medication nurse will administer them.

---

[1] The titles for such persons are quite diverse, ranging from "President" to "Executive Director" to "Hospital Administrator." Quite commonly, such a person (and immediate subordinates) will have received a master's degree in business, public health, or directly in hospital administration. On very rare occasion, the person will be an M.D. In smaller hospitals, the person may have no postgraduate training.

[2] Titles will correspond to that of the "president," for example, "Vice-Administrator" or "Assistant Administrator."

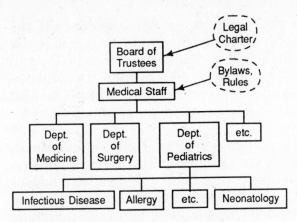

Figure 8.2 Organization of medical staff.

within the "management" organization all have a contract with the hospital; the hospital reviews their performance, pays their wages or salary, and can fire them. By contrast, doctors admitted to the medical staff have no similar relationships with the hospital. In general, they receive no income directly from the hospital, their performance is subject to a very different and weaker review process, and with few exceptions, they cannot be "fired."[3] Once a doctor gains admission to a medical staff, removing that privilege turns out to be exceedingly difficult.[4]

By law, most of the activities carried out with, on, and for patients within the hospital must be done either by or under the direction of a licensed physician. Thus, the doctor "writes orders" in the patients' charts that direct virtually the entire flow of activities for each patient. These orders create "demands" for activities within the hospital, which the hospital organization "supplies."

---

[3] Some of the line managers are doctors in some hospitals, such as the director of the laboratory, the X-ray division, or the emergency room. In other hospitals, these functions will be handled by a separate firm (composed of relevant doctors), with whom the hospital has a contract to perform the management functions. Thus, for example, a radiologist may be a member of the medical staff of the hospital as well as an employee, or may be a member of the medical staff as well as president of a separate corporation that contracts with the hospital to manage the radiology functions. Typically, doctors performing such a management function will retain a separate relationship with their patients, sending them separate bills for the interpretation of an X-ray.

[4] The doctor may have privileges revoked for serious medical mistakes or misbehavior, but even this has proven exceedingly difficult in the past. See Chapter 14, describing the medical legal system, for further discussion.

If the patient receives physical therapy, either the therapist will come to the patient's room or the patient will be delivered (walking or by wheelchair) to the PT unit. If an X-ray is to be taken, a similar process occurs. If a laboratory test is needed, the phlebotomist will come to the patient's room and draw a blood sample. Meals are brought from the kitchen to the floor, and given to the patient there.

All of the specialists performing these activities (pharmacist, therapist, phlebotomist, X-ray technician, food service delivery) report organizationally to their own "boss" in their own department (pharmacy, X-ray, etc.), who reports up the line eventually to the relevant vice president. The activity of all of these hospital departments interacts on the floor unit, focusing around the patients. In addition to directly supervising the nurses in such units, the charge nurse also acts much as does a traffic cop, managing the flow of all of these diverse actors as they interact with the patients.

Remarkably, this complicated interaction and organizational description omits the one person who initiates all of this activity: the doctor who admitted the patient to the hospital. The doctor is not an employee of the hospital. The doctor has no "boss" up the line. The one person upon whom this entire activity depends has only a weak and ambiguous organizational tie to the hospital—the "medical staff."

The hospital medical staff has its own organizational chart, and by-laws of operation. The staff is divided by medical specialty: medicine, pediatrics, obstetrics and gynecology, and so on. These departments may have divisions as well, reflecting subspecialization of the doctors, if the hospital is sufficiently large: orthopedic surgery, neurosurgery, cardiac surgery, and general surgery, for example, within a department of surgery. Figure 8.2 shows a typical organization of the medical staff.

Doctors receive admission to this medical staff by application to the hospital, nominally to the board of trustees, who are responsible for the overall activity of the hospital. The board invariably delegates the responsibility for this decision *de facto* to the existing medical staff, which will commonly have a "Credentials Committee" established to review the applications of potential new members of the staff. Their report, voted on by the full staff, provides the basis for the board's decision. Thus, at least *de facto*, if not *de jure*, the medical staff, like the board, is self-replicating.

An important difference emerges between the "line management" and employees of the hospital, and the medical staff. Those

those orders require a blood test, the phlebotomist appears on the unit, draws blood, and takes it to the laboratory, where it will be analyzed by equipment purchased with the knowledge that such orders will be written. If the patient goes to surgery, the surgeon performing the surgery will literally give direct orders to the nurses and technicians (employees of the hospital) who assist in the operation, even though the surgeon is not their "boss" organizationally. The drugs the patient receives come only after the doctor writes a prescription, and similarly with any therapy, X-rays, and even the diet that the hospital will offer to the patient. The doctor captains the ship, ordering when to start the engines, which direction to go, and how fast to move, and the hospital can do little but respond, no matter what orders the doctor gives. The doctor, independent of the hospital for income and supervision, nevertheless directs virtually all of the activity, and hence the use of resources, within the hospital. Figure 8.3 shows these relationships (albeit vaguely).

The hospital is really two separate organizations—the line management and the medical staff—which serve the roles of supply and demand in the "market" of hospital care.[5] The hospital is really a "job shop," where each product is unique (although often similar to others), and the hospital is set up to provide inputs to the craftsmen who direct the output of this job shop, the doctors on the medical staff. The patient has two distinct contracts before this can all happen,[6] one with the hospital, and the other with the doctor. In the contract with the hospital, the patient promises to pay for care, and the hospital promises to provide necessary medical care under the direction of the patient's physician. In the contract with the doctor, the patient promises to pay for care, and the doctor promises to provide care as needed and to supervise the activities of the hospital.

This type of arrangement has apparent disadvantages, notably that cost control may be difficult if not impossible to achieve in this type of organization, but apparent advantages also emerge. Most notably, in the unpredictable environment of treating illness, the patient's health may be better served by having a doctor in control

---

[5] This viewpoint, which forms the basis for a very useful analysis of the hospital organization, appears in two articles by Jeffrey Harris, M.D., Ph.D. (Harris, 1977, 1979). As mentioned earlier, Dr. Harris is trained in internal medicine and economics.

[6] Here, we use the word *contract* to include both written documents and oral agreements.

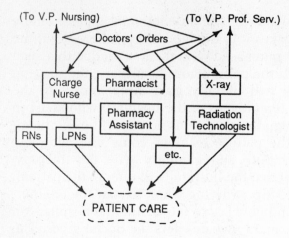

Figure 8.3 The doctor directs the activity.

of resources who is independent of the hospital. In addition, the patient may be able to monitor the physician's activity better than the hospital, since the patient possesses unique knowledge about the doctor's success (i.e., how well does the patient feel?).

WHO IS THE RESIDUAL CLAIMANT?

The hospital described above can and often does "make a profit" in the sense that its revenues exceed its costs. What becomes of that profit, and who makes the decisions about its allocation? This question has haunted analysts of the hospital industry for decades. Some models say that the doctors have seized control of the hospital, and operate it to enhance their own profits.[7] Others say that the hospital administrator (or hospital board) operates the hospital to increase its own happiness, shifting the hospital's resource uses accordingly.[8] Some say that the hospital uses its profits to raise wages of employees above normal levels, in effect being captured by "the nurses."[9] The legal charter of the hospital as a not-for-profit entity at least hints that "profits" will be returned to patients in the

---

[7] This view is most carefully developed by Pauly and Michael Redisch, 1973, and by Pauly, 1980.

[8] This view of the hospital administrator as "owner" first appeared in Newhouse, 1970.

[9] Martin Feldstein's (1971) study of hospital costs developed this theme.

form of lower prices, although no serious analyst has adopted this view in writing.

Most likely, each of the proposed answers to this problem is never wholly correct and never wholly wrong. *Everybody* associated with the hospital has a finger in the pie. Who gets the bigger slice will vary from place to place and time to time, and the study of such questions is probably better suited to modern political science than to economics. Let us look at each of the relevant actors and their roles in control of the hospital resources.

The Doctors as Residual Claimants

In its most simple form, the doctors-win-it-all (Pauly-Redisch) model of the hospital simply says that the doctors on the medical staff "milk" all of the hospital's profits into their own bucket, by having the hospital perform (and pay for) activities that promote the profitability of the doctors' own firms.[10] This model has obvious face appeal, given the central role of doctors in directing the use of hospital resources.

The difficulties in fully accepting this model arise from two directions. First, why don't the doctors just directly own the hospital, and thus eliminate the ambiguity arising from the existing situation? Indeed, in past years, doctors *did* own a majority all of hospitals in the United States.[11] One answer to that might be that the not-for-profit law confers tax advantages on the hospital that can be passed on to the doctors, enhancing their profits.

One study (Sloan and Vraciu, 1983) estimated that for-profit hospitals pay about $15 per bed-day in income taxes, avoided by their nonprofit competitors. In addition, because of regulatory distinctions drawn, the payments from Medicare and Medicaid to nonprofit hospitals were sufficiently higher on a per-bed-day basis to add another $15–$20 per day in profitability to the nonprofit hospital. For a 300-bed hospital operating at average utilization rates, these differences could add over $2 million per year to the hospital's profits. By the Pauly-Redisch model, the $2 million is available for the doctors as added profits.

---

[10] We discussed the physician-firm as an economic entity, and how this differs from the situation of physicians as individual "workers" in those firms, in Chapter 7.

[11] Steinwald and Neuhauser, 1970. The number of for-profit hospitals peaked in the first quarter of this century at about 2,500, representing over half of all hospitals in the early part of the century.

The difficulty with this view, and the second major problem with the Pauly-Redisch model, is that on the whole, it treats the medical staff as essentially homogeneous, operating with unified goals as a single entity. It assumes that the doctors divide up the hospital's profits in even proportion to their own amount of work.

The reality differs considerably from this simple view, which proponents of the "doctor-gets-it-all" school readily recognize. Specialists of one area—say, infectious disease—have very different ways to enhance their own profitability, say, than pediatricians or heart surgeons. While all of the heart surgeons may have a single voice recommending how the hospital, say, should build a laminar-air flow surgical suite to cut down on infection during long operations, the pediatricians will likely prefer some other approach, such as providing a clinical psychologist for testing children's developmental skills. The Pauly-Redisch model does not tell us how the hospital will resolve such conflicting claims. The central difficulty is that, to a considerable degree, the model does not confront the question of choice between conflicting objectives of various medical staff members.

This conflict also circles back to the first question, the alleged financial superiority of the tax-free model. Lower taxes (and higher profits for the hospital) create higher profits for the doctors only if the hospital decides to spend those profits in particular ways. The battle of control within the hospital may itself be costly, so that the hospital dissipates some or all of the gain. For example, the choices made to appease both the pediatricians and the heart surgeons may end up costing so much that neither group gains anything from the nonprofit structure. Pauly and Redisch discuss this problem some, but leave unresolved the question of just how the doctors will cooperate, if at all.

The Administrator as Residual Claimant (Organizational Utility Function)

Another view of the hospital proposes a "decision maker" within the hospital, whom we can either dub "the administrator" or "the board." This person is said to control the hospital in such a way as to maximize its own utility, much in the fashion of a consumer (see Chapter 4). The prototype model (Newhouse, 1970b) describes the decision maker as gaining utility from quantity and quality of output. Rather than having a budget constraint (as would a consumer), the hospital faces a market constraint (the demand curve for its services) and a production constraint (the technical ability to combine inputs into producing output). In the usual fashion of a maximizing decision maker facing constraints, the hospital decision maker trades off quality and quantity in such a way as to maximize utility.[12]

This model suffers from a problem similar to the second problem described for the Pauly-Redisch model: Where does "the utility function" come from? If we accept the existence of such a central decision maker with a stable utility function, then this model is very helpful.

This approach has considerable appeal, and can help illuminate the types of decision making that must occur in a not-for-profit hospital, as we shall see later in this chapter. Unfortunately, little has emerged since the model originally appeared in 1970 to help understand just how "the decision maker" seized control of the organization, or why the utility function "looks" the way it does (e.g., having quality and quantity as its arguments).

Employees or Patients as Residual Claimants (Higher Wages or Lower Prices)

Other views of the hospital have suggested that the hospital deliberately pays its employees more than "market" wages, either because it wants to or because the employees have forced it to. The hospital might do this either because of a powerful employees' union (or other strong bargaining position) or because the administrator "wanted to."

A corresponding view is that the hospital returns any profits to patients in the form of lower prices, simply by not collecting economic profits that it might have achieved. If one takes the legal form of not-for-profit at its face, this might be interpreted as the "desired" choice of the legislators who establish the not-for-profit structure, but this belief holds water only if one does not consider the other possible "residual claimants" such as the medical staff or the administrator. It also is silent on the question of quality choice, and thus does not really help much in understanding hospital behavior. Of course, if the decision maker's utility function in-

---

[12] In Chapter 16, discussing hospital cost regulation, we will see this model developed more fully. If you want, you may peek ahead to get a flavor of how this decision-making process works.

cludes quantity of output (as Newhouse, 1970, modeled it), then lower prices offer one way to achieve this. Since patients will respond to lower prices by demanding more care, lower prices might also directly improve "administrator" well-being.

The For-Profit Hospital: Shareholders as Residual Claimants

The for-profit hospital has other claimants to the profits of the hospital—namely, the owners. Legally, of course, they are entitled to all of the profits. However, in a market populated with nonprofit hospitals, they may have to make concessions in a for-profit hospital—for example, to the medical staff—in order to induce them to bring patients into the hospital. These concessions may make the for-profit hospital appear much like the not-for-profit hospital in structure, organization, equipment, and even management style. (See Box 8.1.)

A Synthesis: The Production of Organizational Power

These conflicting views of hospital ownership and control have a common thread: Each takes *one* of the relevant actors in the hospital, and asks how the hospital would perform if that one actor or group "captured" the hospital for its own benefit. The Pauly-Redisch model can be viewed as a physician-capture model; the Newhouse model as an administrator-capture model. We could also consider employee-capture models, patient-capture models, or others.

The problem of allocating the resources of the not-for-profit hospital closely resembles the problem of a regulator in any economic market allocating rights to economic goods to which the legal title is ambiguous. Such problems abound in our society, and regulation invariably is used to try to solve the problem when the government has failed to establish clear and complete property rights: Air pollution regulation, control of groundwater pumping in western states, imposition of a 12-mile (or 100-mile) fishing limit off our continental borders, right-of-way rules for drivers, and non-smoking sections in public buildings are all examples. The not-for-profit hospital routinely produces economic profits, and something akin to "the regulator" in society must decide their allocation, since the hospital's structure and by-laws do not clearly establish a residual claimant.

In problems of regulation, the allocation of economic goods takes place in a political market, where each of the actors tries to generate political power to persuade the legislators and regulators to favor them. Each interested party *must* enter the fray, or face being pushed out of the picture. Each organizes its resources to control the political-regulatory process, and "pushes" the system as hard as seems economically desirable. In the end, each should push only until the incremental effort yields a return that just matches the effort. In "equilibrium," each party should have equal power *on the margin* to alter the process. While one group or another may have "pushed" the process considerably in its favor, eventually other parties push back hard enough so the system appears to have reached equilibrium. At this point, everybody pushes hard, but all appear at a standstill, like two giant sumo wrestlers locked in battle.

The determinants of how this all works out include (1) the resources at the disposal of each group, (2) the rules of the game, and (3) the organizational-management capabilities of each group *in this type of contest,* or (in other words) their political savvy. The "rules of the game" can be particularly important. Sometimes, they dominate the decision, and most of the other resources fall by the wayside. In other cases, political savvy proves most important, and in still others, the brute strength of resources.

In the not-for-profit hospital, both society-wide legal rules and the particular by-laws and rules of the hospital and its medical staff set the rules of the game. So long as these rules remain stable, the resources and political savvy of each relevant group will determine who appears to dominate the game. If the doctors dominate, then the hospital will appear much like a Pauly-Redisch hospital. If the administrator dominates, then a Newhouse-like hospital will emerge, with that decision maker's preferences. A strong labor union could make the nursing staff appear dominant, and so forth.

In fact, so long as the "rules" remain the same, and so long as the resources of each participating group remain stable, a hospital will *look like* a hospital with a single decision maker who has a stable utility function, much like Newhouse's original model of the hospital. However, the "regulatory contest" view tells us (at least in principle) where the utility function comes from. It blends the interests of each relevant person and pressure group within the hospital, the blend being richer for those who entered the fray with the greatest resources and political savvy. The apparent utility function of such a hospital, of course, could include a variety of

Box 8.1 THE PERFORMANCE OF FOR-PROFIT VERSUS NOT-FOR-PROFIT HOSPITALS

One study of the effects of for-profit ownership compared 80 "matched pairs" of investor-owned ("chain") hospitals and matching not-for-profit hospitals (Watt et al., 1986) using 1980 data. The cost data showed

	INVESTOR-OWNED	NOT-FOR-PROFIT
per admission	$1529	$1453
per day	$ 233	$ 218

These differences were not statistically significant. They did find significant differences in the costs of drugs and supplies (the for-profit hospitals used more).

The for-profit hospital "marked up" its costs more than the not-for-profit hospitals, yielding higher revenue and higher net revenue (profits). After adjusting for differences in income and property taxes paid by the investor-owned hospitals ($14 per day) the study found the charges per day as

	INVESTOR-OWNED	NOT-FOR-PROFIT
Charges per day	$257	$234
"Profit margin"	9%	4%

Another study (Sloan and Vraciu, 1983) of hospitals in Florida found no differences between the two types of hospitals on a wide variety of dimensions. They compared nonteaching hospitals under 400 beds in size in Florida, using 1980 data. In that state, a third of all the hospital beds are owned by for-profit hospitals, strikingly higher than in most other states. Their study used 52 not-for-profit hospitals and 60 for-profit hospitals, matched by size category and location. Their findings on "profitability" show

Investor-owned

  Chain—4.8%

  Independent—5.4%

Not-for-profit—5.6%

One substantial difference between the two studies is the amount of taxes they detected. Sloan and Vraciu also calculated direct income taxes at about $14 per day, just as did Watt et al. However, they found that for-profit hospitals paid $20 per day in "hidden taxes" because of the way Medicare and Medicaid pay for care differently to each type of hospital. These rules paid only "allowable costs," which are less than patient charges for both types of hospitals. However, the greater discount Medicare and Medicaid imposed on the for-profit hospitals ($20 per total patient day more) almost exactly accounts for the discrepancy in charges per day found in the Watt et al. study. The distinction is this: The for-profit hospitals list higher prices than the not-for-profit hospitals, but Medicare and Medicaid don't pay as much. On balance, the two types of hospitals seem indistinguishable.

Sloan and Vraciu also compared the list of facilities and services the hospitals indicated that they offered. They concluded that "the not-for-profits are more likely to offer such 'profitable' services as open-heart surgery, cardiac catheterization, and CT scanning, but are also more likely to offer an 'unprofitable' one like premature nursery." On other dimensions, the hospitals seemed difficult to distinguish between.

interests, including high-quality care, fancy machinery for diagnosis, subsidized obstetrical service, pleasant working environment and high wages for the nurses, total numbers of patients served, the quality of the meals, or any endless set of things.

We should *expect* hospitals to appear different from each other in the nonprofit world, precisely because we should expect that this "contest" to control the economic profits of the hospital will stabilize differently in different hospitals. Hospitals owned by the Roman Catholic church will surely include an obstetric service, but not an abortion service. Hospitals with a particularly "savvy" cardiac surgeon will emphasize capabilities and services to benefit that doctor's patients and profits. The broad message from this view of hospital control is that there is no common message. We should not expect any single group to "capture" the organization consistently, because the rules of the game and the capabilities of the players will differ from setting to setting.

In some important cases, this approach significantly deviates from a "decision maker's utility function" approach to modeling the hospital. The approach outlined here raises an alert about changes in the legal and regulatory structure that will alter the rules of the "game" within the hospital, and thus alter the outcomes. Each time this happens, the apparent "utility function" of the hospital will shift. For example, a labor-relations board ruling decertifying a nurses' union will shift the balance away from them. A legal ruling strengthening the hospital's ability to dismiss a doctor from the medical staff shifts the balance away from doctors. A statewide regulatory system decision to limit wage increases to 3 percent next year will clearly reduce the bargaining power of employees. Each time such an event occurs, the "utility function" of every affected hospital will shift, potentially creating systematic changes in its economic behavior.

Within a stable legal system, however, changes that occur from hospital to hospital should be more sporadic, rather than systematic, as individual leaders of various factions retire, die, move, or emerge. While individual hospitals will change their behavior from time to time, responding to such changes in the balance of power, no systematic changes should appear. In such an environment, it is probably safe to think about hospitals (at least collectively) as behaving according to a model relying on a single decision maker with a stable utility function. This approach, for example, assists in understanding how hospitals will respond to some changes in cost control regulation (but not others!) as we will see in Chapter 16.

HOSPITAL COSTS

The mix of particular activities selected by a hospital to undertake, the mix of inputs it chooses to produce that mix of outputs, and the cost of those inputs all come together to form "the costs" of the hospital. Given all of these choices, one can think about expanding the scale of the hospital, holding constant its output mix (newborn deliveries, cancer therapy, psychological counseling, substance abuse, open heart surgery, . . .) and the cost of its inputs (R.N. wages, interest on loans, the cost of electricity, . . .). As the scale of this hospital/job-shop increases, we can expect its costs per unit to change as well. How this happens forms the basis for studies of "economies of scale" in the hospital industry.

"Economies of scale" studies ask the simple question, "Do costs per unit of 'output' rise or fall as the scale of the organization expands?" A long series of studies of hospitals in the United States (and elsewhere) have sought to answer this question. The resulting set of studies allows you to select almost any answer you wish. Some show costs rising with output (diseconomies of scale), others show costs falling with output (economies of scale), and some show approximately constant average costs over a broad range of outputs.[13] In each case, the "output" of the hospital is measured by the number of patients treated or the number of patient days produced.

To undertake such studies appropriately, one must in concept compare the costs of two hospitals that are "identical" except for

---

[13] The earliest studies of hospital costs included Lave and Lave (1970). A recent review of this literature (Cowing, Holtman, and Powers) appeared in 1983. Except for a few subsequent studies, this review is relatively comprehensive, and provides a good starting point for the person wishing further detail on the various approaches and statistical problems associated with estimating hospital cost functions.

size, or else "control for" their differences in scope and complexity somehow. This may prove to be an impossible task.

As hospitals expand in size, they invariably expand in the scope of their activities as well. This may take place in subtle ways. For example, a larger hospital may be able to justify a larger diagnostic laboratory that not only does more blood tests, but tests for more different items within each blood sample automatically. Doubling the size of the surgical service may make it feasible to build and staff a postsurgical intensive care unit. A larger obstetric service may make possible the construction of a newborn intensive care unit for critically ill babies. Some of such differences will appear in the "list" of capabilities of the hospital, but others will not.

We should remember that the hospital is really a highly specialized "job shop," where each case treated is unique, at least in some ways. There are similarities in types of cases, but even the most finely tuned classification systems available show large variation in the costs of treating patients within a single hospital for a single type of illness or injury, as the accompanying box shows for one city's experience. (See Box 8.2.)

Another important distinction between hospitals concerns the types of patient they serve. Some "community" hospitals specialized in providing routine care to people living in the nearby vicinity. Others specialize in complex care, with referral patterns spanning a wide area, often with patients from more than one state. These referral hospitals will invariably end up with patients who are, on average, sicker than those in community hospitals, even after holding the patients' primary diagnoses constant. For example, patients hospitalized for "chronic obstructive pulmonary disease" in a local hospital are probably "routine" cases; those who are transferred to or deliberately travel to specialty care in "referral" hospitals probably have other complications that make their treatment more difficult.

Statistical methods to "hold constant" the case mix of hospitals when they try to calculate cost functions invariably cannot measure these types of differences, yet they clearly exist and are important. One study, for example, found that for a given set of diagnoses, it cost about 10 percent more per patient to treat Medicaid patients than comparable patients with private insurance (Epstein, Stern, and Weissman, 1990), probably because Medicaid patients have more advanced states of illness when they reach the hospital. All of these factors make accurate estimation of hospital cost functions quite difficult, at best.

Hospitals will also differ on how well they are equipped to

Box 8.2 VARIABILITY IN LENGTH OF STAY WITHIN DRGS

Some data from hospital admissions in Rochester, N.Y., will show how variable the resource use can be within a single diagnostic group. The table below shows the average length of stay and the standard deviation for each of several common illnesses or treatments. These were selected not because they show a lot of variability, but rather because they seem fairly routine and predictable, compared with other diagnoses. These deviations are partly predictable (see Chapter 3), but in a considerable part, they represent the effects of different problems of different patients, emphasizing the "job shop" nature of the hospital.

DIAGNOSIS-RELATED GROUP (DRG)	AVERAGE	SD	RANGE
Otitis media and upper respiratory infections	2.7	1.7	1–10
Inguinal hernia, ages 18–69 without complications	2.0	1.2	1–19
Appendectomy without complications, age <70	3.6	1.6	1–15
Fracture of femur	15.4	12.9	2–54
Prostate removal, transurethral, age <70 without complications	5.3	1.7	3–14
Normal delivery without complications	2.5	0.9	1–24
Caesarean section delivery without complications	4.5	1.9	1–19

Source: Calculations by author from claims data from Blue Cross-Blue Shield of Greater Rochester.

---

handle unusual events, further hindering comparison of their costs. Some of this "standby capacity" is easy to observe (how many beds are usually available in case of a large disaster like an airplane crash or office building fire?). Other differences will not be so apparent in a study of hospitals' costs (such as the capability of conducting a complicated and rare laboratory test). Sometimes the same "capability" means very different things. A very powerful magnetic resonance imaging (MRI) machine can cost two or three times what a small one costs, but both will appear as "MRI" in a list of the hospital's capabilities.

With these caveats, we can turn to the evidence on economies of scale. Although a number of studies exist that have estimated the behavior of costs as scale changes, few have solved the knotty methodological problems well. One recent attempt (Grannemann, Brown, and Pauly, 1986) has data and analysis methods that deal with many of the problems, and has the added advantage that it jointly estimated the costs of both inpatient and hospital outpatient department activity (including emergency rooms). But even they have not been fully successful in controlling for output mix, the problem discussed above.

Their findings show that marginal costs of inpatient care rise both with the number of discharges and the number of patient days the hospital produces. They estimate the marginal cost of a *discharge* at $533 (in 1981 dollars) for low-volume hospitals, $880 for medium-volume, and $1084 for higher-volume hospitals. Similarly, the marginal cost of a day increased from $168 to $237 and then slightly fell to $231 for the hospitals of these respective size groups. The differences in the cost of a "discharge" seem too large for the authors to believe that the differences represent efficiency of operation. As they conclude, "The variables included in this cost function do not fully capture the differences. . . . [and] unmeasured differences in case mix and service content differences may be responsible." Another way to say this is that bigger hospitals are more complex and offer a different "quality" of care than smaller hospitals, in ways that are too difficult to control for using existing data.

This study also finds that in the operation of emergency departments, persistent economies of scale exist. The obvious question is why we don't see ever-larger (and fewer) of such departments, particularly in cities large enough to support multiple hospitals. The answer—and the key policy issue—is that their merger (to produce costs savings) is not feasible, because of an uncounted cost.

Hospitals (including emergency rooms) produce services that require the presence of the patient (and often, the patient's family). Taking travel costs of patients into account, hospitals will always have naturally limited markets, and the question of economies of scale will not matter much. Or put another way, if we counted the cost of transporting the patient to the hospital, every hospital would have eventual diseconomies of scale. In the relevant time frame for transportation of serious emergencies, where sometimes a few minutes can make considerable difference in outcomes, these diseconomies of scale could rapidly appear.

Costs of patient care are not all that change with hospital size. The "quality" of a hospital can also change importantly. One obvious way that quality changes reflects the usual effect of special-ization—larger hospitals can afford to employ specialists to do more jobs, and this, at least to some degree, can improve quality. To paraphrase Adam Smith, "Division of labor is limited by the size of the hospital." Larger hospitals can also afford to have more "standby" equipment that gets used only rarely; as the scale of the hospital increases, the effective demand for such equipment also increases, eventually making the purchase and operation of such equipment economically desirable. Thus, for people with rare diseases, their chances of getting treated by the most sophisticated (if not the best) methods increase if they enter a large hospital.

Just as hospital cost can decline with increasing size, and then finally increase (the classic U-shaped average cost curve), so can specialization lead to increasing and finally decreasing quality of care. Most people have heard of some "horror story" in hospitals where the care had become so fragmented, with numerous spe-cialists each caring for some part of a patient's problem but nobody worrying about the "whole" patient, that major medical mishaps occur. Just as too many workers in the vineyard can trample the grapes, and too many cooks can spoil the broth, too much special-ization in the delivery of health care can produce a poor quality of care.

In one important area, size (or more particularly, experience) seems associated with an advantage in quality—the area of sur-gical mortality. As Box 8.3 summarizes, a body of important work demonstrates that surgical mortality differs considerably for hospi-tals performing many versus few instances of a given surgical procedure, but it remains uncertain whether this is mostly due to numerous referrals to high-quality hospitals or because the hospi-tals "learn" by doing more procedures, and hence have better outcomes.

LONG-RUN VERSUS SHORT-RUN COSTS

A much more interesting question about hospital costs is how much those costs vary as the patient load varies. If a hospital's costs and patient load change together closely, then most of the costs are *variable costs*. If a hospital's costs remain pretty much the same whether it is half full or 90 percent occupied, then much of its cost structure is *fixed*.

In the very shortest of short runs (say, an hour or a day), almost

Box 8.3 **PRACTICE MAKES PERFECT?**

Several studies have investigated the relationship between hospitals' experience in performing surgery and their outcomes (as expressed by surgical mortality). These results show a strong relationship between experience and better outcomes (lower mortality). Practice, while not necessarily making things "perfect," seems to move outcomes in that direction.

The most prominent of these studies have been undertaken by economist Harold Luft and a series of colleagues. The first study (Luft, Bunker, and Enthoven, 1979) used outcomes for 12 operations in nearly 1500 hospitals. For complex procedures like coronary artery bypass grafts (CABG), open heart surgery, vascular surgery, and even more simple procedures like transurethral resection of the prostate (TURP), hospitals doing 200 or more of any of these procedures had mortality rates 25 to 40 percent lower than low-volume hospitals. Some procedures such as hip replacement showed a flattening of the mortality curve at about 100 procedures, and other procedures (e.g., cholecystectomy) showed no apparent gains in outcomes with the frequency of operations.

Critics of the first study suggested alternative explanations for the relationship, including the possibility that hospitals of superior quality attracted more patients, so that the relationship was not the one originally proposed (experience in a procedure improves outcomes), but rather what one would expect from a market that contained good information. Luft (1980) called this the "referral effect." There are also questions of "spillover effects"—for instance, whether broader surgical experience improves outcomes on similar procedures. A second study by Luft (1980) identified the importance of spillover effects in some areas (various types of vascular surgery, for example), but continued to find only "own-procedure" effects for some activities such as CABG. Luft was not able to identify clearly the causal relationships involved, but the data suggest that for some of the procedures, the referral effect explained an important part of the observed association.

In the most recent of the series, Hughes, Hunt, and Luft (1987) examined the outcomes for over 0.5 million patients from 757 hospitals, all of whom had received one of 10 specific surgical procedures. As with most of the other studies (Luft's and others cited in Hughes, Hunt, and Luft), the now-familiar relationship between surgical volume and improved mortality appeared again. In this study, however, a "low-volume-doctor" effect also appeared, in somewhat of a contrast to the previous studies. They found that a hospital was more likely to have poor outcomes, other things being equal, when low-volume doctors did a higher proportion of a hospital's surgery.

The literature is now clouded on the issue of doctor-specific effects, but fairly consistent on hospital-specific effects. Hospitals doing more CABG, vascular surgery, and so forth seem to experience better outcomes. Still incompletely resolved is whether this is a "referral effect" or a "practice-makes-perfect" effect.

all of a hospital's costs are fixed, possibly with the exception of a few supplies (food, medicines, etc.). In the very longest of long runs, all of a hospital's costs are variable, because the entire cost structure, including the decision about rebuilding the hospital, is open for question. However, as Lord John Maynard Keynes once said, "In the long run, we're all dead."[14] How much do a hospital's costs fluctuate as its patient load goes up and down? This question matters considerably right now, because (for a variety of reasons we will discuss in later chapters), hospital use has fallen considerably over the past several years. Should we expect hospital costs to fall in parallel? That depends on the short-run and long-run cost structure.

Recent studies of this question suggest that in a meaningful "short run," defined as one in which the capital stock of the hospital does not vary, about 70 percent of the hospital's costs are fixed (Friedman and Pauly, 1981). To see why this occurs, one need only look at the typical hospital's operations.

Hospitals have "wings" or units dedicated to various types of patient care, commonly with 25 to 40 patient beds. This "unit" must be staffed with nurses, janitors, aides, and so on 24 hours a day, seven days a week, 52 weeks a year. In the course of normal day-to-day and even month-to-month fluctuations, whether the unit is half full (say, 20 patients with 40 beds available) or is 90 percent full (36 patients), the same levels of staff will probably be on duty. It would be difficult to adjust the staffing levels even to accommodate monthly swings in hospital use, because of costs of laying off staff and then retraining or rehiring them. If the unit becomes *systematically* half full, the administrator may feel comfortable reducing the staffing level some, but probably not in direct proportion to the patient load changes.

If a hospital's occupancy rate falls considerably, it can eventually reduce its labor costs by closing down one or more units, and moving the patient into another wing of the hospital (which also presumably had lost some patients). This makes the labor costs variable, but the building and beds continue to sit there, and their costs to the hospital don't change. Only by eventually eliminating new bed construction (what would otherwise occur) can the long-run costs of hospitals be reduced completely in parallel with the reduced hospital use.

---

[14] Economist Rodney T. Smith adds this rejoinder: "Yes, but the short run determines our standard of living."

The importance of the fixed-cost versus variable-cost question appears in many policy questions. States with hospital price regulation need to decide how to change the allowed price if a hospital's occupancy changes. In Medicare's program for paying for hospital care, the payments are based on historical costs of the hospital. If patient loads in hospitals systematically decline through time, then the average costs of the hospital will rise, and a fixed level of payment from Medicare will place the hospital in a serious financial squeeze.

THE HOSPITAL'S "COST CURVE"

We are now in a position to consider what "the cost curve" of a hospital looks like. Begin with the variable costs, those that change as the rate of output of the hospital changes. Usually, economists draw a cost curve like Figure 8.4, where the rate of output appears on the horizontal axis, and the vertical axis has cost/output as its scale. (This can represent price, average cost, or marginal cost.) In most productive activities (and we should not expect the hospital to differ), average variable costs will initially fall as output rises, so the average variable cost curve in such a diagram, labeled AVC here, has a traditional U-shape. The logic behind this stems from the common finding that inputs in the productive process eventually have diminishing marginal productivity (see Chapter 3 appendix). The same phenomena produce an average cost curve that first falls, and then rises.

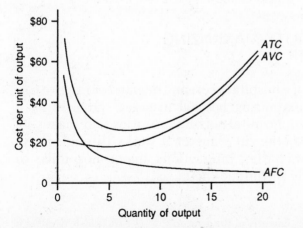

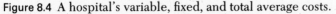

Figure 8.4 A hospital's variable, fixed, and total average costs.

are truly independent of the rate of output, *average fixed costs* appear in this diagram as a curve like *AFC* in Figure 8.4.[15] The average total costs, which must equal average revenue for the not-for-profit hospital, are the sum of the average variable cost and average fixed cost. Figure 8.4 shows this as the curve labeled *ATC*. At each level of output, *ATC* has a height equal to *AFC* + *AVC*.

This figure is drawn to reflect the empirical finding that much of a hospital's costs are fixed in the short run.

THE DEMAND CURVE FACING A SINGLE HOSPITAL

The hospital also faces its own demand curve. For a single hospital within a geographic area, the hospital's demand curve and the market demand curve (the added-up demand curves of all of the individuals within the market) are identical. The same idea holds true in a multihospital market, with an important complication: When we "add up" the patients' demands to create the demand curve for St. Elsewhere Hospital, General Hospital, or their other competitors, we need to have a way to determine which patients in the market will choose which hospital.

We explore in Chapter 9 the mechanisms by which this process takes place, but the process is somewhat similar to that discussed for physicians in Chapter 7. For now, we need only note that the demand curve confronting a single hospital will slope downward (like the market demand curve), and that it will almost certainly be more elastic (quantity demanded more sensitive to price) than the market demand curve. In a market with classic competition, the hospital's demand curve becomes infinitely elastic, but this characterization of hospital markets seems extremely unlikely.

REVISITING THE UTILITY-MAXIMIZING HOSPITAL MANAGER

With the tools in hand of a hospital's cost and demand curve, we can return to the "utility-maximizing hospital manager" problem for a useful example of how a hospital sets its priorities and makes its economic choices. A section in Chapter 9 works through these ideas more formally, including interactions among hospitals, so this discussion will only introduce the ideas.

---

[15] The equation producing this curve is of the form $AFC \times Q = K$, which is called a "rectangular hyperbola."

Suppose the hospital manager cares only about two things—quantity and quality of care. Thus, the hospital decision maker has a utility function of the form $U = U(N, Q)$ where N = the number of patients treated and Q = some measure of quality. The hospital can produce any level of quality it desires, but more quality costs more. Thus, the average cost curve of the hospital depends upon the quality it offers.

The not-for-profit "rules" require that the hospital's total revenue equal its total costs, or (on a per-patient basis), average revenue = average cost. The hospital can achieve exactly "zero profit" by setting a price $P(N, Q) = AC(N, Q)$. It will exactly "clear the market" if the quantity demanded for the quality it selects just equals the quantity supplied. This occurs whenever the demand curve and the average cost curve intersect each other.

If other things "matter" in the utility function, the same sort of logic can yield an analysis of how much of such things will be produced, since we can judge how much patients value them, and how much they cost to produce. Indeed, "quality" in the previous discussion could stand for anything that affected the utility function of a hospital director, including nurses' salaries, doctors' profits, or thicker carpets on the administrator's floor. Of course, patients may not place much marginal value on these dimensions of the hospital's output, in which case the demand curves of the hospital's services won't shift much (if at all) as this dimension of quality varies.

The final idea in this discussion introduces the idea of "cross-subsidies" or "charge shifting" in hospitals. The previous discussion acts as if the hospital had only one output—for instance, patient days—and only a single pricing decision to make. In actuality, the hospital has multiple outputs, and a large number of prices. Thus, it has the opportunity to exploit its market power on one area (e.g., surgical patients) to support another activity creating "utility" in another area (e.g., obstetric services). Thus, the ratio of costs to charges in the typical not-for-profit hospital can differ greatly across activities, and even those ratios can differ greatly across hospitals.

SUMMARY

In this chapter, we explored the hospital as an economic entity. The typical hospital in the United States has a not-for-profit legal status, which eliminates the usual corporate shareholder as a residual claimant. As a result, the hospital's profits must necessarily go somewhere else. Researchers have offered various models to help understand the behavior of the hospital, which generally presume that one or another person (or group) within the hospital has "captured" the hospital, and uses the hospital's economic profits to further their own ends. In the doctor-capture model, the hospital's profits are passed indirectly, and possibly inefficiently, to doctors. In the manager-utility function model, some central decision maker has captured the hospital, and guides the hospital's behavior in a utility-maximizing way. Employee-capture models and patient-capture models also seem possible, although not commonly considered.

A synthesis of these models probably better serves our understanding of the hospital. With no clearly defined legal "rights" to the hospital's profits, something else must serve the same functions as property rights might have, if they existed. This becomes essentially a problem in applied politics. The hospital's by-laws and charter define the "rules of the game" for the participants, who then seek to maximize their own gains, in a contest with others in the same arena. We can expect that these participants will eventually reach a standstill in their struggle, and the hospital will take on the appearance of having a single decision-maker utility function. However, this "apparent utility function" can shift if anything shifts in the political structure of the hospital, including external legal changes, the death of an important participant, or a change in the methods by which the hospital is paid. Whenever any such event shifts the balance of power in the hospital, its apparent utility function will change, and so will the hospital's economic behavior.

PROBLEMS

1. What is the distinguishing feature of the "nonprofit" organizational form as compared with a "for-profit" organization? (Hint: What happens to any profits earned by the nonprofit hospital?)

2. Who controls most of the flow of resources within the standard not-for-profit hospital, and what mechanisms of control does the hospital president have over these people?

3. Suppose Hospital A has 500 patients per day (average) and average costs of $600 per day. Hospital B has 250 patients per day (average) and average cost of $500 per day. Hospital C has 100 patients per day (average) and average costs per day of $550. What can you say (if anything) about the most efficient size of hospital from these data?

4. Some studies show that the average mortality from specific surgical procedures differs depending on the frequency that those procedures are undertaken annually by the hospital. Thus, for example, a hospital doing 200 open-heart cases annually will probably have lower mortality rates than one doing 20 cases a year.

   (a) Based on these data, where would you rather have open-heart surgery done, other things equal?

   (b) If some hospitals achieved reputations for being "very expert" in a particular procedure, do you think that they might actually end up with worse mortality rates? If so, how could this happen, and how would you want to interpret the mortality data? What would you need to be sure you interpreted such data correctly?

Chapter
9

Hospitals in the Marketplace

*T*his chapter discusses the behavior of hospitals in interaction with each other, their patients, and the doctors in their communities. While we will rely in general on the model of a not-for-profit hospital as developed in Chapter 8, nothing really differs greatly in this chapter if the hospital is a for-profit entity, and probably not much if it is operated by a local government (e.g., Cook County Hospital in Chicago). In part this is because the "utility function" model of the hospital readily incorporates "profits" as the source of utility, and because government agencies probably operate quite similarly to not-for-profit agencies in many important dimensions.

Hospitals operate in a variety of markets. They must attract physicians to their staffs, since physicians admit patients to the hospital. They must also attract paying patients. Many steps the hospital takes to do one of these also helps the other, but in some ways, the two types of "competition" differ, particularly under current funding arrangements. Hospitals also operate in factor markets as buyers, most notably in the market for labor. Some of this labor is quite general in nature (e.g., janitorial staff), and some quite specific to the hospital (laboratory technician). This chapter will explore how each of these markets relates to the hospital and to its eventual survival and growth.

HOSPITALS AND THE MARKET FOR MEDICAL STAFF

A primary problem for a hospital is to attract a medical staff, since by law, only doctors may prescribe much of the treatment for patients that hospitals provide, and only doctors can provide some of that treatment (e.g., surgery). Without doctors, a hospital cannot function. In the U.S. health care system, however, doctors typically function as independent economic entities. Thus, the hospital does not "hire" doctors, but rather "attracts" them.

This system of organizing doctors and hospitals does not occur in all other countries. In Great Britain and Germany, for example, "community" doctors typically have no ability to admit patients to hospitals, but rather refer hospitalized patients to separate doctors on the hospital staffs who treat patients, and then return them to their community doctors. (See Chapter 17.) In that setting, the hospital hires the staff doctors directly.

In the type of market organization found in the United States, the problem of attracting doctors creates a type of competition separate from what one would normally find in an industry, because the doctor is somewhat like an employee and somewhat like a customer. Much economic research has made clear that doctors are an important "input" in the production of medical care in the hospital, and can be analyzed in the same way as other types of labor (RNs, technicians, janitors, etc.) even though they are not "on payroll." However, since the hospital and doctor do not directly exchange money (payroll), the hospital must do something else in order to attract doctors. The solution has typically been for the hospital to compete, where necessary, by providing the doctor with facilities and services that make the doctor's practice more profitable.

As should be apparent, the question of how a hospital attracts doctors to its staff is closely tied to the problem discussed in the previous chapter of "dividing up the profits" of the hospital. The model of the hospital developed by Pauly and Redisch focuses on this problem of attracting doctors quite intensively, and ends up characterizing the hospital as if it were "owned" by the doctors. Of course, the hospital has other things it must or wants to do as well as attract doctors, so we would not expect a complete "capture" of the hospital by doctors, but to the extent that attracting doctors matters (as it must), the hospital will behave at least partly in ways as characterized by the Pauly-Redisch model.

Some types of doctors are intrinsically more attractive to a hospital than others, depending on the hospital's goals. Some doctors are systematic money makers, creating a "cash cow" that helps finance other hospital activities. Other doctors provide prestige, glamour, and headlines for the hospital; they may actually create financial losses for the hospital with their activities. The mix that the hospital tries to attract will depend in part on the goals of the hospital's "director."

A primary way by which hospitals attract doctors to their staffs is to provide the capability for doctors to do things that they cannot do elsewhere. For cardiologists, for example, the provision of a hospital with a cardiac intensive care unit (CICU) makes the hospital more attractive, because it provides a high-tech environment in which to care for seriously sick patients. For heart surgeons, provision of excellent nursing staff for postsurgical patients is important, and in some cases, provision of special operating rooms and equipment is the key attraction. For obstetricians, a good "delivery room" is available in almost any hospital, but having a newborn intensive care unit available in case an infant is born with a complex illness or difficult delivery makes the hospital attractive.

Some features of the hospital, including prompt laboratory work, the provision of good assistance to help the doctor complete the medical record, and the availability of standby emergency equipment (the so-called crash cart for cardiac arrest provides a good example) all increase the general attractiveness of the hospital for doctors.

Of course, the doctor and the hospital stand in somewhat symmetric positions in many ways. Not only does the hospital need doctors, but most doctors need access to at least one hospital in order to carry out their medical practice. Denying some doctors access to hospital medical staffs is tantamount to denying them the ability to earn a living, at least in some specialties. (For some specialties such as psychiatry and dermatology, hospital privileges are not as important as, say, for surgery.) Thus, in towns where only one or two hospitals exist, the hospital may be in a stronger bargaining position with the doctor than it would be in a town with many hospitals. Similarly, hospitals in regions that are intrinsically attractive to doctors need not provide as many "goodies" to the doctors as those hospitals in less desirable regions.

In every hospital, the "bargain" struck between doctors and hospitals will differ, depending on the strengths of each in the marketplace and their uniqueness as a medical resource. Some hospitals, for example, require every doctor on their medical staff to provide some time in the emergency room as a "quid pro quo" for membership to the medical staff.[1] In other hospitals, by con-

---

[1] *Quid pro quo* is Latin, literally translated as "this for that." Doctors, like lawyers, are fond of using phrases from ancient languages. Economists do this too, with phrases like "ceteris paribus" (all other things being equal) and "mutatis mutandis" (moving all those things that will be moved). Doctors tend to rely more on Greece for obscure words and phrases, such as *iatrogenesis, cardiac,* and the like, probably because of the importance of early Greeks in the development of medicine, led by the famous Hippocrates.

trast, the hospital not only does not exact such a "payment" but it provides office space for the doctor—sometimes at a very low or zero charge—to see patients within the hospital.

Depending on how hospital care is paid for, it is easy to see how competition for doctors can become quite costly. Particularly in an environment where health insurance pays for most if not all of the costs of most hospitalizations, it may well be that competition among hospitals mostly manifests itself in a technology-intensive war to attract physicians. As we shall see later in this chapter, this appears to be what happened in the United States until quite recently.

The growth in some types of service has been quite remarkable. By 1986, 20 percent of American hospitals offered cardiac catheterization services, and 13 percent (1 out of every 8 hospitals) offered open heart surgery. Sixty percent had a CT scanner. The fraction of hospitals with an emergency department rose in the last decade from 88 percent to 95 percent. "Low-tech" services have also proliferated. Home care units increased from 7 percent to 35 percent of hospitals over the last decade. Outpatient alcohol-chemical dependency units increased to 13 percent of hospitals, where only a few specialized hospitals offered these services in 1970. Box 9.1 shows some examples of the type of information available to show what services any particular hospital offers.

This proliferation of new services into smaller and smaller hospitals (thereby increasing the fraction of hospitals with such services) coincides well with the observed diffusion of specialized doctors into smaller communities, as Chapter 7 described.

HOSPITALS AND PATIENTS

Hospitals must also attract patients, since the patients bring with them the revenue the hospital needs to pay its costs. Some of the same things that attract doctors to a hospital also attract patients, but sometimes different things matter, and sometimes the problem of attracting patients conflicts with the problem of attracting doctors. For patients, of course, a reputation of providing quality medical care is important, but other things sometimes loom large in patients' perceptions of a hospital. The quality of food, the friendliness of the nursing staff, and even to some degree, the price of the hospital, its proximity to patients and family, and its general neatness and cleanliness can all affect the desirability of a hospital to patients.

Box 9.1   INFORMATION ABOUT HOSPITALS

Some standard data sources provide a considerable amount of basic information about most hospitals in the United States. The most widespread information about individual hospitals appears in an annual publication of the American Hospital Association, describing (for each hospital responding to a survey form they send out) some key facts about each hospital, including its size, occupancy rate, employees, costs, and a list of "facilities" that it offers. This annual report is published as a supplement to *Hospitals, The Journal of the American Hospital Association*, and should be available in the library of almost every hospital in the United States. This supplement is called *The Guide Issue*, or more formally, *Guide to American Hospitals and Health Care Facilities*.

The following entries describe the hospitals in Rochester, N.Y., as an example of the types of data available in this resource. The "control" variable indicates the form of ownership, with numbers in the 20s the most common (not-for-profit), 23 being non-church-affiliated, 21 being church-affiliated. (The name of Rochester St. Mary's Hospital of the Sisters of Charity provides a reasonable clue of church affiliation also.) A "service" code of 10 indicates a short-term general hospital, the typical designation of "hospitals" as we normally think of them. Subsequent numeric data are self-explanatory. (Readers should be alert that the average occupancy rates for hospitals in this city exceed those in many areas of the country at present. For all community hospitals in the United States in 1986, the average occupancy had fallen to 64 percent, down from 75 percent in 1976.)

The "facilities codes" listed under the hospital's name and address provide a description of how extensive the scope of services is. The length of the list for Strong Memorial Hospital, for example, is typical of medical-school-affiliated hospitals, offering specialized expertise not found in some other facilities.

The codes that describe these hospitals are somewhat obscure, but the entire set appears in the front of the AHA *Guide Issue*. The codes of L and S for "stay" mean "long" and "short." The various codes listed under the hospitals' addresses (e.g., F1 7 12 14 . . .for Highland Hospital) represent different facilities the hospitals offer: For example F1 = general inpatient care for AIDS/ARC patients; F7 = birthing room; F12 = angioplasty (a procedure to clean out clogged arteries of the heart nonsurgically); F14 = Emergency Department, and so on. One can learn a considerable amount about the size, cost, and capabilities of a specific hospital from the information provided in this directory. (Two hospitals did not complete the survey forms this year, one of the hazards of relying on voluntary reporting. The AHA does not have comparable powers of persuasion to those of the IRS.)

★ American Hospital Association (AHA) membership
□ Joint Commission on Accreditation of Healthcare Organizations (JCAHO) accreditation
+ American Osteopathic Hospital Association (AOHA) membership
○ American Osteopathic Association (AOA) accreditation
△ Commission on Accreditation of Rehabilitation Facilities (CARF) accreditation Control codes 61, 63, 64, 71, 72 and 73 indicate hospitals listed by AOHA, but not registered by AHA. For definition of numerical codes, see page A6

ROCHESTER—Monroe County

Hospital, Address, Telephone, Administrator, Approval and Facility Codes, Health Care System	Control	Service	Stay	Facilities	Beds	Admissions	Census	Occupancy (percent)	Bassinets	Births	Total	Payroll	Personnel
▣ GENESEE HOSPITAL, 224 Alexander St., Zip 14607; tel. 716/263-6000; Paul W. Hanson, President (Total facility includes 40 beds in nursing home-type unit) A1a 2 3 5 8 9 10 F1 5 7 9 11 12 13 14 30 33 34 35 37 39 40 41 43 44 45 46 47 48 49 50 51 52 54 56 60 61 62 63 64 66 67 68 69 70 71 73 74 75 76.	23	10	S	TF	421	14571	375	89.1	41	2782	96979	51974	1894
				H	381	14554	335	—	41	2782	96142	51329	1863
▣ HIGHLAND HOSPITAL OF ROCHESTER, 1000 South Ave., Zip 14620; tel. 716/473-2200; Michael J. Weidner, President A1a 2 3 5 8 9 10 F1 7 12 14 23 26 27 33 34 35 37 38 39 42 49 50 51 52 53 54 56 61 62 64 66 67 68 69 71 72 73 75 77	23	10	S		261	10132	206	78.9	44	3033	52551	26719	1206
□ MONROE COMMUNITY HOSPITAL (Aged & Chronically Ill), 435 E. Henrietta Rd., Zip 14620; tel. 716/274-7100; Jane A. Mahoney, Executive Director (Total facility includes 566 beds in nursing home-type unit) A1a 3 5 9 10 F1 6 13 20 21 23 24 29 39 42 45 58 60 61 62 63 66 67 68 69 71 72 73 74 75 76	13	49	S	TF	605	953	590	97.4	0	0	31404	15591	755
				H	39	489	33	—	0	0	6851	3404	422
▣ PARK RIDGE HOSPITAL, 1555 Long Pond Rd., Zip 14626; tel. 716/723-7000; Timothy R. McCormick, President A1a 2 9 10 F1 4 9 13 14 27 28 33 34 35 39 42 44 54 56 57 58 59 61 62 63 64 66 67 68 69 71 72 73 75 76	23	10	S		234	7764	211	90.2	0	0	42896	19029	763
▣ ROCHESTER GENERAL HOSPITAL, 1425 Portland Ave., Zip 14621; tel. 716/338-4000; Arthur E. Liebert, President A1a 2 3 5 8 9 10 F1 7 9 10 12 13 14 21 26 27 29 30 33 34 35 36 37 39 40 42 44 45 46 48 49 50 51 52 53 54 56 58 59 61 62 63 64 66 68 69 71 72 73 74 75 76 77	23	10	S		517	19575	449	86.8	18	2233	125286	61149	2481
▣ ROCHESTER PSYCHIATRIC CENTER, 1600 South Ave., Zip 14620; tel. 716/473-3230; Martin H. von Holden, Executive Director (Nonreporting) A1a 3 5 10	12	22	L		805	—	—	—	—	—	—	—	—
▣ ROCHESTER ST. MARY'S HOSPITAL OF THE SISTERS OF CHARITY, 89 Genesee St., Zip 14611; tel. 716/464-3000; Patrick Madden, President (Nonreporting) A1a 2 3 5 8 9 10 S1855	21	10	S		201	—	—	—	—	—	—	—	—
▣ STRONG MEMORIAL HOSPITAL OF THE UNIVERSITY OF ROCHESTER, 601 Elmwood Ave., Zip 14642; tel. 716/275-2644; Paul F. Griner M.D., General Director A1a 3 5 8 9 F1 3 5 6 7 8 9 10 12 13 14 16 18 20 27 28 29 30 33 34 35 36 37 38 39 40 41 42 43 44 45 46 47 48 49 50 51 52 53 54 55 56 57 58 59 61 62 63 64 65 66 67 68 69 70 71 72 73 74 75 76	23	10	S		702	24988	622	86.6	40	3721	178438	81583	3438

"Price" matters only a little, if at all, to many patients, because their health insurance offers to pay for any costs billed by the hospital. From the patient's point of view with such an insurance policy in hand, the price of care is zero in many ways, as Chapter 4 discussed.[2] For patients like this, the hospital will mostly compete on the basis of the quality of nursing care, the food and so on. With these types of patients, doing things that attract doctors to the medical staff and things that attract patients to the hospital often work in tandem. Moreover, since "the insurance pays for it," decisions to raise the hospital's quality have no apparent cost to the hospital or the patient.

Some patients have insurance that pays only a specific amount per day of hospitalization, and others (about 13 percent of the U.S. population under age 65) have no insurance at all. For these people, price does matter in the choice of hospital. (In addition, some new forms of health insurance create something akin to price competiton for hospitals. We discuss this momentarily.) For these types of patients, the things a hospital might do to attract doctors (e.g., more elaborate standby equipment in case of cardiac arrest, more numerous and more specialized intensive care units, etc.) may also prove attractive to patients, but they also raise the cost (and hence, the price) of the hospital. For patients who pay for quality "on the margin" (for example, those who have an insurance policy that pays a fixed dollar amount per day of hospitalization, no matter what the cost of the hospital), an inherent conflict arises between attracting doctors and attracting patients.

Until quite recently, the large majority of patients in United States hospitals had health insurance that essentially insulated them from hospital insurance bills, thus making the world more like that of "fully insured" patients than "uninsured" patients. In "the good old days" for U.S. hospitals, Medicare essentially paid dollar for dollar any costs incurred by the hospital,[3] most Blue Cross insurance plans paid nearly full-dollar,[4] and much commercial insurance also paid with similar generosity. For this reason,

---

[2] An important distinction: Most insurance policies will pay only for "semiprivate" rooms, i.e., rooms shared by two patients. Thus, if the patient selects a private room, often at a considerably higher price, the insurance policy will usually pay only the amount that would be paid for a semiprivate room. On this dimension of "quality" the patient bears the full incremental cost.

[3] This was not literally true, but a reasonable approximation for our discussion here.

[4] Again, not literally true, but close enough here for discussion purposes.

most hospitals apparently adopted a strategy of competing on quality, both for doctors and patients. One direct piece of evidence to
support this claim comes from a study by Pauly (1978), comparing
the costs of various hospitals with the makeup of their medical
staffs. As the above discussion suggests, hospitals with "fancy"
doctors (i.e., highly specialized doctors requiring specialized,
rather than general-purpose equipment and staff) will also have
"fancy" costs. Pauly's study shows a strong connection between
the medical staff composition and a hospital's costs. Such a study
cannot tell for sure which causes which, but logic dictates that the
hospital cannot attract a highly specialized surgeon to its staff (for
example) without at least the promise to develop the equipment
and support staff necessary to support the surgeon's work.

Returning to the question of the modes of competition between
hospitals, even with this extensive insurance coverage, it is clear
that price matters, at least to some degree. Remember that some
people (about one in every ten persons who enters the hospital)
have no insurance, and that for them, hospital care is *very* expensive on average, and that others have insurance that only pays a
lump sum per day, thus making these persons "price conscious"
as well. One study attempted to estimate the demand curve confronting individual hospitals to determine just how much price
mattered. This study (Feldman and Dowd, 1986) used data from
a single metropolitan area (Minneapolis–St. Paul) in 1984, and
estimated the demand curves facing 31 hospitals in that area, using
techniques that depended upon a standard economic idea—namely,
that the optimum "markup" for a firm to set its price above cost in
order to maximize profits is inversely related to the demand elasticity confronting the firm.[5] With estimates of cost and price,
Feldman and Dowd inferred the demand elasticity, which stands
as a proxy for the degree of competitiveness. (See also the appendix to this chapter for a more detailed discussion.)

Under one commonly invoked model of market pricing, the
price elasticity confronting a noncolluding firm in the market is
approximately the market demand elasticity divided by the share
of the market held by the firm. For example, if the market demand
elasticity is −1, and the firm has 10 percent of the market, then its
demand curve should have an elasticity of −10. This arises under a

---

[5] This is the so-called Lerner index; Lerner (1934) first showed that the optimum markup
for a monopolist is found by setting price such that the relative markup
$\lambda = (P - MC)/P = -1/\eta$, where η is the demand elasticity and λ is the Lerner index.

model of oligopoly pricing set forth by the French economist Cournot; it presumes that firm i sets its price under the belief that no other firm in the market will change its output if firm i changes its output. While this "story" may seem unrealistic, it has some desirable properties.[6] Since there were 34 total hospitals in the Minneapolis–St. Paul area (with no cost data for 3, leading to the estimates for 31), and the demand for hospital care at the market level in that city should be about the same as that found in the RAND HIS ($\eta = -0.15$ for hospital care), the demand elasticity confronting any single hospital should be about $-0.15 \div (1/34) \approx -5.1$ if the hospitals were all in the same market, and if they behaved in a simple Cournot-like way.

In fact, the estimates found by Feldman and Dowd using this approach did not differ considerably from that target. The price elasticity they inferred for private-pay patients was approximately -4; for Blue Cross patients (who have better coverage), approximately -2.3. However, using another completely different approach (directly estimating the demand facing each hospital as a function of its own price), they found price elasticities near -1, suggesting more monopoly power. These estimates represent a large share of the total literature on this topic,[7] so the matter cannot be considered conclusively settled. Nevertheless, these estimates do suggest that hospitals have some market power, confirming the desirability of using something other than a pure competitive model for analyzing the behavior of hospitals. In the next section, we can turn to what amounts to the beginnings of a model that describes how hospitals determine their price and quality jointly, an important problem in the U.S. hospital setting, with considerable implications for public policy.

A MODEL OF EQUILIBRIUM QUALITY AND PRICE

A model of how the choice is made to pick quality and cost of a hospital emerges (at least in part) from Newhouse's discussion (1970) of the decision making of a hospital.[8] That model character-

---

[6] The Cournot model has the advantage that as the number of firms rises toward infinity, the pricing rules of each firm approach those of perfect competition. More precisely, as $n \to \infty$ each firm's share approaches zero, and the markup (Lerner index) becomes very small.

[7] One hundred percent of that known to the author of this textbook.

[8] This discussion follows directly from the study, and some of the figures that follow are quite similar to Newhouse's figures.

izes the hospital as having a single utility function, describing how much the hospital balances its various goals. In his simple model, the hospital desires only two things—size and quality—but the ideas generalize to other dimensions of hospital choice. We will begin with this model, and then discuss how various forces in the market, notably competition and insurance, will alter the outcomes. The appendix to this chapter derives the solution to this problem formally.

We assume that the hospital utility function includes two characteristics—"quantity" (denoted N for number of days) and quality per day of care (denoted S for service). The hospital attempts to maximize the utility function $U(N,S)$. The patients' willingness to pay for the hospital service is represented by the inverse demand curve facing the hospital $P(N,S)$, where the willingness to pay P decreases with total quantity N (as is customary with demand curves) and increases with the quality of care S.[9] The not-for-profit hospital also faces a breakeven constraint specifying that revenue equals cost ($P(N,S) \times N = C(N,S)$). We presume that patients respond to quality and price in the normal fashion. We can assume that costs increase both with more quality S and more quantity N.[10]

If we were to draw the demand curves for different qualities (Figure 9.1a), then each demand curve would slope downward, but the demand curves (willingness to pay) would be higher at higher qualities. *Remember that in this diagram, everything else is held constant except quantity (N), quality (S) and price-cost. Specifically, the insurance coverage of the patients and the quality and output of other hospitals is held constant.* In this way, the demand curve facing a single hospital is "stable."

Similarly, if we were to draw the average cost curves of the hospital for different levels of quality, the cost curves would stack up, each with the characteristic U-shape of average cost curves (see discussion in Chapter 6 on physician-firms), so that higher quality costs more at any given level of output. These curves are shown in Figure 9.1b as approximately parallel, but nothing requires this in general.

Now combine these two sides of the market, as in Figure 9.2: In Figure 9.2, indicators for the level of quality (e.g., S_1, S_2, S_3, are omitted to avoid clutter, but we can take D_1 to mean $D_1(S_1)$, and so forth. In general, the demand curve for a specific quality (say S_1)

---

[9] In formal terms of the calculus, $\partial dP/\partial N < 0$ and $\partial P/\partial S > 0$.

[10] Again, this means formally that $\partial C/\partial N > 0$ and $\partial C/\partial S > 0$.

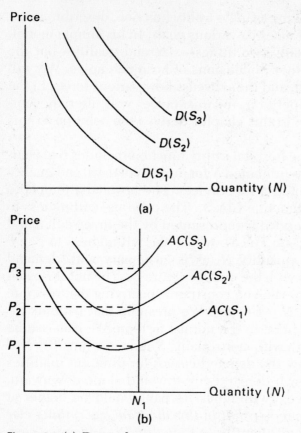

Figure 9.1 (a) Demand curves and (b) average cost curves of the hospital for different levels of quality.

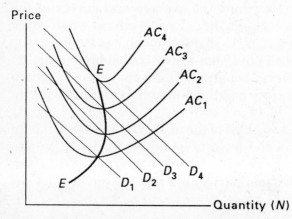

Figure 9.2 Equilibrium combinations of quality and quantity.

must intersect the corresponding average cost curve at either two points, one point (just tangent), or never. Points where the demand and cost curves intersect are important, because they show the combinations of price, cost, quality, and quantity that keep the hospital in equilibrium, because it can charge a price equal to its average cost for that output and quality, and at that price, quantity demanded will just equal quantity supplied. In other words, these points of intersection are equilibrium points. If no intersections occur, that choice of quality is not feasible, because it would always cost the hospital more to produce that level of quality than patients would be willing to pay, no matter what the level of output.

The next thing to note is that when two intersections occur, the one to the lower right is the best one for the hospital, because it has more output, and by assumption, hospitals would prefer to produce more hospital care, so long as quality doesn't suffer. Thus, the upper left of any intersections of a quality-specific demand curve and average cost curve don't matter.

We can find the set of all possible equilibrium combinations of quality and quantity using these tools. First, pick any level of quality, say, S_1, and draw the corresponding average cost and demand curves AC_1 and D_1. The lower right of the two intersections is one possible equilibrium choice for the hospital. Now repeat that process for some other quality level S_2, and again for S_3, and all other possible combinations. Figure 9.2 shows all possible such combinations of cost and demand curves, tracing out the entire set of quantity–quality points that jointly satisfy both the demand conditions facing the hospital and its zero profit constraint. This collection of points is the line EE, a continuous line so long as quality can vary continuously. In general, this collection of points can slope either downward or upward, but only the downward-sloping portion should matter, since on the upward-sloping portion of that set of points, the hospital could increase both N and S, which (by our assumption) are both "goods" to the hospital decision maker. In other words, if the hospital inadvertently found itself at some quality level on the upward-sloping portion of the EE curve, it could increase both quantity and quality and still be in equilibrium.

We can pick the "best" of all possible equilibrium points for the hospital by returning to the ideas set forth in Chapter 4 on the demand for medical care, by using the concept of an "indifference curve." Since we have assumed that the hospital decision maker

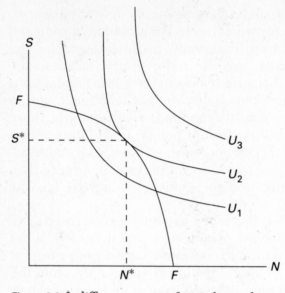

Figure 9.3 Indifference curves for quality and quantity.

has a utility function that increases with two "goods" (N and S), we can draw a set of indifference curves for such a decision maker, just as we did for the individual in Chapter 4. Figure 9.3 shows such indifference curves. The problem of finding the best possible combination of N and S is found by taking information from the EE curve in Figure 9.2 and redrawing it into Figure 9.3. We can do this, since each point on the EE curve describes a unique combination of quality (S) and quantity (N). Thus, the points on the EE curve can map into an opportunity-possibilities frontier FF for the hospital, with the optimum mix of N and S defined by the familiar tangency to an indifference curve of the hospital decision maker at N* and S*.

INSURANCE AND COMPETITION IN THE HOSPITAL'S DECISION

The next step in understanding hospital decision making is to ask what various market (economic) events would do to the location and slope of the hospital's demand curve, since for all of the discussion surrounding Figures 9.1–9.3, "everything else" was held constant, including the quality and quantity of other hospitals, and the scope and style of the insurance plans of the patients using the hospital. If we can learn what happens to demand curves in response to various stimuli, we can at least begin to infer what happens to the hospital's best choices for output and quality, and hence for the cost per hospital day.

All of the previous discussion presumes that the hospital is a monopolist, or more specifically, that the family of demand curves facing the hospital (one for each level of quality) does not change as the pricing, quality, or output of any other hospital changes. In some small towns with only one hospital, this accurately describes the world. In many cities, however, the relevant market is shared by a few hospitals—more than one, but not so many that we can safely presume a purely competitive market. A more complete analysis of this problem appears in Phelps and Sened, 1990.

Other Hospitals' Quality and Output Changes

What happens when one hospital changes its quality, and all other hospitals make no response? We can begin the analysis by picking one specific hospital (say, St. Elsewhere Hospital), and asking what affects its demand curve's location and slope (elasticity). Suppose that another hospital in the region (say, General Hospital) increases its quality—for example, by adding a new intensive care unit, or increasing the average level of training of its nursing staff. All of the demand curves facing St. Elsewhere Hospital (for different levels of quality) will shift. Demand for higher quality confronting St. Elsewhere will fall, because of the increased quality of General Hospital. Demand for lower-quality service at St. Elsewhere may rise, because the higher quality (and price) at General Hospital may drive some of its patients elsewhere. On net, this will tip the EE curve confronting St. Elsewhere Hospital so that the downward-sloping portion is flatter in Figure 9.2, by rotating the EE curve counterclockwise. In Figure 9.3, the opportunity-possibilities frontier FF will also flatten out, making the most desired point one of higher output (more N) and lower quality (smaller S). This suggests that hospitals will tend to specialize in different styles of output (different qualities) depending on their tastes and preferences, but the ideas described here suggest that the market will not generally behave explosively.[11]

---

[11] An explosive market would have the following characteristic: If Hospital A increased its quality (perhaps by mistake), then Hospital B's optimal response would also be to increase quality. This would in turn cause Hospital A to increase quality more. . . . Soon all of the resources in the universe would be sucked into the black hole of hospital quality.

Entry by a new hospital does the same thing to St. Elsewhere's demand curves, with the shift being greater for those qualities where the new entrant specializes. Thus, if a lower-quality hospital enters, St. Elsewhere's demand curves will shift most for low quality, and not much for high quality. If the new entrant has precisely the same quality as St. Elsewhere, then all demand curves would shift similarly, and the entire *EE* curve (Figure 9.2) and corresponding *FF* curve (Figure 9.3) would just shift inward.

The ultimate equilibrium appears as a variant on the standard monopolistic pricing market analysis. As noted, entry by other hospitals will cause some of St. Elsewhere's patients to shift to the new entrant. How long can this continue? If the demand curve shifts so far for a given level of quality that it no longer touches the *AC* curve for that quality level, then St. Elsewhere can no longer produce at that quality. If the same thing happens for all possible qualities that St. Elsewhere might produce, it will not be able to cover its costs of production at any quality level, and would be forced out of business.

In general, for hospitals with access to the same technology, any entering hospital would face the same problem as St. Elsewhere faces—that is, attracting enough customers to have demand curves crossing *AC* curves for at least one point. The new entrant could not in fact enter unless it had some tangencies or intersections of demand and *AC* curves at one or more qualities. The limiting case occurs when each hospital in the market has its demand curves *at all levels of quality* just touching their *AC* curves at one point—that is, just tangent to the *AC* curves. This obviously will occur on the left side of the *AC* curve, where *AC* is declining and the hospital's capacity is underutilized. This is the classic excess capacity story of Chamberlin's (1962) monopolistic competition. Figure 9.4 shows St. Elsewhere's opportunity set in this equilibrium, with the *EE* curve just a collection of tangencies of demand curves to *AC* curves for various levels of quality. Any further entry will cause St. Elsewhere's market opportunities to collapse. If every hospital in the community has similar *EE* curves, then the market is stable, with no incentives for any hospital to enter or exit.

Changing the Insurance Coverage of Patients

We can also think about what would happen if hospital insurance coverage changed for the patients in the market. As a thought experiment, consider what would happen if more patients sud-

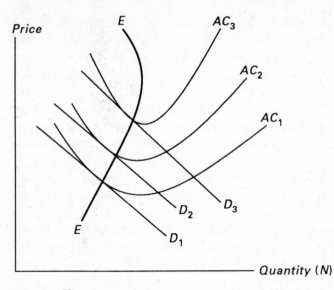

Figure 9.4 Excess capacity.

denly became insured with standard "coinsurance"-type insurance. We know from Chapter 4 that such insurance rotates demand
curves clockwise around the point where the demand curve inter-
sects the quantity axis. In other words, the demand curves would
become steeper, and, in general, be further to the right than previously. This would cause the EE curve to shift outward to the right
in Figure 9.2, cause the production possibilities frontier FF to
expand outward in Figure 9.3, and lead to a higher equilibrium
quality and quantity for the hospital. This would happen for *all*
hospitals in the market, because the expansion of insurance would
cause the demand curves for all hospitals to rotate outward, rather
than having an increase on one hospital's demand come at the
expense of another (as happens when one hospital increases its
quality unilaterally). Thus—and this should really come as no
surprise—we reach the conclusion that better insurance not only
increases the quantity of care demanded, but it also increases
average quality, and hence average cost.

HOW DOCTORS AND HOSPITALS INTERACT: "GOODIES" FOR THE DOCTOR

This same type of model can help us understand how hospitals and
doctors interact, and how hospitals "compete" with each other for
medical staff. To begin, consider a doctor in a monopolistically

competitive market with n physician-firms total, each initially with "typical" demand curves like D_n and cost curves like AC_1 in Figure 9.5. (This figure shows the same situation as Figure 7.1, where the idea of monopolistic competition among doctors was developed, but this one is more complex; AC_1 and D_n represent the situation shown in Figure 7.1.) Now suppose that the hospital used some of its surplus in a way that reduced doctors' AC curves. (The hospital has numerous ways it could do this: provision of specialized surgical equipment that the doctor can use at no charge, rental of in-hospital office space at reduced or no charge, provision of interns and residents to "cover" the doctor's patients on nights and weekends, etc.) This would shift the AC curve downward, to something like AC_2 or AC_3. If the costs for each doctor shift as low as AC_3 because of the hospital's largesse, then the existing doctors are no better off, because the costs will have shifted so low that entry will occur. (This takes place when the AC curve is just tangent to the demand curve D_{n+1} that each doctor would have with $n + 1$ doctors in town, shown as AC_3; indeed, AC_3 was chosen so that it is just tangent to D_{n+1}.) However, if the hospital picks some intermediate amount of largesse to bestow on its doctors, it will lower their costs to AC_2, which will not support another doctor in town, and hence will not induce entry. At this point, the doctors can begin to price monopolistically (see Box 6.1 on monopolistic pricing), setting $MR_n = MC_2$, the marginal cost

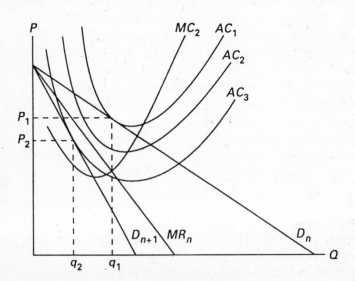

Figure 9.5 Monopolistically competitive markets.

curve corresponding to AC_2, and the price will be P_1. (If entry occurred, the price would fall to P_2 in the monopolistically competitive environment.) This is drawn in Figure 9.5 as the same price that had prevailed before (at the tangency of D_n and AC_1), but this just represents another deliberate choice in drawing Figure 9.5. The price could rise or fall from P_1 depending on whether the AC shift was above or below AC_2. The key point is that the doctor receives a direct benefit, because costs fall for the physician-firm, and the optimal pricing response *must* make the physician-firm more profitable. Of course, it's also important that the hospital not "overdo it" to the point where entry is provoked (i.e., reducing costs at or below AC_3).

HOW DOCTORS AND HOSPITALS INTERACT: PATIENTS FOR THE HOSPITAL

The story is not yet complete, for we must reach an understanding of why the hospital would agree to do something to cut the physician-firm's costs as described just previously and shown in Figure 9.5. One can ask why the doctor becomes a "stakeholder" in the hospital, or in the model of the not-for-profit hospital developed in the previous chapter, ask why the physician-firm's profits enter the utility function of the hospital "director." One obvious answer is that the doctor can provide something the hospital badly needs: patients. Patients cannot "admit" themselves into a hospital—a doctor who has "admitting privileges" in the hospital must admit each patient. Thus, the very location of the demand curve for each hospital depends on how many doctors are on its staff, and how many patients the doctor admits to that hospital.

We can now see how this model reaches closure: Hospitals compete for doctors by doing things that make their physician-firms more profitable; doctors respond by admitting patients to the hospital, thereby shifting out the demand curve (or, more properly, demand curves for all levels of quality) for the hospital. As the earlier discussion showed, when the demand curves shift out, the EE curve expands outward, and the hospital can accomplish more of its desired goals. The doctors who bring the patients to the hospital share somewhat in the fruits of this expanded EE curve, but not completely.

This market seems generally to have a stable equilibrium, where hospitals compete for doctors (and hence patients) by providing cost-reducing "benefits" to its medical staff (Phelps and Sened, 1990). Doctors in return bring patients to the hospital. In a market where doctors can move freely from one community to another, and where hospital staffs are "open" to new doctors, the equilibrium will have the characteristics described here, with doctors getting some gains, but not all of the possible economic returns that their patients provide for the hospital.

COMPETITION—"OLD STYLE" VERSUS "NEW STYLE"

This discussion of hospital and doctor behavior leads naturally to a more extended discussion of how hospitals compete. Of course, in addition to competing for doctors, they must also compete for patients, since patients' preferences will have at least some effect on decisions of where they get hospitalized.[12] The next section explores how this competition takes place, and how it has changed over the most recent decade.

We might (loosely) characterize the "old-style" hospital market in the United States as one in which most of the people going to the hospital had insurance that covered most of their expenses, so that demand curves would be quite inelastic. (Some of Feldman and Dowd's estimates confirm this assertion.) In general, we could even draw the following tentative conclusion: With "old-style" insurance, the hospital's problem of attracting patients would be solved mostly in concert with its problem of attracting doctors— namely, by increasing the quality of the hospital.

In this type of market, increasing the number of hospitals could have an unusual effect on costs and prices. Since more hospitals would increase the bargaining power of doctors relatively, we might expect that areas with more hospitals (more competition) would have to compete more intensively for doctors. In this case, competition would be cost-increasing by driving every hospital to a higher quality level in order to attract doctors (and patients).

An alternative type of insurance (and market equilibrium that would result) occurs if a large fraction of the insured population

---

[12] Some doctors have admitting privileges at several hospitals, and admit their patients to that hospital most preferred by the patient. Of course, some patients also choose doctors on the basis of where they have admitting privileges, which makes the attractiveness of the hospital the key issue.

has incentives or administrative mechanisms that induce a high degree of cost-consciousness on the part of patients. We could call this a "new-style" insurance.

Importantly for understanding how hospital-doctor-patient markets work, we have seen something of a natural experiment in hospital insurance over the past several years that provides some insight into hospital behavior. The first step came when Medicare (which pays for about a quarter of all hospital admissions) switched from a pay-dollar-for-dollar system to a very different system called the Prospective Payment System (PPS), which pays a flat amount to each hospital for a given category of admissions.[13] For now, all we need to know is that it suddenly put into the decision-making process a cost-consciousness that had been notably absent previously.

During the same period, particularly in a few states, cost-conscious insurance spread through other sectors of the health care market as well. Some private insurance plans adopted some very aggressive shopping-for-deals strategies, and even some state Medicaid programs for low-income people began to seek competitive bids for hospital care. If you will, the health care buyers began to play "Monte Hall" with health care providers, particularly with hospitals.[14] The insurers began to negotiate with hospitals and doctors, saying, "We'll send you all of the patients enrolled in our insurance plans if you will give us a good price." The insurers enforced this with their subscribers by paying 100 percent of the hospital bill for "approved" providers, but a lower amount for others.[15]

The results of this change, which began in the early 1980s, create the possibility that the economic environment might have shifted between the "old style" and this new cost-conscious style. In particular, whereas competition might have been cost-enhancing in the old style (due to the pursuit of doctors and competition for patients with increasing quality), competition under the "new style" is much more likely to create cost reductions.

A perfect testbed for this type of analysis appeared in California during the mid 1980s, because the change in Medicare pricing

---

[13] This change in Medicare pricing is discussed in much more detail in Chapter 12.

[14] Monte Hall is the host of a TV game show called "Let's Make a Deal."

[15] These insurance plans, called "Preferred Provider Organizations," are discussed in more detail in Chapter 11.

coincided with a large increase in the market share of cost-conscious private health insurance as well as a shift to competitive bidding for the state's Medicaid hospital business. While in the "old style," competition indeed was cost-increasing for California hospitals; in the new-style era, the opposite appears to be happening more and more. In one recent analysis (Melnick and Zwanziger, 1988), the role of competition appeared quite dramatically. In markets characterized by low competition,[16] hospital costs (adjusted for general inflation) increased by only 1 percent in the 1983–1985 period, a historically very low rate. In markets characterized by high competition, inflation-adjusted hospital costs *decreased* by over 11 percent during this same period. This decrease represents a very unusual event in the history of U.S. hospital costs, and suggests (as Melnick and Zwanziger describe it) that "pro-competition policies are having dramatic and potentially far-reaching effects on the nature of hospital competition, leading to increased competition based on price."

ENTRY AND EXIT: THE PIVOTAL ROLE OF FOR-PROFIT HOSPITALS

One of the characteristics of a market economy widely admired by economists, if not others, is the ability of the market forces to generate investment in a hurry when new demand emerges. Unusually high economic returns create incentives to enter an industry. Not-for-profit hospitals, by contrast, may not respond as swiftly to changes in demand. For example, if a community grows very rapidly, or more particularly, if demand grows for a service in some community, one might expect that the first line of response could well be a for-profit hospital.

The same characteristic of rapid entry in the face of rising profits may cause for-profit firms to exit an industry faster than nonprofit firms when profitability falls (i.e., demand falls).

It is difficult to characterize market responses to changes in demand fully in the hospital industry, because (1) some states forbid or severely curtail the presence of for-profit hospitals, and

---

[16] The measure of competition was the Hirshman-Hirfindahl index, $HHI = \Sigma (1/s_i$ are the market shares of each hospital. If each seller is of the same size, the HHI equals the number of sellers, which is of course 1 in a monopoly. The U.S. Department of Justice uses the HHI as a test of whether monopoly power is present, particularly when deciding whether to contest mergers. They have used this measure even in questions about hospital mergers.

(2) governments sometimes subsidize or create hospitals where they would not otherwise exist. However, the growth of for-profit hospitals in regions of the country that grew rapidly over the past several decades lends credence to the idea that the for-profit response is faster than not-for-profits are capable of producing. The casual observation that for-profit hospitals have grown most rapidly in California, Florida, and Texas supports this idea, since these areas had rapidly growing populations, and even more rapidly growing demand for hospital care, since they are popular retirement states with many elderly people, who systematically use much more than the average amount of hospital care. In one early study of the role of for-profit hospitals, Steinwald and Neuhauser (1970) indeed found that for-profit growth was correlated positively with population growth more than the not-for-profits.

Demand for hospital services is now shrinking, particularly in California and other areas where "competitive" health strategies have been more prominent, but also due to the growth of substitutes, particularly ambulatory surgery. The model of a "fast response" for-profit sector suggests that the for-profit hospitals will be first to move out of the industry. However, no details have yet emerged on the pattern of shrinkage in these areas. (See Box 9.2.)

Box 9.2   THE DEMISE OF DEMAND FOR THE AMERICAN HOSPITAL

Aggregate hospital statistics reveal a striking decline in the use of hospitals in the United States, beginning in 1981. The total number of inpatient days peaked at 280 million in 1981, and has fallen precipitously since then to 230 million in 1986, and probably as low as 200 million in 1988. For any other industry, this fall in the quantity demanded would create financial ruin for many firms, surely accompanied by considerable exit from the industry. In the U.S. hospital industry, however, there have been fewer than 100 hospital closures during this same period out of a total of almost 6000 hospitals.

One consequence of the not-for-profit structure of many hospitals, coupled with the support of cost-paying insurance (and regulations, as we shall see in Chapter 16) has maintained most hospitals' existence, while the average cost per day has increased rapidly. (This is in part due to the substantial component of fixed costs in the structure of hospitals.)

The causes of this decline in demand are numerous. An important part of the change occurred because of the change in the way insurance plans, most notably Medicare, pay for hospital care. However, another important part of the decline in demand for inpatient days probably arises from a change in the technology of medical care. Most notably, a large fraction of surgical interventions now takes place without overnight hospitalization of the patient.

The decline in demand is shown vividly by comparing the occupancy rate of U.S. hospitals. By bed-size category, the data show

GROUP SIZE	1976	1986	PERCENT DECLINE
6–24	48%	32%	33%
25–49	56%	39%	31%
50–99	64%	51%	20%
100–199	71%	59%	17%
200–299	77%	65%	15%
300–399	79%	68%	14%
400–499	80%	70%	12%
500+	81%	75%	7%

The smallest hospitals have lost the largest fraction of their patients. One obvious explanation for this is the rise of outpatient surgery, which is used for the relatively simple surgical cases—that is, those most likely to be carried out in a smaller hospital.

The growth in outpatient surgery has continued rapidly just over the past few years. In 1983, about 24 percent of all hospital surgery was on an outpatient basis (i.e., the patient did not stay overnight). By 1986, that had increased to 40 percent and by 1990 to 47 percent. A decade earlier, outpatient surgery constituted a trivial fraction of all operative procedures. The increase in outpatient surgery has been even greater in rural hospitals than in urban hospitals. These data probably understate the total rate of evaporation of demand for inpatient hospital days, since some "outpatient surgery" is performed in doctors' offices now, just as it is now performed in the "outpatient surgery" suite of a hospital.

THE HOSPITAL IN LABOR MARKETS

We now turn ourselves to a different problem confronting the hospital—its market environment for buying its inputs, such as equipment, supplies, and labor. The demand for the services of a hospital in turn creates a *derived* demand for the inputs that a hospital uses, including those of capital (buildings, equipment) and various types of labor. These derived demand curves slope downward, just as do the demand curves for the final product. Sometimes, these demand curves are called *factor* demand curves, with reference to the phrase "factors of production." Demands for input factors will depend in part on the quality of the hospital's service and the mix of complexity of patients' problems. Since larger hospitals seem to specialize in more complex patients, it would seem natural that they demand both more total staff per patient and a more highly trained staff. Box 9.3 shows some relevant data for U.S. hospitals in the aggregate.

Box 9.3 AGGREGATE DATA

The American Hospital Association also publishes an annual report summarizing key facts about the hospital industry. These data originally appeared in *Hospitals* (see Box 9.1 on data about hospitals), but the American Hospital Association changed to separate publications, *AHA Hospital Statistics* and *AHA Guide* in 1971. Most hospitals have a complete collection of the latter in their library, as well as an extensive collection of *Hospitals,* including the annual *AHA Guide.*

The aggregate data are shown in various tables, providing a good "snapshot" of the U.S. hospital industry for the year in question. In addition, some of the tables show time-trends for key variables.

As an example of the types of data available and the inferences that can be drawn from these data, look at the following part of one of their tables, reproduced from the *Hospital Statistics,* Table 3, for 1987, showing personnel and expenses for short-term general hospitals in the United States by size category. This table shows the personnel of the hospital, including specific indicators of how many RNs (Registered Nurses, the most highly trained nursing category) and LPNs (Licensed Practical Nurses, a less highly trained category) per 100 "adjusted census" (The AHA adjusts to account for outpatient activity of the hospital, so "adjusted census" accounts for outpatient visits as well as actual inpatients.)

	FULL TIME EQUIV. PERSONNEL PER 100 ADJUSTED CENSUS			EXPENSES			APPARENT AVERAGE
	TOTAL	RN	LPN	LABOR/ DAY	TOTAL/ DAY	TOTAL/ STAY	LENGTH OF STAY
All	367	85	25	$230	$410	$2934	7.15
6–24 beds	395	77	30	$207	$353	$1582	4.48
25–49	357	66	35	$180	$329	$1808	5.49
50–99	323	67	32	$174	$322	$2119	6.58
100–199	333	76	29	$194	$350	$2464	7.04
200–299	350	84	26	$221	$397	$2795	7.04
300–399	370	90	23	$242	$431	$3134	7.27
400–499	381	91	22	$251	$434	$3297	7.60
500 or more	409	94	21	$269	$468	$3906	8.34

Several things stand out in these data. First, total costs clearly increase with size, almost surely due to increased complexity of the case mix. Only the very smallest hospitals show apparent diseconomies of scale, since average costs per day fall until the 50–99 size category is reached. The apparent average length of stay (in days) appears in the last column, calculated as the ratio of costs per stay to costs per day. Note that the ALOS increases systematically with hospital size, almost surely an indicator of increased complexity of cases. Another direct indicator of complexity is the hospital's choice of personnel. Total personnel per 100 census increases with hospital size (except for the smallest hospitals, which show their diseconomies of scale here as well as in total costs).

Also, the mix of RNs to LPNs increases steadily with hospital size, in response to the more complicated patient mix confronting the larger hospitals. For example, in the largest size category, 82 percent of the nurses have RN training, where only about two-thirds of the nurses do in small hospitals.

Hospitals compete for these inputs, just as do firms in any other industry. In some cases, they compete only against other hospitals—for instance, for very specialized forms of labor. In other cases, they compete against a very broad spectrum of the economy—for instance, for secretaries, janitors, food service workers, and so on. In the former case, the wages paid within the industry are determined in part by the industry itself. In the latter case, the wages of workers are almost certainly determined in broader markets, where the hospital has no pivotal role at all.

To see how and why this works, think about two generic types of labor, one specialized to the hospital sector, and the other used broadly across all industries.[17] As a shorthand notation, we might call these types of labor "nurses" and "janitors," understanding that the labels should not connote anything about skill level, but rather the extent to which the occupation crosses the bounds of many industries. Thus, "nurse" will represent nurses, various types of technicians, therapists, medical records librarians, and so on, and "janitor" will represent janitors, food service workers, accountants, computer programmers, lawyers, and so on. These prototypes form extreme cases, whereas much of the labor working in hospitals fits "in between" these cases. For example, nurses work not only in hospitals, but also in doctors' offices, in public health settings, and as private practice nurse practitioners. The fundamental difference between these two types of labor is whether the supply curve confronting the hospital industry in a particular community is likely to be upward sloping or flat. The distinction has the following importance: When the supply curve is upward sloping, then if the hospital attempts to expand its output or quality (by hiring more "nurse" labor), it will have to pay an increasingly higher wage, and if it contracts its demands for "nurses" then the equilibrium wage will fall. However, in the

---

[17] The same discussion would apply equally to other types of inputs for the hospital, such as equipment and supplies, which might be quite "special" for the hospital, and might be wholly "generic."

market for "janitors," changes in the hospital's demands have no effect on the market wage, because "janitors" can find ready employment elsewhere. Thus, if the hospital tries to expand its output or quality by hiring more "janitors," it can do so without driving up the wage rate. The reason is that the pool of people available for hiring includes not only those janitors currently unemployed, but also all those working in all other industries. By the same token, if the hospital reduces its demand for janitors, those laid off can find ready work elsewhere, so the wage rate will not fall.

Figure 9.6 portrays these two cases, with the wage rates for nurses shown in Figure 9.6a at two levels of demand, and the (unvarying) wage rate for janitors in Figure 9.6b.

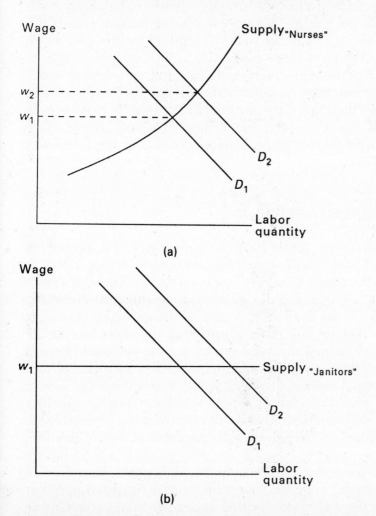

Figure 9.6 Wage rates for (a) specialized and (b) unspecialized hospital workers.

What might cause the demand for labor to shift outward like this? The most obvious answer is that anything causing a shift in the demand for hospital care will also shift the demand for labor used in the hospital. A key example would be provided by the introduction of Medicare in 1965, creating a large increase in the amount of medical care (including hospital care) demanded by the elderly. That shift in demand for hospital care also created a shift in the *derived* demand for labor of all sorts. The shift in the demand for nurses would create an upward pressure on the wages of nurses, because the supply of nurses (at least in the short run) is upward sloping, or "inelastic." This in turn will cause the costs of the hospital to rise. As a first-order approximation, the change in the costs of a hospital (or any other firm) behave according to the following rule: The percent change in costs equals the percent change in the factor price multiplied by the cost share (%ΔCost = %ΔWage $\times$ Cost share). Thus, for example, if nurses constitute half of the cost structure of a hospital, and nurses' wages rise by 10 percent, then hospital costs will rise by 5 percent. The increase is less than that if the hospital has the ability to substitute other forms of labor (or equipment, capital, or any other productive input) for nurses as the wage of nurses rises, but the first-order effect is as stated.

NURSING "SHORTAGES"

Hospital markets frequently have the appearance of a shortage of some types of labor, most commonly, registered nurses (RNs), as distinct from the generic "nurses" we used earlier. These phenomena have appeared repeatedly in the U.S. health care literature, most recently during the 1980s. A true "shortage" in an economic sense means that something is constricting the wage rate or the supply beyond normal market forces, because we would expect that whenever the quantity demanded exceeded the quantity supplied, the normal market response would cause the wage rate to rise, thereby eliciting a larger supply while causing the quantity demanded to fall at the same time. In equililbrium, of course, the quantity supplied and demanded are the same, at the equilibrium wage rate.

From the perspective of competitive markets, therefore, a "nursing shortage" suggests that something is constricting either the supply or the wage rate. Neither appears to exist in the United

States. People can decide to enter nursing freely, with a large number of schools offering training to become a nurse. The traditional training program for much of this century was a hospital-oriented school of nursing, offering a three-year training program leading to a degree in nursing. Many of these schools were operated in close affiliation with large hositals. More recently, two-year community colleges began to offer a two-year program leading to an Associates of Arts (A.A.) degree in nursing, after which the student could take state licensure examinations, and become a legally registered nurse (RN). Some colleges offer a four-year course of study leading to a Bachelor of Science in nursing, and graduate study in nursing is now available at some universities, providing M.S. and Ph.D. degrees. At the M.S. level, the training is often specialized (for example, in intensive care nursing) and the Ph.D. programs mostly prepare nurses for teaching or high-level administrative positions. Nevertheless, the widespread availability of these programs suggests strongly that there is no artificial constriction on the supply of nurses. We would anticipate a competitive labor market from the supply side, at least on the decision to enter nursing.

Monopsonistic Markets

One alternative explanation that has appeared cyclically along with nursing "shortages" is that hospitals have market power in the labor markets for nurses, and are responding in an appropriate way to that market power. Note that the presence of market power is not illegal; colluding to obtain that power is illegal. However, by the very nature of the hospital, whereby it has market power on the product side (i.e., faces a downward-sloping demand curve for its final product), we could well expect also that it has market power on the supply side, at least for types of labor that are specialized to the health care industry. It is precisely this type of labor that we called "nurses" generically earlier in this chapter.

To see what happens when a hospital confronts an upward-sloping supply curve for labor, look at Figure 9.7. This curve shows the supply of labor (adding up all of the hours of supply from individual participants) and the hospital's derived demand for labor, sloping downward in the usual fashion. The third curve in Figure 9.7 is the "marginal factor cost" curve, so called because it shows the hospital how much its factor payments (total wages to "nurses") rise as it tries to expand its use of "nurses." The *MFC*

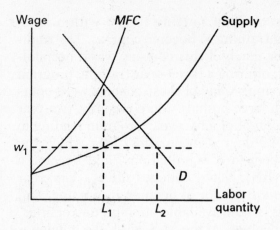

Figure 9.7 Hospital confronting an upward-sloping supply curve for labor.

curve rises faster than the supply curve because of the market power of the hospital in the labor market; it has *monopsony* power.[18]

A hospital in this setting will choose the "right" amount of labor by finding the point where the *MFC* curve crosses its demand curve, at the point labeled L_1 in Figure 9.7. The "going wage" for the market will then be set at the point where the market supply curve and the vertical line L_1 intersect, at the wage rate w_1. Of course, at that wage rate, the hospital would gladly hire L_2 amount of labor, but at that wage rate, a competitive labor market will supply only L_1. The gap between L_1 and L_2 represents the "shortages" of RNs confronting U.S. hospitals.

One recent study of the market for nurses in Utah found an apparently considerable degree of potential monopsony power in the market for nurses (Booton and Lane, 1985). Three firms controlled 26 of the state's hospitals, and one firm controlled over half of the hospital market in Salt Lake City. Their study estimated that a 10 percent increase in unfilled vacancies in nursing by hospitals would lead to over 4 percent *lower*, not higher, wages. This is precisely what would happen in a monopsonistic market if the gap between L_1 and L_2 represents "unfilled vacancies."

---

[18] Again, we should emphasize that nothing is illegal or immoral about this. It is just a fact of life, like gravity, or sticky fingers after eating a jelly doughnut.

Unions and Bilateral Monopoly?

A further extension to this model considers the potential role of increased unionization in nursing. The "classic" (but seldom fulfilled) model of a union portrays the union as achieving monopoly (not monopsony) power in the labor market, thereby driving up wages and driving down employment. Several difficulties emerge if one attempts to hang this model onto the nursing market. First, in order to drive wages up, the number of nurses actually hired would have to fall. (See Figure 9.8: A successful monopolization of the market for labor by a union would drive the wage from w_1 to w_2, and employment would fall from L_1 to L_2.) This would create the appearance of some nurses seeking work but being unable to find it (or rationing via the labor union's rules), since at the wage rate w_2, a quantity L_3 of hours would willingly be supplied to the market by nurses. This would also raise the economic returns to nursing, creating an increased demand for training in nursing. By contrast, however, enrollment in nursing programs has fallen steadily during the 1980s.

If a successful unionization did take place, creating monopoly power in the labor market, it would also likely confront the monopsony power described previously. This would create a classic "bilateral monopoly" with an indeterminant solution to the quantity and price in the labor market.

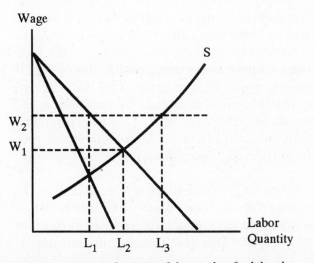

Figure 9.8 Monopolization of the market for labor by a union.

SUMMARY

Hospitals face demand curves for their services that shift inward or outward, depending on the price and quality of competing hospitals. Only if the hospital is a monopolist (one hospital) does the demand curve for a hospital match the demand curve for the market. Any hospital also can choose its quality, and the demand it confronts (think of it as willingness to pay for the hospital's services) will shift upward as the quality increases. In general, we can state that demand curves for a hospital will shift upward (outward) whenever

· A competing hospital's price increases.
· A competing hospital's quality falls.
· The extent of hospital insurance held by patients increases.

Hospitals also must decide about their scope of services. In part, the interaction between doctors and hospitals hinges on these types of decisions. When hospitals want to attract more patients, they need doctors to do this; to attract more doctors, they need to provide special facilities that assist the doctors' practices. Thus, the "scope of services" decisions importantly determine much of the overall success of a hospital.

We can think of hospitals as confronting a family of demand curves, with a different curve for each quality of care that the hospital might produce. Each level of quality also has associated with it a different cost curve, so there is a family of average cost curves just as there is a family of demand curves. The not-for-profit rules of the hospital require that the hospital set a price so that revenue just covers costs, which means that it must operate where demand curves intersect average cost curves. The set of all such intersections, called the *EE* curve in the text, provides the set of all feasible choices of quality, quantity, and price. The hospital can select which of these feasible choices is best by comparing the preferences of the fictitious "director" of the hospital (using a utility function) with the available choices. One choice will provide the highest utility, and this becomes the point actually chosen by the hospital.

Hospitals use resources to produce their services, leading to derived demand curves for inputs. The choice of inputs is determined by the technology of production available (the production function) and the costs of inputs. One particularly interesting part of the hospital's decision is that it may have *monopsony* power in some labor markets, such that they affect the price as they alter their demands. In these cases, hospitals may set their wage offers to some staff (e.g., nurses) in a way that creates the appearance of a permanent shortage.

PROBLEMS

1. Using graphs like Figures 9.2 and 9.3, carefully explain how a hospital decides what level of quality to produce.

2. Using graphs like Figures 9.2 and 9.3, show what happens to the quantity and quality of a not-for-profit hospital that has been a monopolist, and suddenly has a new competitor in town.

3. What necessarily happens to average quality in a hospital market when the generosity of insurance coverage increases? How does this relate, if at all, to the data appearing in Chapter 2 on increases in spending through time in the hospital sector?

4. If you looked in the Want Ads section of your local newspaper, would you expect to find more advertisements run by hospitals in your town seeking nurses or janitors? Similarly, among ads for nurses, would you expect the ads to focus more on "general" nurses or specialized nurses like those with training to work in intensive care units? Why?

APPENDIX

The Hospital's Quality and Quantity Decision

One way to determine the best choice of quality and quantity is to use a technique of the calculus developed by Lagrange. This technique finds the best choice to maximize utility subject to the constraint that revenue does not exceed cost (the limitation of the not-for-profit form of organization). The Lagrangian formulation of this problem says to maximize the function L, which is the sum of the objective function and λ times the constraint (rewritten to equal zero). Then the technique takes the partial derivatives of L with respect to each variable, including the new variable λ, and sets each equal to zero. The resulting equations define the best mix of output and quality:

$$L = U(N, S) + \lambda(P(N, S) \times N - C(N, S)) \tag{1a}$$

Define the derivatives by subscripts, so (for example) $\partial U/\partial N = U_N$, and similarly for U_S, and C_N and C_S. The conditions to maximize the hospital's utility require that:

$$U_N + \lambda[P(N, S) \times (1 + 1/\eta) - C_N] = 0 \qquad (2a)$$

$$U_S + \lambda[P_S N - C_S] = 0 \qquad (2b)$$

$$P = C/N = AC \qquad (2c)$$

where η is the own-price elasticity of demand confronting the hospital (holding service quality constant). Reorganizing equations (2a) and (2b) to solve each for λ, gives

$$U_N/[(C_N - P(1 + 1/\eta)] = \lambda \qquad (3a)$$

$$U_S/(C_S - P_S N) = \lambda \qquad (3b)$$

In this familiar form, the decision rules emerging from the maximization say to set the ratio of marginal utility to marginal cost equal for both of the choices N and S. In equation (3a), the net cost of N is marginal cost minus marginal revenue. In equlilbrium, MC exceeds MR, which in turn requires expanding production past the monopoly level (where $MR = MC$). The hospital "spends" potential profit on the expansion of sales. In one corner solution where $U_N = 0$, the firm simply picks the monopoly price, using its market power as a "cash cow" to augment quality.[19] So long as C is nondecreasing in N and normal demand conditions prevail (so marginal revenue falls as N rises), the net cost of N increases with larger N.

A corresponding condition holds for service intensity ("quality"); the marginal cost to the hospital is $C_S - P_S N$, the logical equivalent of the difference between marginal cost and marginal revenue, and again, the hospital "spends" potential profits on more quality. So long as cost is nondecreasing in S and demand is increasing in S, the "net cost" to the hospital of S also increases with S.

Since the "costs" of both N and S increase as their own scale increases (at least in the relevant range of behavior), the hospital faces decreasing returns to scale in production of utility in both N and S. Thus, the opportunity-possibility curve is concave to the origin for the relevant choice set, as shown in the figures in the body of the chapter.

---

[19] This might correspond closely to the Pauly-Redisch (1973) model, where "service" was selected to create more profits for doctors. Conversely, a Baumol-like "sales maximization" model would have $U_S = 0$, and would milk the quality dimension in order to expand sales.

Policy Applications of the Model

The model we have here has some interesting policy applications as well. We can solve equation (2a) for the equilibrium price, and then make use of that result in further discussion:

$$P = \left[C_N - \frac{U_N}{\lambda} \right] \frac{\eta}{(1 + \eta)} \tag{3c}$$

This is similar to the standard "markup pricing" of a monopolist, except for the term U_N/λ, which represents (scaled to dollars by λ) the utility benefits to the hospital director from additional output.

This characterization of the optimal price helps explain two particularly interesting questions in the economics of hospital behavior. The first returns to the analysis by Feldman and Dowd described in the text. They used an equation like (3c) to estimate the demand elasticity facing a hospital. To do that, they solved their version of equation (3c) for η, which is

$$\eta = \frac{-P}{(P - MC + U_N/\lambda)}$$

The only difference between their work and this approach is that they assumed the hospital sought to maximize profits alone, so that $U_N = 0$. Thus, the calculated $\eta = -P/(P - MC)$, as one would for a profit-maximizing monopolist. The estimates they derived are the larger-valued elasticities reported in the text. The direct estimates of demand curves gave values of η near -1, which Feldman and Dowd appeared not to believe. In part, this disbelief arises from the well-known monopolist's rule that one never willingly operates in the region where demand is inelastic (i.e., $|\eta| < 1$) because marginal revenue is negative: The monopolist can always make more money by raising price and reducing output if demand is inelastic. However, as equation (3c) shows clearly, if the utility function of the hospital sufficiently emphasizes quantity of output, then the hospital will willingly operate in the realm of inelastic demand, and indeed might even willingly charge a negative price (bribe people to use the service) under some situations.

The second issue arising from this analysis concerns the oft-studied question of "cost shifting" in hospitals. Hospitals all have prices that they use when sending bills to every patient. However, many customers—usually big insurance organizations or government insurance plans—do not pay "charges," but rather something less, often an approximation to the actual average costs of the hospital. Groups that continue to pay billed charges often complain about the arrangement, and are particularly worried that when (say) Medicare or Medicaid reduces its payment to hospitals, the hospital merely shifts its costs to the charge-paying customers by charging higher prices. (The hospital keeps sending the bills with higher charges to the government plans as well, but the government plans ignore them and send back their own calculations of "costs.")

The model of behavior developed here extends directly to the question of cost shifting. Suppose there are J groups of patients, each with demand elasticity $\eta_1, \ldots \eta_j, \ldots \eta_J$, that the hospital doesn't really care more about providing service to one group than another, and that each costs the same to serve. Thus, overall costs $C\left(\sum_{j=1}^{J} N_j\right)$ are just the cost of providing the total amount of care, and total revenue is $\sum_{j=1}^{J} P_j N_j$. The optimal strategy for the hospital sets a different price for each group, using the rule

$$P_j = \left[\frac{dC}{dN} - \frac{U_N}{\lambda}\right] \frac{\eta_j}{(1 + \eta_j)}$$

Thus, although it costs the same to produce care for each group, the groups can end up paying different prices. The smaller the demand elasticity, the higher the price.

Dranove (1988) has used this type of model to show what should happen in a hospital that faces an arbitrary reduction in payments by a government program. We can think about a government program with a fixed price per day (or stay) as one where the demand elasticity is very high, indeed, infinitely high. Dranove shows what happens to the price paid by "charge-paying" patients with smaller demand elasticities. The worst fears of the charge-paying customers are "confirmed" in his model: Lower Medicare or Medicaid payments lead to higher prices for everybody else. Further, he actually estimated the response of hospitals in Illinois to a substantial reduction in that state's Medicaid payments, and found that for each $1 reduction in Medicaid income, $.5 was recouped from charge-paying customers.

Chapter
10

The Demand for
Health Insurance

*T*he world around us generates risks of innumerable sorts. Fire can damage or destroy homes. Thieves can steal automobiles, or careless drivers can crash them. Vandals can smash plate glass windows. Any person or business can purchase insurance against the financial consequences of these risks. Similarly, illness can make it impossible for people to go to work, and they will lose income. Disability insurance offsets the financial loss. Even the financial consequences of death can be insured against, so that family (or others) can replace at least the financial loss occurring when a person dies.

Many persons purchase insurance against these and other risks. Almost every homeowner carries fire insurance, and most automobile owners carry collision insurance. Life insurance is commonplace. Yet perhaps the most pervasive of all forms of insurance is that insuring against the costs of medical care. Nearly 90 percent of the people in the United States under age 65 have some form of health insurance, and everybody over age 65 is protected through Medicare. In this chapter, we explore the demand for health insurance, both from a conceptual point of view and with some concrete evidence on the importance of various economic forces in affecting people's demand for health insurance.

We must think about the demand for health insurance together with the demand for medical care—the two cannot be separated meaningfully. Of course, this raises the obvious question of whether it is easier to think first about the demand for health insurance or the demand for medical care. We have solved that problem by thinking (in Chapters 4 and 5) about the effects of insurance on medical care, without any particular concern about why a person might have a particular insurance plan. We can now return to the question of how a consumer selects that insurance, taking into account how the insurance (whatever it might be) will affect the demand for medical care.

THE SOURCE OF UNCERTAINTY

The fundamental uncertainty driving the demand for health insurance arises not because of any financial events, but rather because of the random nature of health and illness. The response of a consumer who becomes sick—to seek cure for the illness with appropriate medical care—creates a financial risk. Health insurance insures against this *derived* risk.

In a hypothetical world where no medical care existed, people still might buy insurance against the risks of poor health, but it would differ considerably from the type of health insurance we see now. Without medical care, each person's stock of health would be a unique and irreplaceable object, similar to an original painting by Picasso or the original copy of the United States Constitution. The loss could be tragic, but no amount of money can replace these unique items. Money, however, might help in a different way. With sufficient money, a person might purchase a near-substitute that would create almost as much utility as the original object. Of course, that money might be used to buy another Picasso, or it might be used to buy a new car. Since substitutes exist for almost everything, money will surely help offset the loss of even unique objects for which no markets exist. Life insurance will not replace the loss of a person, but the money does substitute (for example) for the earning power of a working person who dies.[1]

With health insurance, we typically seek something else. We know from Chapters 4 and 5 that people use more medical care as their illness severity increases. While nothing *compels* a sick person to seek medical care, it is a rational act, so long as the care has some positive benefit in improving health and doesn't cost too much. The expenses associated with that medical care create the financial risk against which health insurance protects us. Some

---

[1] Cook and Graham, 1977, discuss the insurance of irreplaceable objects.

people complain sardonically that "health insurance" is nothing of the sort, because it doesn't insure our health. Of course, this is correct, but meaningless. Our society simply does not possess the technology to insure health. We must accept the second-best alternative of insuring against the financial risks associated with buying medical care.

WHY PEOPLE DISLIKE RISKY EVENTS

People seem to dislike risk. The pervasive purchase of insurance of many types offers concrete evidence of this dislike. People willingly (and often) pay insurance companies far more than the *average* loss they confront, in order to eliminate the chance of really risky (large) losses. We can describe people who behave this way as *risk averse*.[2]

Risk aversion arises quite naturally from a simple assumption about people's utility functions. Recall earlier that we had described a utility function $U(X, H)$ for individuals, and we had said that more X or more H created more utility. In other words, the *marginal utility* of either X or H is positive.[3] Since income (I) can be used to purchase X or medical care, which can increase H, we can also say that the marginal utility of income is positive.[4] The basic idea here is that a person has a fixed and stable set of preferences, so that once we know how much X and H the person has, we know his or her level of utility. A process of rational decision making also tells us that once we know a person's income and the prices for X and m, we can also determine the person's utility.[5]

Risk aversion arises from a simple additional assumption—namely, that the marginal utility of income, while positive, gets smaller and smaller as a person's income gets larger. In other

---

[2] Some people simultaneously buy insurance and gamble. This sort of behavior creates a real puzzle. At this point, we will simply ignore such contradictory behavior.

[3] In the language of the calculus, $U_X = \partial U/\partial X > 0$ and $U_H = \partial U/\partial H > 0$.

[4] This link between income and utility allows us to redefine $U = U(X, H)$ into a comparable "indirect" utility function $V = V(I, p_X, p_m)$. Utility increases with income (holding prices constant), so $\partial V/\partial I > 0$. Utility also falls as prices rise, holding nominal income constant, so $\partial V/\partial p_m < 0$, for example.

[5] For a detailed description of this process, including the direct measurement of consumer welfare from quantities, prices, and income, see McKenzie (1983) or McKenzie and Pearce (1976).

words, if we were to plot a person's utility against his or her total income (which, recall, corresponds to increasing ability to buy X and m), the diagram would look like Figure 10.1. The person's utility would always increase as income increased, (since "more" is "better"), but the graph of utility versus income would always flatten out more and more. In Figure 10.1, this is shown by the two tangent lines at incomes I_1 and I_2. The slope of the tangent line shows the marginal utility at that level of income. At I_2, the slope is flatter, equivalent to saying that the marginal utility of income is smaller.

A person with a utility function shaped like this, it turns out, is *risk averse*, and will always prefer a less risky situation to a more risky situation, other things being equal. This is called *diminishing marginal utility*, and this idea is central to the question of why people buy insurance.

THE RISK-AVERSE DECISION MAKER

We can explore briefly the nature of risk aversion by considering a very simple gamble. Suppose the person with the utility function shown in Figure 10.1 starts out with an income I_2, but knows that some externally generated risk (over which the person has no control) may reduce this year's income to I_1. If this risky event occurs with probability f, then the statistical *expected* income of this person is $E(I) = fI_1 + (1 - f)I_2 = I^*$. Now look at Figure 10.2, where, to make things simple, we can select a particular value

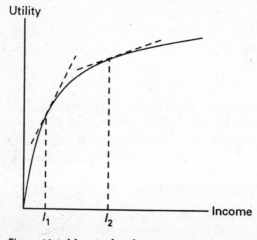

Figure 10.1 Marginal utility at two income levels.

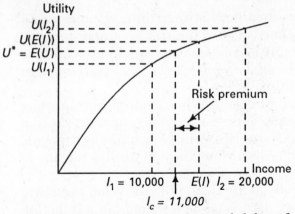

Figure 10.2 Expected utility when probability of I_1 = 0.4 and probability of I_2 = 0.6.

for f, say, f = 0.4. If I_2 = \$20,000 and I_1 = \$10,000, then $E(I)$ = (0.4 × 10,000) + (0.6 × 20,000) = \$16,000.

What is the expected *utility* of a person confronting this risky gamble on income? Using the utility function in Figure 10.2 tells us the correct answer. If the income level I_2 occurs, then the utility associated with that, $U(I_2)$, measures the person's level of happiness. If I_1 occurs, then similarly $U(I_1)$ is the right measure. The *expected utility* for the person with this risky income is $fU(I_1) + (1 - f)U(I_2)$. In the specific case we have used, $E(U)$ = [0.4 × U(10,000)] +[0.6 × U(20,000)] = $E(U)$.

Because of diminishing marginal utility, U(20,000) is not simply two times U(10,000), but rather is smaller. The *expected utility* $E(U)$ of this risky income lies 60 percent of the way between U(10,000) and U(20,000) on the vertical axis of Figure 10.2 (the 60 percent figure coming from the probability f = 0.6 that I_2 will occur).

Since this particular gamble has an average income (expected income) of \$16,000, we could also find the utility associated with \$16,000, shown as U(16,000) in the figure. *Note that U(16,000) exceeds the expected utility [$E(U)$] of the risky gamble.* The expected utility of the gamble is lower than the utility of the average income.

We could find the *certain* income I_c that would create utility of $E(U)$ by reading the utility-income graph the other way—that is, moving across the graph at $U = E(U)$ until coming to the utility-income curve, and then dropping down to the income axis to find the corresponding income. This "certainty equivalent" income I_c

is less than \$16,000. The difference between the certainty equivalent and the average income is called the *risk premium*. It represents the maximum that a risk-averse person would be willing to pay to avoid this risk, *if he or she made decisions in such a way as to maximize expected utility.*

This model of consumer behavior when confronting uncertain (risky) financial events stands at the heart of the economist's way of thinking about such decisions. Economists presume that people act to maximize expected utility. When they do so, they buy insurance against risky events.[6]

To restate things slightly, if the Gone With the Wind Insurance Company came up to this risk-averse person and offered the following deal, this risk-averse person would say yes:

> Each year, you give me your paycheck, whether it is \$10,000 or \$20,000. In turn, every year, we will give you a certain income that is larger than I_c.

How far can I_c get below $E(I)$ (here, \$16,000)? That depends on how fast the person's marginal utility of income diminishes as income rises. Intuitively, the more tightly bent the graph of utility versus income (like Figures 10.1 or 10.2), the more the person dislikes risk. The straighter this graph is, the more "neutral" the person is about risk. A perfectly risk-neutral person has a utility function plotted versus income that is just a straight line.

The more the person dislikes risk, the bigger the gap between I_c and $E(I)$. It can be shown that the risk premium $[(E(I) - I_c)]$ a person will willingly pay to avoid a risky gamble is directly proportional to the variability of the gamble (the variance, in statistical terms) and to a specific measure of how rapidly marginal utility declines as income increases.[7]

---

[6] *Caveat emptor!* Some important work, mostly arising out of the field of psychology, but some out of economics, shows that people don't behave according to this model in all settings. However, no clearly competing alternative has yet emerged that can fully explain all of the anomalies either, and the expected utility model does predict much of the key behavior we see in the purchase of health insurance.

[7] For the calculus freak: Define the second derivative of the utility function as $d^2U(I)/dI^2 = U''$ and the first derivative as $dU(I)/dI = U'$. Now define the ratio $r(I) = -U''/U'$. John Pratt (1964) has proven that the risk premium a person will pay when confronting a risky gamble is approximately $0.5 \times r(I) \times \sigma^2$, where σ^2 is the variance of the risky income distribution.

A related measure is the *relative risk-aversion measure* $r^*(I) = Ir(I)$. This is simply the income elasticity of the marginal utility of income, i.e., the percentage change in marginal utility associated with a 1 percent change in income.

The welfare gain that people get from an insurance policy is simply any difference between the risk premium they would be willing to pay and the amount the insurance company charges them for risk-bearing. Here, we need to be very careful about definitions of terms. The risk-bearing amount charged by the insurance company is any amount above and beyond the amount of benefits that the insurance company can expect to pay. We will define these terms more precisely below.

Of course, events involving illness are much more complicated than this simple risk, but the idea of risk aversion remains the same. If a person's utility function is defined in terms of X and H and is "stable" in the sense that the *function* itself doesn't change as health changes, then the ideas of this simple risk-aversion example carry through to the complicated world of health insurance. We can now proceed to consider the particular problem of selecting a health insurance policy.

CHOOSING THE INSURANCE POLICY

We had defined in Chapter 4 the most simple of all health insurance policies, one with a consumer coinsurance rate C that paid $(1 - C)$ percent of all the consumer's medical bills, leaving the consumer to pay C percent. While many insurance policies are more complicated than this, this simple insurance plan captures the essence of many real-world plans, and also allows us to explore the issues associated with selecting the insurance. In this simple world, the consumer's decision problem is to find the value of C that maximizes expected utility.

To think about this problem clearly, we need to begin with some common statistical definitions. Box 3.1 provides the most basic definitions of mean and variance that we will need; if this material seems unfamiliar after review, any beginning text in statistics will probably provide useful additional help. If an insurance company intends to stay in business, it must charge the consumer an insurance *premium* that at least covers the *expected benefits* it will pay out plus any administrative expenses. Suppose the insurer knows precisely the *distribution* of medical expenses that a person confronted for the coming year, although neither the insurer nor the consumer knows actual expenses that will arise.

The insurance contract says that the insurer will pay $(1 - C)p_m m$ if the consumer buys m units of medical care at a price p_m each. (Again, we assume only one type of medical care for simplicity in discussion. The idea readily generalizes to many types of care.) Suppose that there are N different amounts of medical care that the consumer *might* buy during the year (say, each corresponding to one of N different illnesses the consumer might acquire), and that each one of these would occur with probability $f_i (i = 1, \ldots N)$. Then the *expected benefit payment* from the insurance company to the consumer is

$$E(B) = \sum_{i=1}^{N} f_i (1 - C) p_m m_i$$

or more simply, $(1 - C)p_m m^*$, where m^* is the expected (average) quantity of care.

Herein lies the most complicated part of the problem: As we learned in Chapters 4 and 5, the amount of m that people select depends upon the coinsurance of their health insurance plan. Thus, the insurance company cannot blithely assume that m^* is the same, no matter which insurance plan (which coinsurance) the consumer chooses. The insurance plan is chosen in advance (at the beginning of the year, say), but that choice affects all subsequent choices of medical care. When the consumer actually gets sick or injured, the medical care demanded will depend upon the coinsurance C previously chosen.

The dependence of m on C has sometimes been described as "moral hazard," but (as Pauly, 1968, has pointed out) this behavior has nothing to do with morals, and it isn't even hazardous, in the sense that it is predictable. (The RAND HIS results provide the type of information one needs to make the right calculations.) "Moral hazard" is really just a predictable response of a rational consumer to the reduction of a price. In this case, the insurance plan causes the out-of-pocket price to fall at the time medical care is purchased. This price response by consumers is somewhat of an unwanted side effect of insuring against the risks of health loss by paying for part or all of the medical care people buy when they become sick.

The effects of the insurance coverage on demand for care feed back on the demand for insurance itself. Recall the discussion of the demand curve for medical care, or more particularly, the "value curve"—called the "inverse demand curve." The demand curve (and the inverse demand curve) slopes downward, so the marginal value of a particular amount of m consumed falls as the total amount of m rises.

Since health insurance reduces the price of medical care, it induces people to buy some care that creates less marginal value (as measured by the inverse demand curve) than it actually costs to provide the care.[8] The induced demand due to the health insurance coverage creates a *welfare loss* in the market for medical care. The insurance policy breaks the link between the costs of care and the price charged for it, since the health insurance is paid for no matter which illness the person actually gets, and no matter what amount of medical care the person buys.

This welfare loss from buying more medical care offsets the welfare gains that consumers get by reducing the financial risks. The choice of the best coinsurance rate C balances these two ideas—reduction of financial risk versus the effects of increasing demand for care (Zeckhauser, 1970).

A Specific Example

We can explore these ideas better by thinking of a *very* simple world in which only two illnesses might occur (with probability f_1 and f_2). Since probabilities must add up to 1, "not getting sick" has probability $(1 - f_1 - f_2)$ in this simple world. Think now about a specific insurance policy that the consumer might select, say with $C = 0.2$. Then (as in Chapter 4), the demand curves for medical care will depend on the particular illness that actually occurs. If illness level 1 occurs, the demand curve is D_1, and similarly D_2 for illness level 2. For illness 1, the insurance plan induces the consumer to buy m_2 of care, where an uninsured consumer would buy m_1. The welfare loss generated by this purchase is the triangle labeled A in Figure 10.3. Similarly, if illness 2 occurs, demand is m_4, versus m_3 for a consumer without insurance, and the welfare loss is shown as triangle B.

The medical spending associated with the risk of illness is a distribution that (here) has outcomes m_2 (with probability f_1) and m_4 (with probability f_2). Thus, the expected insurance benefit is $p_m(f_1 m_2 + f_2 m_4)(1 - C)$. To place things in very concrete terms, if $C = 0.2$, $p_m = \$500$ per hospital day, $m_2 = 4$ days in the hospital, $m_4 = 9$ days in the hospital, $f_1 = 0.3$ and $f_2 = 0.1$, then the expected benefit to be paid by the insurance company is $500[(0.3 \times 4) + (0.1 \times 9)] \times (0.8) = \$840$.

---

[8] This is the same type of welfare loss that was discussed in the section on "variations" in medical care in Chapter 5.

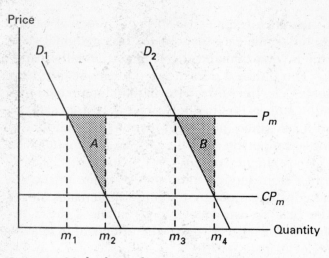

Figure 10.3 Medical spending associated with two levels of illness.

The total insurance premium will be the expected benefit (the $840) plus any "loading fee" for risk bearing. Insurance companies commonly compute a loading fee as a percent of the expected benefit (as a subsequent section discusses). Suppose that loading fee is 10 percent. Then the actual premium charged will be $924, of which $840 is expected benefit, and $84 is the fee for risk bearing and administrative costs of the insurer.

The net welfare gain to the consumer depends on two things—the amount of the risk premium (willingness to pay for risk reduction), and the size of the triangles A and B. An appendix works out a specific computation for this problem, but for now, we can just assume the results. Suppose the triangles A and B have areas equivalent to $200 each. Then the *expected* welfare loss in purchasing medical care is $(0.3 \times \$200) + (0.1 \times \$200) = \$80$. (Recall that illness 1 and hence welfare loss A happens with probability $f_1 = 0.3$ and illness 2 with welfare loss B happens with probability $f_2 = 0.1$ in our simple world.)

We would have to know the consumer's utility function to derive the maximum risk premium for this risk. (See the discussion regarding Figure 10.2 about risk premiums to remind yourself of what this term means.) Suppose this risk premium were $220.[9]

[9] In the simple problem presented here, the variance of the uninsured risk equals $1,372,500. For a risk aversion parameter of −0.0003, a "moderately high" level of risk aversion, the risk premium is approximately $220. See footnote 7 for the formula.

This means, of course, that the consumer would be willing to pay $220 (risk premium) *plus* $840 (expected benefits) = $1060 for the insurance policy, minus the $80 in "moral hazard" welfare losses, for a net of $980. If the insurance company actually charges $924 then the consumer gains $980 − $924 = $56 in welfare from the risk reduction.

The consumer who wishes to maximize expected utility will (in concept) think about all the possible coinsurance rates that the insurance company might sell, and will pick the one value for C that gives the greatest net gain. Each value of C selected requires the same balancing: Lower C reduces risk more, but creates bigger welfare losses like triangles A and B. Bigger values of C create less risk reduction, but don't generate as much welfare loss in purchasing medical care. We will revisit this example shortly to discuss the effect of income taxes on this decision and the net well-being from having the insurance policy.

How the Medical Care Demand Elasticity Affects the Demand for Insurance

A key idea in the health insurance problem is that the expected "welfare loss" ($80 in the example) depends directly on the elasticity of demand for medical care. If the demand curve were very inelastic (say, $\eta = -.05$), then there would be very little change in demand due to the insurance coverage, and the expected welfare loss would be much smaller, say, $20. Conversely, if the demand elasticity were much larger, the welfare loss would also be much larger.[10]

The welfare loss arising from the purchase of insurance tells us something particular about the link between medical care demand and health insurance demand: *The more price responsive (price elastic) the demand for medical care is, the less desirable it is to insure against that risk with "normal" types of health insurance.*

The reason for this is that large price responsiveness (elastic demand) creates bigger welfare losses in the demand for care, for

---

[10] Specifically, the welfare loss for any illness is measured by the size of the triangle $(1/2)\Delta p \Delta m$, where Δp is the change in effective price due to insurance and Δm is the change in demand due to that change in price. For a simple insurance policy paying $(1 - C)p$ for each unit of m purchased, $\Delta p = (1 - C)p$. We can find Δm from the demand elasticity; $(\Delta m/m) = \eta(\Delta p/p)$. Thus the percent change in m is $(1 - C)\eta$. A little algebra makes it easy to prove (see Problems at the end of the chapter) that the welfare loss for any illness i is approximately $-\eta(1 - C)^2 pm_i/2$.

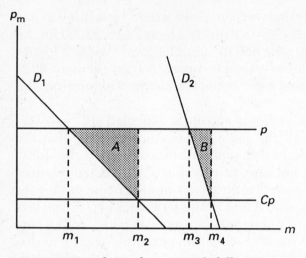

Figure 10.4 Two demand curves with different price responsiveness.

any given level of insurance coverage. Figure 10.4 shows this for two demand curves with different price responsiveness. Demand curve D_2 is not very price responsive (very steep demand curve), so if an insurance policy is acquired with copayment rate C, the increase in demand on demand curve D_2 is only to quantity m_4 from m_3 without insurance. The welfare loss is the small triangle B. However, if the demand curve is much more elastic (price responsive), like D_1, then the consumption with the insurance policy rises to m_2 from m_1, and the welfare loss is the triangle A, obviously much larger than B. Indeed, if demand for medical care were perfectly price insensitive, then more complete insurance would be a better idea.

PATTERNS OF INSURANCE COVERAGE

This relatively complicated model of consumer behavior (the model that consumers act to maximize expected utility when confronting risky events) contains several clear predictions about empirical regularities that we should observe in the demand for insurance. It says that (1) demand for insurance should be higher, the more financial risk (variance) confronting the consumer; and (2) demand for insurance should be less, the more price-elastic the demand for the type of medical care being insured. How well do these ideas stand up in practice?

Table 10.1   PATTERNS OF INSURANCE COVERAGE

Type of health care	Variance of risk	Demand elasticity (RAND HIS)	Percent of people under 65 insured
Hospital care	highest	−0.15	80%
Surgical and in-hospital medical	high	−0.15	78%
Outpatient doctor	medium	−0.3	40%–50%
Dental	low	−0.4	40%

One simple test asks what portion of the population carries health insurance against specific risks such as hospital care, surgical procedures, dental care, psychiatric care, and so on. It turns out that the patterns of coverage conform well to the expected utility model, as it applies to health insurance.

As shown in Table 10.1, the type of coverage most commonly held is for hospital care, which is the type of care creating the greatest financial risk (highest variance) and which has the smallest elasticity of demand. In-hospital surgical and doctor expenses rank next on both accounts, and has the next-highest extent of coverage. Finally, at the bottom of the list, dental care creates the smallest financial risk, and has the largest elasticity of demand, and by far the lowest frequency of coverage. Indeed, only within the last decade has dental insurance enjoyed any popularity, and that (as we will see below) probably only because of favorable tax treatment for health insurance.

The Price of Insurance

We normally expect that demand for a good or service falls as its price increases. With an insurance policy, we must be careful to define the price appropriately. The price is *not* simply the premium paid, because that premium includes the average expense of something that the consumer would have to pay anyway. The price of insurance is just any markup above those expected benefits that the insurance company adds. To return to the previous discussion, suppose the expected benefits are $E(B) = (1 - C)p_m m^*$. Now define the insurance premium (the amount actually paid by the consumer each year) as

$$R = (1 + L)(1 - C)p_m m^*$$

The price of insurance is L, the "loading fee" of the insurance company above expected benefits. If $L = 0$, the insurance is "free" in the sense that there is no charge for risk bearing or administration of the insurance plan.[11] Demand for health insurance behaves in response to this "price" just as other goods and services do to their own price—the higher the price, the less insurance will be demanded. In the case of our simple insurance policy with a coinsurance rate C, this simply means that at higher loading fees (bigger L), the consumer will select a bigger coinsurance rate (C), and the portion paid by the insurance company $(1 - C)$ will be smaller.

If the health insurance policy has a deductible in it (see Chapter 4), the same idea holds: Bigger loading fees cause the consumer to select bigger deductibles.

HOW THE PRICE OF INSURANCE DIFFERS FOR DIFFERENT PEOPLE

Most health insurance in the United States (and in many other nations as well) is sold not to individuals or families, but to large groups. Group insurance offers two advantages over individual insurance: scale economies and avoidance of adverse selection.

Economies of Scale

Many of the costs of developing and selling an insurance policy are nearly the same, whether there are ten or ten thousand people enrolled in the plan. Group insurance sales take advantage of this economy of scale. Group insurance purchases also eliminate some of the activities often involved in the sale of nongroup insurance, such as gathering and analyzing information about the health history of persons buying the insurance. Insurance companies can do this because of the role of groups in eliminating adverse selection.

Avoiding Adverse Selection

The second major gain with group insurance is that the insurance company can avoid the problem of adverse selection. With individual insurance policies, the consumer (for once!) has an informa-

---

[11] Insurance that is priced so that $R = E(B)$ is called "actuarially fair" insurance. (Actuaries are the people insurance companies hire to predict the benefits they will pay with a specific insurance plan.) Of course, no insurance company can afford to sell "actuarially fair insurance," because insurance companies use real resources to conduct their operations.

tional advantage over the seller. Consumers know better than the seller about their particular health conditions (or propensities to become sick). Thus, the insurance company can set the premium for the insurance policy only based on averages for people "like" the prospective customer. The customer may know, however, that cancer or high blood pressure runs in the family. Indeed, the insurance customer may even be planning to have an operation (like a hernia repair), and may seek an insurance policy just to pay for that special event. This problem is known as adverse selection, because the insurance company's view of the world is indeed biased unfavorably. Insurance companies sometimes attempt to deal with this problem by requiring people to give a health history before enrolling them in a health insurance plan. Sometimes they will exclude coverage for previously existing health conditions (such as cancer). Sometimes they will even give a physical examination to the person. Even with such detailed information gathering (which adds to the selling expense), the insurance company cannot in concept know as much about the health of the person as the person does. Thus, they always face the problem of adverse selection. The insurer may be able to control it somewhat through careful insurance practices, but it will always persist to some extent.

Group insurance solves this problem by selling insurance to a group of people assembled for some purpose other than buying insurance. For example, an employment group (all of the people working for a particular company) probably have their share of good and bad health risks, but the average should be pretty average. Thus, the insurance company can sell insurance without having to go through the trouble of examining people, taking histories, and so on that are common with individual insurance policies.

The economies of scale associated with group insurance are considerable, and the loading fees that result reflect these gains. At the most basic level, we can see the importance of group insurance in reducing the price of insurance by looking at the aggregate data for the past several years. Table 10.2 shows the premiums paid by consumers, the benefits paid by the insurance company, and the ratio of these two, which equals $(1 + L)$. Although the numbers vary from year to year (depending on how well the actuaries forecast the benefit payments), the group policies always have a much lower loading fee (averaging about 15 to 20 percent in most years), compared with nongroup policies that have an average loading fee of 65 to 100 percent or more. As a consequence, one of the key

Table 10.2 PREMIUMS, BENEFITS, AND LOADING FEES FOR
GROUP AND NONGROUP INSURANCE OF COMMERCIAL
INSURANCE COMPANIES FOR HOSPITALS AND
MEDICAL/SURGICAL INSURANCE

	Group insurance			Nongroup insurance		
	Premium[a]	Benefit[a]	Ratio	Premium[a]	Benefit[a]	Ratio
1970	6.181	6.043	1.02	2.293	1.090	2.10
1975	13.090	11.607	1.13	3.070	1.557	1.97
1980	31.524	25.89	1.22	4.909	2.970	1.65
1985	59.530	44.340	1.34	7.886	4.696	1.68
1990[b]	81	68	1.2	12.2	7.2	1.7

[a] In billions of dollars.
[b] Estimated.

factors affecting demand for health insurance in this country is
access to group insurance. When a person belongs to some group
(such as working for a firm that provides insurance as a benefit), the
chances are much greater that the person will have insurance
coverage.

Larger groups have bigger economies of scale than smaller
groups, so the price of insurance should be even lower in big
groups than in small groups. Health insurance sources describe
the typical loading fee by group size as shown in Table 10.3.

Several studies of demand for insurance have shown that the
insurance coverage selected in larger groups is systematically
greater than in smaller groups (Phelps, 1973, 1976; Goldstein and
Pauly, 1976). In fact, the size of a work group appears to be one of
the best predictors of the amount of insurance any family in the
United States will have.

GROUP DEMAND VERSUS INDIVIDUAL DEMAND
FOR INSURANCE

Since group insurance plays such an important role in health
insurance, it seems worth exploring the question of how a group
might select the insurance plan that everybody in the group will
use. Historically, most groups offered only one plan for all members. More groups now offer a small menu of plans from which the
employee can select one plan, typically with the employer contrib-
uting an equal dollar amount no matter which plan is chosen. Even

Table 10.3   TYPICAL LOADING FEES BY GROUP SIZE

Number of employees	Loading fee (as percent of benefits)
Individual policies	60%–80%
Small groups (1–10)	30%–40%
Moderate groups (11–100)	20%–30%
Medium groups (100–200)	15%–20%
Large groups (201–1000)	8%–15%
Very large groups (over 1000)	5%–8%
Overall for all plans (weighted average)	15%–25% (varies annually)

now, only a small fraction of workers have more than one insurance plan from which to choose, although (by law) many now also have the option of enrolling in a Health Maintenance Organization (HMO), which we will discuss in Chapter 11.

Employers commonly offer health insurance as a fringe benefit for workers, usually paying for part of and sometimes for the entire premium. "Well," one might say, "that would make it crazy for the person not to get as much insurance as possible, since it's free, with the employer paying for it." However, as with lunches, there is no such thing as a free health insurance policy. Strong economic forces in the labor market make it quite likely that the economic *incidence* of the health insurance premium falls on the worker, not the employer (Mitchell and Phelps, 1976). Every dollar paid for health insurance premiums is a dollar not paid in wages. This means that the workers want to be careful in their selection of insurance, because they really are paying for it, even if the employer is making the payments on paper.

The employer must decide which policy to offer as a fringe benefit for all workers in the firm. Which policy will be most attractive to the workers?[12] Think about a fictitious process whereby workers voted on an insurance policy. To keep things relatively simple, suppose that the only dimension of the insurance policy being considered were the coinsurance rate C. Then each possible insurance plan (with some specific coinsurance rate)

---

[12] It is in the interest of the employer to offer the most attractive package available. Otherwise, the employer will just be throwing away money that will not help attract the quality and quantity of workers desired.

would have an insurance premium R associated with it, so $R = R(C)$. Suppose the employer asked for a showing of hands on who wanted a low-cost plan with $C = 0.5$. A few people might find this best (thinking about it as we have previously discussed), but most would probably prefer a more generous plan. How about a plan with $C = 0.4$? More hands will appear in the air. Finally, some specific plan will command a majority, and this will be the plan selected. In fact, because of the way majority rule voting works, the plan selected will match the preferences of the *median voter*, the worker (if you will) who casts the tie-breaking vote. Workers do not really vote in this way, but the idea that the preferences of the median voter (median worker) dominate the decision about employer group insurance is still useful. Goldstein and Pauly (1976) have used this process to study demand for employer-group insurance, and find that the ideas explain a good deal about the patterns of insurance coverage found in firms throughout the country.

HOW INCOME TAXES SUBSIDIZE HEALTH INSURANCE

The other important part of the "price of insurance" for most health insurance purchases in the United States is the income tax system. We had noted earlier that employers pay for a significant fraction of all the health insurance premiums of workers in the United States. Over the past several decades, payments by the employer toward the health insurance premium have increased from about two-thirds of the total premium to over 80 percent (Phelps, 1986). Since (as noted above) the worker really ends up paying for this premium anyway in terms of lower wages, why go through the extra step of having the employer pay for it? The answer lies hidden in the bowels of the Internal Revenue Code: Employer payments for health insurance are not taxed as income to the employee, but remain a legitimate deduction for the employer. Thus, these employer premiums escape the tax system. As a result, they make health insurance cheaper than any other good or service the employee might buy, because the health insurance is purchased with "before-tax" dollars. Relative to everything the employee might purchase with after-tax dollars, health insurance costs only $(1 - t)$ as much, where t is the employee's marginal tax rate. Rather than paying $R = (1 + L)(1 - C)p_m m^*$ for an insurance policy (with after-tax dollars), the employee (and the group, when thinking about the best policy to choose) can buy health insurance for an effective premium of $(1 - t)R = (1 - t)(1 + L)(1 - C)p_m m$. Thus, the *effective* price of insurance is not L, but rather is $L - t(1 + L)$. In other words, the rate L is still paid, but the tax effect provides a subsidy of $t(1 + L)$. This makes insurance very low cost indeed! Since the marginal tax rate (t) for the median worker in many firms would be about 0.25 to 0.35 over recent years, and since actual loading fees (L) from insurance companies are about 0.1 to 0.3 for large groups, many people will find themselves in the position where the net price of insurance is negative! (For example, if $t = 0.25$ and $L = 0.2$, then $L - t(1 + L) = 0.2 - 0.25 \times 1.2 = 0.2 - 0.3 = -0.1$.)

What puts any limit at all on the demand for insurance in this setting? One limit, of course, is the group insurance choice. The group plan is a compromise over the interests of many workers, some of whom surely have different preferences than others. The heterogeneous nature of the group (for example) with respect to age also puts an eventual lid on the demand. Since every worker in the group pays the average premium (over all workers), irrespective of age or health habits, younger workers will find an extremely generous health insurance plan too expensive for their preferences, even with the tax subsidy. Since they "vote" on the plan, they help constrain the choice.

The other natural limit for health insurance, of course, is the welfare loss generated by extending coverage too far. As C approaches zero (full coverage), the welfare loss caused by overpurchases of medical care increases further and further. Even with a tax subsidy for insurance, it is probably preferable to quit before full coverage is reached for most people.

Revisiting the Ficticious Insurance Purchase Example

Earlier in the chapter, we had a simple example of a consumer deciding whether or not to purchase an insurance policy against the simple risk that a four-day hospitalization might occur (with probability 0.3) and that a nine-day hospitalization might occur (with probability 0.1). There, for a moderately risk-averse consumer, we found that the net gain from purchasing the insurance policy was $56 after all was taken into account. (Go back and review that example, in the section called "A Specific Example" including Figure 10.3 earlier in this chapter. We will use it again here.)

Now consider what happens if the same insurance policy is acquired through a group insurance plan, and the consumer's marginal tax rate is 0.3 ($t = 0.3$). The consumer gains all of the benefit previously stated, but also the tax liability confronting this consumer drops by 0.3 times the insurance premium—that is, $0.3 \times \$924 = \$277$. This "tax benefit" swamps the overall gain from the insurance policy alone (\$56 net). Indeed, it could readily tip the decision about whether or not to purchase the insurance.

To see how this could happen, suppose that an otherwise identical consumer was not as risk averse as the first consumer we described, and had a risk premium (willingness to pay to avoid risks) of only \$70, rather than the \$220 in the original example. Then *without the tax subsidy*, the net benefit of buying the insurance would be negative, because the costs (\$924 premium and the welfare loss triangles A and B, with expected value of \$80) exceed the willingness to pay for the insurance (\$840 expected benefit plus \$70 risk premium = \$910). In a world with no tax subsidy, this person would not buy this insurance policy. (She might well buy one with larger coinsurance.) However, adding in the "tax benefit" of \$277, we can see that this insurance policy is now attractive, and the smart consumer will choose it. This will expand the amount of insurance in force, and will increase the overall demand for medical care.

Note that the same thing can happen as the marginal tax rate changes. Even our second "mildly risk-averse consumer" would not take this insurance policy if the marginal tax rate were smaller. For example, if this consumer were in a 10 percent tax bracket, rather than a 30 percent bracket, the "tax benefit" from buying the insurance policy would be only \$92, not \$277, and this would not be enough to offset the other costs of insurance.[13]

EMPIRICAL ESTIMATES OF DEMAND FOR INSURANCE

Studies of demand for insurance have taken two approaches. The first studies the choices actually made by individuals or groups (Phelps, 1973, 1976; Goldstein and Pauly, 1976; Holmer, 1984). In such studies, the differences in income (for example) allow esti-

---

[13] The policy costs the consumer \$924 for the premium plus \$80 in welfare loss, for a total cost of \$1004. The willingness to pay is the expected benefit of \$840 plus the risk premium of \$70 plus the tax benefit of \$92, for a total of \$1002.

mating how demand varies by income. Differences in the group size generate variation in effective price, allowing estimates of price responsiveness. The other approach uses aggregate data over time, estimating how total insurance premiums (for the entire economy) change in response to changes in income, loading fees, tax subsidy, and so forth.

Income Effects

In most studies using individual data, estimated income elasticities are generally positive, but less than 1 for almost any measure of insurance chosen. Measuring the "correct" income is difficult for insurance chosen in a group setting, because the median worker's income presumably is more important than any other worker's income. Aggregate data (such as Phelps, 1986; Long and Scott, 1982; Woodbury, 1983) avoid this problem. Generally the income elasticity of demand for insurance (as measured by premiums) in such studies is above 1, and may be closer to 2.

Price Effects

The effects of the price of insurance on demand turn out to constitute an important question for public policy, because of the way the income tax system subsidizes health insurance purchases. Unfortunately, the range of estimates from the literature is disquietingly large. Again, the source of data seems to determine the magnitude of the estimate. Aggregate data usually provide price elasticity estimates (using variation in the marginal tax rate through time as a price change) in the neighborhood of -1.5 to -2 (Phelps, 1986; Long and Scott, 1982, Woodbury, 1983). Studies using data on individual households have varied some in their findings. When the size of the work group is used to generate price variation, the estimates are also typically large, in the range of -1 (Phelps, 1973, 1976; Goldstein and Pauly, 1976; Ginsberg, 1981). Other studies use differences in the marginal tax across households to determine the effect of price on insurance demand (Taylor and Wilensky, 1983) and get smaller estimates (about -0.2).

Certainly there is clear evidence that the price of insurance matters considerably in demand for insurance. The question of just how much is not as well settled as in the case of demand for medical care itself, because nothing comparable to the RAND HIS exists for studies of demand for insurance. One other piece of evidence arises from the tax cuts initiated by the Reagan administration. Those tax cuts reduced the desirability of extensive insurance, since they reduced the tax subsidy for almost all working persons. (The top marginal tax bracket fell in several steps from 50 percent to 28 percent.) In response, the number of persons with hospital insurance began to fall in the United States in 1983, the first year after the tax cut took place. The peak enrollment occurred in 1982 (188.4 million), and had fallen to 181 million by 1985. This represents the first time since World War II that a systematic decline in the number of insured persons had occurred, and this decline took place in a period of rising employment and income.

THE OVERALL EFFECT OF THE TAX SUBSIDY ON THE HEALTH SECTOR

The cumulative effects of the federal income tax subsidy of health insurance could be very large indeed. The health insurance subsidy produces a secondary effect in the market for medical care, since the increases in insurance coverage in turn increase the demand for medical care. The leverage that this interaction can generate has the potential for substantially altering the shape and size of the health care system.

One recent study (Phelps, 1986) estimated that employer-group health insurance premiums would be only about 55 percent as large today if the tax subsidy were not in effect. Cutting the premiums in half is not as radical a restructuring of insurance coverage as it might seem. Recall (for a simple insurance policy) that $R = (1 + L)(1 - C)p_m m$, and that m falls as C rises. Based on estimates of how m varies with C of about 25 to 30 percentage points (e.g., from 0 to 0.25 or 0.25 to 0.5) would cause premiums to fall by the required amount. In turn, this change in insurance coverage would cause the demand for medical care to fall among those insured persons. Again, the RAND HIS results tell us that demand is about 20 percent lower with $C = 0.25$ than for $C = 0$, and about 10 percent lower for $C = 0.5$ than for $C = 0.25$. Thus, total medical care could decline by some 10 to 20 percent among the persons under age 65 who are insured by this mechanism.

In addition, if private health insurance had (from the beginning) contained higher coinsurance or deductibles (without the tax subsidy), the structure of Medicare would probably reflect that difference as well. Medicare was clearly patterned after private health insurance when it was instituted in 1965, with essentially full coverage for hospital care and a "major medical" type of insurance for doctor services, with a $50 deductible and a 20 percent coinsurance. Thus, private insurance with greater cost sharing probably would have led to public insurance with greater cost sharing as well. In aggregate, it seems possible that the health sector would be at least 10 to 20 percent smaller without the tax subsidy for health insurance. Lest this seem "small," we can translate that difference into something that represents 1 to 2 percent of the gross national product.

"OPTIMAL" INSURANCE

The model used by economists to study demand for insurance (the expected utility model) offers one further insight into the question of how insurance contracts might look without the tax subsidy in place. According to that model (Arrow, 1963), a consumer seeking to maximize expected utility will select a policy with full coverage above a deductible, when the losses are fully independent of the insurance coverage (truly random losses). The size of the deductible increases as the loading fee increases. Arrow also shows that the optimal policy has a coinsurance feature included when the insurance company is also risk averse, as well as the consumer. Taking into account the effect of coinsurance on demand for medical care, Keeler, Buchanan, Rolph, et al. (1988) have estimated the expected utility among a variety of insurance policies, based on the RAND HIS results. The "best" plans all contain a coinsurance rate of about 25 percent, and an initial deductible of $100 to $300. Most "major medical" plans that pay for doctor services out of the hospital have about this structure, whereas most hospital and in-hospital medical-surgical insurance plans cover nearly 100 percent of all costs. The theory of demand for insurance, coupled with the empirical results from the HIS, suggests that consumers would be better off with less complete coverage. Elimination of the income tax subsidy would certainly move people in that direction, and (based on the estimates in Phelps, 1986) the magnitude of change from the full coverage to a major-medical type of plan would correspond with the premium changes predicted to occur.

OTHER MODELS OF DEMAND FOR INSURANCE

The model of expected utility maximization has enjoyed wide popularity among those studying demand for insurance. Nevertheless, the model has some particular defects in terms of precise predictions about how people behave in settings containing uncertainty. The strongest challenges have come from the discipline of psychology, most notably from Kahnemann and Tversky (1979). Their "prospect theory" model sets aside the idea of a stable utility function (one that does not change, for example, as income or health change). Instead, they offer a model where deviations from "today's" world affect behavior. Everything is viewed in terms of "where you are." The model goes on to suggest that people prefer risk (are willing to accept gambles) for degradations in their well-being, but are risk averse with regard to improvements. This type of person, for example, would purchase insurance against losses but also would gamble actively with the prospect of large gains.

Hershey et al. (1984) have studied the health insurance choices of a sample of people, and concluded that their purchases closely correspond to those predicted by the standard expected utility model. However, another study (Marquis and Phelps, 1987) shows some contradictions with the expected utility model in insurance purchases, and another (Marquis and Holmer, 1986) suggests that prospect theory fits the data better.

An analysis of insurance policies people have purchased to supplement the Medicare coverage (Phelps and Reisinger, 1988) shows some inconsistencies with the expected utility model as well. A further discussion of this issue appears in Chapter 12.

Thus, we must leave as ambiguous the complete validity of the expected utility model. However, except for very particular problems and questions, most of the empirical work conducted on the demand for insurance remains valid, and can guide public policy choices about insurance.

SUMMARY

Health insurance offers a way to protect against financial risk. Economists' models of expected utility maximization predict certain patterns of insurance purchases, including the presence of deductibles, copayments, and insurance against the riskiest, rather than commonly occurring events. These models seem to predict actual patterns of demand for health insurance reasonably well, but not perfectly.

Most insurance is sold through groups, most commonly by far, employer work groups. This insurance provides the added benefit to consumers of reducing income taxes. In some cases, the tax benefit is large enough to tip the balance between buying and not buying insurance.

Health insurance creates a subsidy for medical care, so a "welfare loss" emerges each time medical care is purchased, because consumers are induced (by the lower price of care generated by the insurance policy) to buy more medical care than they otherwise would. Indeed, they are "tricked" into buying care that costs them (through the insurance premium) more than it is worth in restoring health. This is an added "cost" of health insurance that doesn't show up on anybody's accounting books, but it is a real cost in any economic sense.

PROBLEMS

1. "People most commonly buy insurance against hospital costs because it is that type of medical care that has the biggest average expense for individuals year in and year out." Comment.

2. What are the two major efficiencies arising out of providing health insurance through employer-related work groups?

3. Since the average loading fee for work-group insurance is about 15 percent to 20 percent, and the average tax subsidy to work group insurance is about 30 percent or more, insurance really comes at a negative price. What forces prevent work-group insurance from exploding in costs?

4. "The single most important health policy choice in the United States over the last four decades has nothing to do with the Department of Health and Human Services, but rather the Internal Revenue Service." Comment.

5. Prove that the welfare loss triangle described in footnote 10 has a value of approximately $-0.5\eta(1 - C)^2 P_m mi$ for medical event i.

APPENDIX

The problem in the text has the following characteristics: Suppose the person confronts an illness risk such as the following:

Probability	Uninsured Demand	Insured Demand
.3	3	4
.1	8	9

Each day of hospital care costs $500. The uninsured risk has an expected value of $500[(0.3 \times 3) + (0.1 \times 8)] = $500 \times 1.7 = 850. The variance of that uninsured risk is $(0.3 \times $1500^2) + (0.1 \times $4000^2) - $850^2 = $1,552,500$.

With the insurance policy, the patient buys more medical care, but pays for only 20 percent of it. The insured risk has an expected value to the patient (the patient's share) of $(0.2 \times $500)[(0.3 \times 4) + (0.1 \times 9)] = $100 \times 2.1 = 210. The variance of this risk is $(0.3 \times $400^2) +$

$(0.1 \times 900^2) - \$210^2 = \$84,900$. Thus, the *change* in risk to the consumer by purchasing this $C = 0.2$ insurance policy is $\$1,552,500 - \$84,900 = \$1,467,600$.

According to Pratt (1964), the risk premium a person would be willing to pay for this reduction in risk is approximately $0.5 \times r \times$ (change in variance). If the risk-aversion parameter $r = -0.0003$, then the risk premium the consumer would be willing to pay for this reduction in risk is $0.0003 \times 0.5 \times \$1,467,600 = \$220$.

Now consider the welfare loss from the extra demand for medical care, the triangles A and B in Figure 10.3. Consider first what happens for illness 1. The change in quantity is 1 day. The change in marginal value is $0.8 \times \$500 = \$400$. The welfare loss triangle for that event is $0.5 \times \$400 \times 1 = \$200$. Similarly, if illness 2 occurs, the change in demand is also 1 day (although there is no reason why it has to be the same), the change in marginal value is again $400, and the welfare loss from that event is also $200. Thus, the expected welfare loss from "moral hazard" is $(0.3 \times \$200) + (0.1 \times \$200) = \$80$.

These data are sufficient to allow calculation of the net welfare gain from purchasing an insurance plan with $C = 0.2$ as described in the text. The risk premium is $220, and the loading fee charged by the insurance company is 10 percent of $840, or $84. Then the net gain from risk reduction is $220 - \$84 = \$136$. The welfare loss from the "moral hazard" is $80, as computed in the previous paragraph. Thus, the net gain from purchasing the insurance is $136 - \$80 = \$56$.

A useful approximation to the welfare loss triangles (see problem 5) has the following computations: The arc-elasticity of demand for the first illness is -0.2143. The change in price is $400. The approximate welfare change is $0.5 \times \$400 \times 4$ days $\times (-0.2143) = \$171$. This occurs with probability .3, for an expected loss of $51. The "exact" calculation shows a welfare loss of $200, rather than $171.

Chapter
11

Health Insurance in the Marketplace

*T*he previous chapter described how and why people demand health insurance. This chapter expands the discussion, addressing some additional issues in health insurance. These issues include (1) how health insurance is supplied; (2) what the role is of not-for-profit and for-profit insurers, and how taxes paid by the latter affect their business; (3) stability of the insurance market; and (4) state regulation of health insurers. We will also explore some innovative new insurance plans that offer different ways of rebalancing the trade-off between risk spreading and "moral hazard" that pervades the choice of a "standard" health insurance plan. These new plans include a wide spectrum of prepayment plans with provider-based controls on utilization and spending, rather than "consumer-based" controls that depend on copayments. Finally, we turn to a persistent question in public policy—what to do (if anything) about uninsured persons—and discuss some common proposals to reduce or eliminate the possibility of a person being uninsured (universal health insurance).

THE SUPPLY OF INSURANCE

Insurance companies perform two primary tasks, the first of which is more obvious to the consumer, and the second more fundamental. The obvious activity is the processing of claims. When a patient incurs an expense for medical care, either the doctor's office bills the insurance carrier directly, or the patient pays the doctor and then submits the bill to the insurance company for reimbursement. The medical expense is compared with the coverage of the person, and a check is issued to the doctor or patient, as appropriate.

The risk-bearing activity is less obvious, but more important. Insurance companies are in the business of bearing and spreading risk. Since consumers dislike risk, if they are able to shed that risk, somebody must acquire it. Insurance companies perform this task.

The simple exchange of risk from one person to another does little to reduce risk in our society, but merely transfers it from one person to another. However, by pooling risks, insurance companies can actually reduce average risk. If the risks are completely independent, the average risk (the risk per person in the group) of an insured group with N members is $1/N$ times the variance that each individual confronts.[1] If the risks are correlated across persons, then the risk can't be reduced as much. Thus (for example) a group of people all living in Southern California can't gain much by pooling their risks against earthquake losses, because if one home gets destroyed by earthquake, probably others will as well.

In health care, the presence of contagious diseases like colds and influenza contribute to the correlation of risks, but overall, the correlations across people are very small, and risk spreading can work quite well.

THE FUNCTIONS OF HEALTH INSURERS

Health insurance companies exist to spread risk. They do so, of course, by paying for part or all of the medical expenses associated with illnesses of the people insured by the company. Thus, the operating costs of insurance companies include the tasks of processing insurance claims and making appropriate payment. These tasks primarily involve people and computers, which are the main inputs, along with buildings, in the production function of health insurance. These costs are paid for by the loading fee charged by the insurance company described in the previous chapter.

The loading fee for the insurance company will rise and fall as its costs rise (wages of workers, for example) and fall (computing costs, for example). The structure of an insurance policy itself will also affect the loading fee. If the insurance plan has complicated provisions that require many human decisions before a payment

---

[1] The appendix to this chapter provides the proof and discussion.

can be made, the loading fee will of necessity be higher. If the insurance is a "standard" plan that can be administered mainly by computers, loading fees will be smaller. The number and complexity of claims filed by insured people similarly will affect loading fees. Insurance plans with large deductibles, for example, will generate fewer claims, since people will not bother filing small claims on the policy. Plans with many small claims, like dental insurance, will probably generate higher loading fees than plans with smaller number of large claims, like a surgical insurance plan.

The other relevant cost of insurance companies is a negative cost—the "cost" of money. Since insurance companies collect the insurance premium at the beginning of a year (or whenever the insurance contract begins), and pay claims throughout the year as they are sent in by patients, they get to hold the money in the interim, earning return on their investment. Competitive forces will generally lead to the pricing of the insurance contract to account for any earnings from the insurance company's investment portfolio. For a typical health insurance contract with a one-year duration, the average time between payment of the premium and payment of the "typical" benefit is six months. Thus, the higher the "real" return on capital for the insurance company, the lower will be its loading fee.[2] Over the past several decades, the real return (net of inflation) for insurance companies has fluctuated between 0 and 10 percent, averaging about 5 percent.[3] Thus, with an average holding period of half a year, the insurance company will earn about 2–3 percent of premiums as investment income. While it is conceptually possible for the premiums collected by the insurance company to be *less* than the average benefit paid out, portfolio earnings will not commonly be sufficient to offset the entire operating costs of the insurer, and loading fees will be positive.

---

[2] The "real" rate of earnings is approximately the actual rate of interest earned minus the rate of inflation. For example, if the insurance company can earn 10 percent on its investment portfolio, and there is a 6 percent inflation in the economy, then the real rate of return is 4 percent.

With health insurance companies, most of the premium is paid out in terms of health care costs, and only a small part (the administrative costs of the insurer) depends on the overall price level in the economy. Since health care prices have increased faster than overall prices for almost all of the past 40 years, health insurance companies really need to concern themselves with the rate of price increase for the bundle of goods and services they pay for in benefits, such as hospital care and physician services, versus the interest rate they receive.

[3] The Department of Commerce gathers data on insurance companies' return on investment.

Not-for-Profit and For-Profit Insurers

As with hospitals, health insurers come in several "flavors." One primary distinction is the form of ownership. For insurance covering hospital care (Blue Cross plans), physician services (Blue Shield plans), and dental care (Delta Dental plans), many large (and in some cases, dominant) insurance carriers are organized on a not-for-profit basis.

The not-for-profit status typically is established in special "enabling acts" in each state, allowing this form of insurer to do business in addition to normal insurance companies. These enabling acts typically provide for the establishment of these insurers, and commonly set up different (less stringent) regulations for them to do business than for commercial insurance companies. The not-for-profit status typically exempts these insurance carriers from the payment of "premium taxes," a sales-tax-like levy that varies from state to state, commonly 1 to 3 percent. Finally, in some states (including, for example, New York), Blue Cross plans within the state are granted preferential treatment in the state's hospital cost control regulations, with the Blues paying hospitals less than their commercial insurance competitors pay.

The Effects of Premium Taxes

In concept, premium taxes paid by one insurer but not another should give one a price advantage. If the insurance market were perfectly competitive and the tax advantage were not dissipated in some fashion, then the Blue plans would have the entire market to themselves. However, for a variety of reasons, this has not happened.

One of the key factors in this market is the long history of the Blue Cross plans offering full coverage for a specified number of hospital days (originally, 30 or 60 days). Similarly, Blue Shield plans typically offered "service benefits" that paid in full for the costs of covered services, usually doctors' services within the hospital. By contrast, commercial insurance companies have tended to offer plans with deductibles, copayments, and fee schedules. Thus, while not perfectly so, the market is somewhat segmented by type of provider.

The purposes behind the full coverage plans offered by the Blues have sparked considerable debate. Just as the profits of a not-for-profit hospital can be redirected to various groups (employees, doctors, or even patients), so can the "profits" of a not-for-profit insurance plan. Several writers have argued that the Blue plans have been captured by doctors, who use the insurance to further their own benefits (Frech, 1979; Kass and Paulter, 1979; Arnould and Eisenstadt, 1980; Eisenstadt and Kennedy, 1981; Greenberg, 1981). Other analysts have argued that the advantages of not-for-profit status have been squandered in higher administrative costs of the Blues (Blair, Ginsberg, and Vogel, 1975). Historians of the Blue Cross movement offer the view that the plans were designed to benefit consumers most of all, and that the Blue plans fiercely held to their style of insurance on the beliefs that it was "best" for the consumer (Anderson, 1975).

If we accept the economists' model of consumer demand for insurance, then it seems clear that the Blue Cross style is inappropriate, however well-meaning the designers of the plans might have been. The early Blue plans, for example, covered 30 days of hospitalization fully, and then no coverage thereafter. This coverage was commonly extended to 60 days per hospitalization, and then 120, in part under competition from "major medical" insurance plans of the commercial carriers, that offered very high coverage limits ($50,000 to $100,000 coverage). The traditional Blue plan is upside down from the point of view of a risk-averse consumer, because it covers first-dollar risks (no deductible) without covering the really large risks. (Recall Arrow's proof that a risk-averse consumer would select a plan with unlimited coverage above a deductible.) These plans have no deductible, but expose the consumer to rare but large risks of extended hospitalization.

For whatever reason, the Blue plans held to their style of full-coverage "service benefits" while the commercial plans typically offered plans more in accord with economic models of consumer demand theory. This may explain in part how the two types of firms coexist, even with a tax advantage for the Blues: One type of firm (the Blues) offers a type of insurance that is conceptually less desirable, but without the burden of the premium tax. The other type of firm (commercials) offers a more desirable type of coverage, but with the burden of the premium tax. Different consumers could readily choose either style, given the price differences that might emerge. In a comment at a conference on this issue in 1978, a vice president of the national Blue Cross Association said, "If the buyer feels that access to care and service benefits are important, and that he is getting value, he opts for the Blue Cross Plan product. If he wants lesser coverage, he opts for a different segment of the market" (Berman, 1978).

Empirical studies of the actual effect of the premium tax paid by the commercial insurance companies on their market share (or the market share of the Blues) have reached quite different results. An early study (Frech and Ginsberg, 1978) found that larger premium taxes paid by commercial insurance plans increased the market share of Blue plans significantly. A later study (Adamache and Sloan, 1983) found absolutely no effect of the premium tax on the Blues' market share, while another study (Sindelar, 1987) found significant effects. This issue, like many empirical questions involving health insurance and health care, is not yet fully resolved.

The Blue Cross "Discount"

Several studies have combined an analysis of the tax advantage of the Blue plans with an analysis of "the discount" they receive from hospitals. On average, Blue Cross plans account for about half of the hospital insurance for persons under age 65, with shares as high as 80 to 90 percent in some regions. Given this size, the question naturally arises as to whether the Blues have used this market position to advantage in their dealings with the hospitals in this area. This is not a standard "monopsony" problem,[4] but rather a problem in negotiated price. The Blue plans assert the right to pay less for hospital care than the list price of the hospitals. (The commercial insurance plans typically pay the list price, as do customers with no insurance.) The Blues defend this arrangement on the grounds that they do not contribute to the bad debt problems of the hospital, because they have "full-service" benefits, and hence should not have to pay for bad debts. Generally, the Blues enforce this arrangement through contracts with hospitals; if the hospital does not sign such an agreement, it gets classified as a nonparticipating hospital in the Blue plan, and patients using that hospital there have less generous benefits than at a participating hospital.

The larger the market share of the Blue plan, the more potential leverage it has with the hospital in achieving a discount. Of course, the market share and the discount interact, since a bigger discount makes the premium lower for the Blue plan. For this reason, the tax

---

[4] In a classic monopsony, the buyer reduces the quantity demanded in order to avoid higher prices. The model depends upon a supply curve showing increasing costs. In the case of health insurance, one would not think of the Blue Cross plans as restricting the quantity of hospital care demanded.

advantage of the Blue plans potentially interacts with "the discount," because of its potential effect on market share.

One analysis of this arrangement (Feldman and Greenberg, 1981) estimated that the "discount" achieved by Blue plans helped them gain larger market shares, but that the market share was not responsible for the discount. By contrast, another study (Adamache and Sloan, 1983) found that the discount was quite important both in reducing the premium for Blue Cross hospital coverage and in establishing their market share. In both cases, the tax advantage of the Blue plans was included in the analysis as a possible determinant of the Blue plans' market share.[5]

If we return briefly to the discussion in Chapter 8 about the hospital's decisions about how to use its profits, the idea of a Blue Cross hospital discount makes more sense. The Blue plans, by obtaining a contractual discount, cut themselves into the decision process of the hospital. They force a partial spending of the hospital's profits as lower prices *to the Blues.* The Blue plan insurers can in turn use the lower hospital price in ways meeting their own organizational objectives. The work of Adamache and Sloan shows that the Blue plans "spend" this advantage (obtained through lower hospital prices) by expanding their market share. This automatically passes along to customers of Blue plans the lower hospital price, by keeping the premium lower than it would otherwise be.

Is the Insurance Market Stable? The Question of Self-Selection

Despite the widespread presence and persistence of health insurance, a lingering question persists, at least in theory, about the intrinsic stability of the health insurance market. The problem hinges on the difference in information held by buyers of insurance (consumers) and sellers of insurance (insurance companies).

---

[5] In econometric jargon, the tax advantage of the Blue plans helps to identify the effect of market share on "the discount." It does this as follows: To the extent that the tax advantage (which is set by law) changes the market share of the Blue plans, one can think of the differences in tax law as a set of experiments conducted for us by various state legislatures. A higher tax on the commercial insurance plans (avoided by the Blues) would potentially create a larger "experimentally induced" market share. This in turn would allow an "experimentally induced" estimate of the effect of market share on the size of the discount the Blue plan could achieve. This is basically the strategy followed by the researchers studying this question.

The buyers know more about their own health than the sellers. Thus, the risk exists that insurance companies will put an insurance plan into the market, using one set of actuarial projections about the costs of insured people, and end up attracting a special subset of the population with unusually high health care costs. Obviously, the insurance company would go broke if this happened repeatedly. This is called the problem of "self-selection" or "adverse selection." It represents yet another place where uncertainty and information intrude considerably into the economic analysis of health care.

Self-selection could manifest itself in many ways. People who believe they have just contracted cancer (say, because of unusual weight loss), couples who plan to have a baby, or people who think it is time to get their hemorrhoids or hernia repaired all have special reasons to buy high-coverage health insurance. They will try to subscribe to an insurance plan with excellent coverage. People with better health risks will obviously not want such people in their insurance pool, because the average costs and hence the premium will be driven up. We might characterize this process as "bad risks chasing good risks." This phrase creates the vision of people racing from group to group in a frenzy of sickly people seeking coverage and healthy people trying to evade sickly people. Of course, this doesn't literally happen, but it is useful to think about the idea in order to understand how and why the insurance market does remain stable.

One reason why this doesn't happen in reality is that the costs of making such changes would probably overwhelm the benefits. If any single group contained only a smattering of high-cost enrollees, it would cost more (in time and money) for other members of the group to shop around for a "better" group than it would be worth. The other and perhaps more important reason is that the primary method of providing health insurance in the United States—the employer work group—goes a long way toward solving the problem of self-selection. Since the employment group is brought together for some purpose other than buying insurance (e.g., building cars, selling groceries), insurers usually need not worry about the problem of bad risks chasing good risks. It is difficult for "bad risks" to attach themselves to a particularly desirable insurance plan, because they must be healthy enough to work, and also possess skills desired by the employer through which the insurance plan is offered.

Avoiding self-selection in insurance plans seems to be an important function of the work group arrangement for selling insurance. A key to making this system work is that the work group must all use the same insurance plan, so the preferences of the "median" worker dominate the choice of plan. Probably for this reason, a minority of employer groups offer a choice of more than one insurance plan to their employees.[6]

Finally, we should note that the issue of self-selection and market stability may matter considerably more in such areas as long-term care insurance than in more traditional hospital insurance. Long-term care insurance is mostly viewed as pertaining to the elderly. If the demand for such insurance mostly arises from retired people, then the mechanism of the work group will not be available. How much this matters remains to be seen, since these markets are only just beginning to form.

COST-CONTROLLING HEALTH INSURANCE PLANS

As we have seen in Chapters 4 and 5, health insurance creates incentives to consumers to increase the amount of medical care they buy. These "demand-side" incentives reduce the attractiveness of health insurance, more so if the underlying demand for medical care exhibits considerable price sensitivity. One way to help control the costs of medical care is to build into the insurance plan a counteracting "supply-side" restriction on the use of medical care. A number of different alternatives have emerged to do just this, including the closed-panel prepaid group, also known as the health maintenance organization (HMO), the independent practice association (IPA), the preferred provider organization (PPO), and many cousins, aunts, and uncles. These plans create the new alphabet soup describing the health care system as it currently evolves.

The Closed-Panel Prepaid Care Plan

The prototype of the "cost-controlling" health care plan is the closed-panel prepaid care plan, now more commonly called a health maintenance organization. The concept originated with the

---

[6] An exception involves the option of a health maintenance organization (HMO), which federal law requires employers to offer as an option to employees in any area where a federally certified HMO exists.

Kaiser Permanente Corporation in the 1940s in Portland, Hono-lulu, and Los Angeles, as a way to provide health care benefits to corporation employees. The idea is very simple: The insurance coverage and the delivery of medical care are integrated into a single organization. The insurance plan hires the doctors (in the pure form, on a straight salary) or contracts with a specific group of doctors to provide care, and either builds its own hospital or con-tracts for the services of a hospital within the community. (The original Kaiser Permanente group plans all had their own hospitals associated with them.) The key idea with such programs is that the providers of care (doctors, hospitals, etc.) have no financial incen-tive to provide extra care to the patients enrolled with the insur-ance plan, because the insurance plan, the hospital, and the doc-tors all share common financial interests.

In many of the "pure HMO" plans, the doctors operate as a separate legal entity (commonly, a large multispecialty group prac-tice) that enters a contract with the HMO-insurance company to provide care for the enrolled patients, agreeing to use the HMO's hospital exclusively. The patients pay a fixed premium for the year, for which all care is provided (sometimes, with small copayments for physician visits). The physician group shares in any extra money remaining at the end of the year. Thus, the group has strong financial incentives to avoid expending resources unnecessarily. They can avoid some of the coordination problems of large groups (see Chapter 7) through rules and supervision, and—impor-tantly—by being able to fire doctors who do not cooperate.

These plans create a supply-side incentive to restrict the amount of medical care rendered. Not surprisingly, the major cost savings generated by this form of care come from reduced hospital-ization of enrolled patients, the one very costly activity over which physicians have almost exclusive control. Extensive studies of HMOs (see Luft, 1981, for a summary) have repeatedly shown their reduced propensity to use the hospital, compared with fee-for-service arrangements with extensive insurance coverage. (Note that the relevant comparison is the rate at which medical resources are used in a full-coverage insurance plan, since HMO patients have no financial risk and an out-of-pocket price at or very near zero.)

These nonexperimental findings were confirmed in the RAND Health Insurance Study, which enrolled some patients (randomly assigned) into an HMO in Seattle. These HMO patients were compared with other HIS enrollees with full-coverage insurance, and also with a set of patients previously enrolled within the HMO. The first comparison (with full-coverage fee-for-service patients) tests the effect of the HMO-incentive system, holding constant the coverage from the patient's point of view (row 1 versus row 3 of Table 11.1). The second comparison (with previously enrolled HMO patients) tests a separate issue—namely, whether the HMOs have been getting an unusually favorable or unfavorable set of patients in their usual enrollment (row 1 versus row 2). This is the "self-selection" problem previously discussed. It has loomed important in the long debate about HMOs, because the previous comparison studies, like those undertaken by Luft and others, always contained the risk that the results achieved by the HMO were due to a favorable patient mix.

Two important results emerge from this study. First (in parallel with the previous nonexperimental results of Luft and others), the HMO plan used considerably less resources than the comparable $C = 0$ fee-for-service plan in the same city (Seattle). (The importance of using the same city as a comparison is heightened by the cross-regional variations in hospitalization shown in the work of Wennberg and others—see Chapter 3).

Second, the RAND HIS studied the health outcomes of people enrolled in the experimental HMO, and compared those with people on the full-coverage fee-for-service plan. As reported by Sloss et al. (1987), at least on a battery of 20 physiologic measures of health status that encompass every major organ system (vision, musculo-skeletal, digestive, etc.), the HMO enrollees fared at least as well as the corresponding fee-for-service enrollees with full coverage ($C = 0$). Thus, at least in this HMO, the reduced use of resources did not lead to reduced health outcomes for the enrollees.

THE ADVERSE SELECTION QUESTION REVISITED

Another RAND HIS comparison of importance is that of the HMO "experimentals" and the HMO "controls." The experimental subjects were enrolled through the RAND HIS in the standard fashion, and thus represent a random selection of people in the region, randomly assigned to the HMO. The "controls" within the HMO were a random sample of people previously enrolled in the HMO under its normal enrollment procedures. Thus, any differences between use of the HMO experimental group and the HMO con-

Table 11.1   HMO HOSPITAL USE AND TOTAL COSTS IN THE RAND HIS

	Percent using hospital	Percent ambulatory any care	Total cost
HMO experiment group	7.1%	87%	$439
HMO controls[a]	6.4%	91%	$469
Fee for service with $C = 0$	11.1%	85%	$609

[a] Previously enrolled in HMO, not as part of RAND HIS.

trol group would be due to self-selection into the HMO from its normal enrollment process. In fact, the differences between the plans are small and insignificant (compare rows 1 and 2 in Table 11.1). This null result increases the confidence that we can place in other nonexperimental comparisons of HMOs and fee-for-service systems. Since there was no apparent selection in favor or against the HMO in its normal enrollment, we can infer that the differences in observed use are due to the institutional arrangement, rather than patient selection.

This result differs somewhat from several alternative approaches to studying the question of patient selection. One study (Eggers and Prihoda, 1982) looked at the Medicare population that had enrolled in HMOs in a demonstration of the HMO concept for Medicare, conducted in three cities. The study looked at the previous year's utilization by the people who decided to enroll in the HMO. They found that among the Medicare population, HMOs had experienced quite favorable self-selection, since those who enrolled had shown unusually low medical care use in the year previous to their decision to switch into the HMO.

In another similar study of preenrollment use by an under-65 population (Jackson-Beeck and Kleinman, 1983) in the Minneapolis–St. Paul area, a similar favorable self-selection seemed to appear. HMO enrollees had used considerably less hospital care in the year prior to enrollment, compared with other people in the region. Overall, the average use was 53 percent lower, but this average is misleading, since older people (who use more hospital care in any setting) were less likely to enroll in the HMO.

A third study of this issue (Buchanan and Cretin, 1986) studied the previous year's medical care use by employees of a large Southern California aerospace company that offered HMO enrollment to its employees. This study, like the previously described studies, found that HMO enrollees had considerably lower health care use in the year before enrollment than those who stayed in the standard insurance plans.

These results contrast sharply with those of the RAND HIS, which showed no self-selection into the HMO in Seattle used in that experiment. Of course, one might just write off the differences as idiosyncratic to the cities in which the studies were conducted. The RAND HIS study (Seattle), the Buchanan-Cretin study (Los Angeles), and the Jackson-Beeck and Kleinman study (Minneapolis–St. Paul area) each involved only one city. Perhaps we just don't have a definitive answer yet. However, differences in research design may explain part of the differences in results.

In the RAND HIS study, the important contrast was between the HMO use of those who had enrolled in the HMO voluntarily versus the HMO use of those who had been experimentally assigned to the HMO. The other studies (Minneapolis–St. Paul and the Medicare study) compared prior-year fee-for-service use for those who enrolled in the HMO when given the opportunity.

One explanation for the apparently favorable self-selection in the HMO in these latter two studies is that at least some of those who chose to stay in the fee-for-service system had ongoing medical problems from the previous year, still under treatment by their doctors. These people would have unusually large expenses in the previous year, but also would be less willing to shift into the HMO because it would require a change of doctor. One part of the Medicare-HMO study supports this belief. Of the three HMOs studied, two had "closed panels" that would require changing doctors, and both of these plans experienced apparently favorable self-selection. The third plan had an "open" panel throughout the community (really an IPA plan, as discussed below), in which almost no enrollee would have to change doctors in order to change insurance plans. In this group, there was no favorable selection for the HMO at all.

On net, the strongest evidence is the controlled trial study from the RAND HIS, but the conflicting evidence from several other studies, showing unusually low prior-year expenses for new HMO enrollees, still casts some shadow of doubt on the matter. We do know this: Utilization results all appear as predicted (that HMOs would have lower total costs, and probably would use hospitals less often). Favorable self-selection into HMOs would produce the same types of results. The issue will probably only be settled by further research, including a better understanding of the importance of closed versus open panel HMOs, one of which requires changing doctors, the other not.

INDEPENDENT PRACTICE ASSOCIATIONS

IPAs represent a sort of halfway house between the pure HMO and a fee-for-service system. Typically, the IPA has a panel of doctors enrolled to provide care, but they may well not be part of the same group of physicians. They commonly have only a part of their practice devoted to the IPA patients, and often treat a substantial majority of their patients in a standard fee-for-service arrangement. They treat patients either on a fee-for-service basis or on a per capita basis (capitation), although the capitation plan is not common.[7] More commonly, the IPA will withhold a portion of the fees due to the doctors within the IPA (say, 20 percent), as a reserve against high costs, paid to the doctors (proportional to their year's billings) if costs are kept sufficiently low overall. As with the pure HMO form, the doctors thus share collectively in any savings from reduced hospitalization or other cost savings.

Since the IPA is not a centralized group, but rather a collection of independent practitioners, the controls over physician behavior are weaker than within a single organizational setting like the HMO. It is more difficult to monitor the behavior of doctors, and each doctor has only ambiguous incentives to create cost savings. The problem is identical to that described earlier for group practices (Newhouse, 1973)—any cost savings generated by one of N doctors pays back to that doctor only $\$1/N$ for each $\$1$ saved.

Studies of the use of hospital care and other medical resources within IPAs (Luft, 1981) confirm what these concepts might lead one to believe: IPAs are not as effective as a pure HMO in reducing the costs of medical care. While individual plans vary considerably, the IPA has generally used hospital care at rates between those of the HMO model and the fee-for-service system. Physician use is typically not much different, perhaps because many physician visits are initiated by patients, and are thus under less control of the IPA or its doctors.

---

[7] The reason why pure capitation is not common is that it imposes a large amount of risk on individual doctors. The doctor would have to take on the role of insurance company, pooling risks across patients. However, the typical doctor's practice is not large enough to take full advantage of the gains from pooling risks for the segment of patients within the IPA, and a pure capitation plan would leave the doctor exposed to potentially large financial risk.

The concept of the IPA is still evolving. Some IPAs now have complex rules for treatment of patients, including the filing of treatment plans for patients with "prior authorization" required before the treatment can proceed, or in some cases, the requirement that a primary care doctor see the patient before a specialist is involved. This "gatekeeper" model is discussed separately below.

PREFERRED PROVIDER ORGANIZATIONS

The PPO model is conceptually closer to the fee-for-service system than the IPA. Basically, the PPO attempts to strike a deal with doctors about the fee they will receive, or hospitals for the per-day costs they will be paid. The PPO offers to bring to the doctor (or hospital) a large number of patients in return for the reduced price. To the extent that the PPO succeeds in obtaining favorable prices, they will have lower total premiums, and hence can attract more patients to enroll in their insurance plan.

The effects on utilization of medical care within the PPO are likely to be quite small, and perhaps even ambiguous in direction. Since the doctor within the PPO receives a smaller amount of money for each patient visit or procedure than the "usual" fee for the same visit or procedure (because of the agreement struck with the PPO), the doctor may reduce the number of visits for patients, spending more time with more lucrative fee-for-service patients. However, others have expressed concern that the PPO model will simply generate more demand inducement by the doctor, who will (in extreme form) cut down on the time spent with each patient (lower quality) and perhaps even try to convince the patient to return for more visits, for each of which the PPO can be billed.

At this point, few empirical studies of the effectiveness of PPOs in controlling health care costs are available. The diverse set of controls and agreements within the PPO family hinders analysis of their results and the concept of the PPO has only recently emerged. It would appear from the simple economics of the model, however, that the PPO would achieve most of its costs savings in reduced fees per visit, rather than changes in the number of physician visits or hospitalizations incurred.

We should note that the line between a PPO and an IPA may be difficult to distinguish. Both types of plans operate with physicians throughout the medical community, rather than a closed panel of providers. Both have at their disposal an array of cost-control devices, including the use of fee schedules, hold-back arrangements (keeping a fraction of the fees due to the doctors in reserve against high hospital spending within the plan), prior authorization, and the like. Within a decade, perhaps a new alphabetical nomenclature will emerge that differs from the current IPA/PPO system.

Legal Questions Regarding PPOs

The types of organization forming or sponsoring PPOs are as diverse as one could imagine, ranging from Blue Shield plans to county medical societies, to hospitals, to commercial insurance companies, and to labor unions. A particular legal difficulty emerges when the PPO is sponsored by a group of physicians, such as a county or state medical society: The arrangement may violate federal and state laws about price fixing and collusion. The classic case was the U.S. Supreme Court decision in *Arizona* v. *Maricopa County Medical Society,* which struck down as illegal a PPO sponsored by the medical society in the Phoenix region. Many observers believe that this ruling effectively precludes physician-sponsored PPOs, but other variations may eventually permit such plans (Greaney and Sindelar, 1987). The concern, of course, is that the doctors may use the PPO fee-setting arrangement to their own benefit—for example, by suppressing competition.

SECOND OPINION PLANS

One way that insurance plans can control the use of hospital care for enrolled patients is to use "second opinion" programs. These programs pay for the visit of a patient to a second (or sometimes third or fourth) doctor to determine how widely a first doctor's recommendations for treatment are shared. Almost always, the second opinion deals with a decision to send the patient to surgery. A second doctor sees the patient, knowing that any surgery performed will be by another doctor. Thus, in concept, the second opinion provides information to the patient about the desirability of treatment without the financial conflict (if any) created by the fee-for-service system. This has the potential for removing the asymmetry in information present in many medical encounters.

Some patients seek second opinions on their own account. In one study of "doctor shopping," (Olsen, Kane, and Kasteler, 1976) almost one in five "doctor shopping" episodes by patients was intended to verify a diagnosis by the first doctor. The goal of second opinion plans is to reduce the financial cost of such actions, offering to pay for the cost of the visit to a second doctor.

Private insurance second opinion programs have achieved only moderate success, if any, in controlling medical care costs. One reason may be that the insurance coverage for second opinion plans changed only a few people's use of the idea (although there is no evidence on this point). However, one study shows the potential for more successful use of this concept, particularly if made a mandatory feature of the insurance plan.

An experiment with mandatory second opinions occurred in the Massachusetts Medicaid program (Gertman et al., 1980). In that program, Medicaid patients for whom certain surgical procedures were recommended were required to obtain a second opinion before proceeding to surgery. The patients were not denied surgery even if the second opinion contradicted the first recommendation for surgery, but they were required to seek the second opinion. If a conflicting opinion appeared, the patients could seek a third opinion as well. The results of this program show that mandatory second opinion plans have more capability for success than the spotty success of voluntary plans has suggested. Overall, about 10 percent of the surgery was avoided in the program, either because it was not undertaken, or because an alternative treatment was recommended and undertaken.

GATEKEEPER MODELS

Another model of insurance that has been attempted on occasion, with only limited success, is the "gatekeeper" model. In this arrangement, each patient has a primary care doctor who is responsible directly or indirectly for all patient care costs. In effect, each doctor has a "quota" of funds to use for her patient population. If a doctor refers the patient to expensive specialists who use many resources or hospitalize patients often (or if the doctor herself hospitalizes the patients often), the "quota" of funds runs out, and the doctor becomes partly responsible for subsequent costs. The most prominent of these experiments was the SafeCo Insurance Company plan for its employees in Seattle. In that plan, no measurable cost savings emerged from the arrangement (Moore, Martin, and Richardson, 1983) and it was eventually dismantled. The failure of this plan to control costs is not too surprising in retrospect—the amount of money for which the doctor was "at risk" was quite limited, producing only very small incentives for changes in practice style.

The failure of the SafeCo plan reminds us vividly of a central feature of insurance markets: Without sufficient risk pooling, one cannot "eliminate" risk, but merely transfer it from one person or corporation to another. In the SafeCo plan, the doctors confronted only a small portion of the risk arising from patients' illnesses and their treatment; the insurance plan continued to carry almost all of that risk in the larger pool (the entire plan). Indeed, one might ask how much risk a single physician practice would want to absorb. As with other individuals, physician-entrepreneurs are probably risk averse, so getting them to agree to take on a large risk (sufficiently large to make them change their style of medical practice) may be impossible, without compensating them in some way (say, with a lump-sum payment each year). The intrinsic problem with these arrangements, it appears, is that the scale of a typical physician-firm is too small to spread the risk of patients' expenses well. Given the small scale, if the insurance plan puts a lot of risk (incentive to control costs) onto the doctor's shoulders, the doctor will not want to participate. If the risk to the doctor is sufficiently reduced to induce the doctor to participate, it probably will not provide enough incentive to alter behavior. The SafeCo experience seems to fit this model reasonably well.

HOW COMMON ARE COST-CONTROL MEASURES?

Despite the apparent attractiveness of various cost-control measures, they are not yet used uniformly by health plans across the country. A recent survey of employer-based health insurance plans (HIAA, 1990) studied use of various cost-containment strategies in "conventional" health insurance plans (i.e., not HMOs, IPAs, or PPOs, which by definition employ cost-containment strategies). By 1989, the last year of their data, 65 percent of all "conventional" plans were using preadmission certification for hospitalization (i.e., the requirement that for at least some elective surgery, the doctor obtain agreement from the plan that the hospitalization was necessary). Over half (54 percent) used some form of utilization review (checking length of stay and other hospital use by doctors), 44 percent required mandatory second opinions for at least some surgeries, and 39 percent used some form of case management (tracking patients during their hospital and posthospital treatment). These data point out, more than anything, how blurred the distinction has become between "managed care" plans and "conventional" plans; even the latter group now seems firmly if not uniformly in the business of cost management.

PLURALISTIC PROVISION OF HEALTH CARE

The market force of these "cost-conscious" insurance plans appears to be increasing considerably. The experiences of recent years suggest that these insurance plans will prove very attractive to many insurance groups. In California, for example, the growth of PPOs, IPAs, and HMOs has been almost explosive during the first half of the 1980s. According to one recent survey, 80 percent of the physicians in metropolitan areas of California had signed contracts with at least one PPO, and 60 percent of rural physicians had done so. Enrollment grew in PPOs from 0.25 million subscribers in 1983 to over 6 million by mid-1986 (Melnick and Zwanziger, 1988), with national enrollment for such plans estimated between 10 and 15 million total. In addition, HMO enrollment in California had increased to over 6 million by the same time period. Thus, at least in this one state, "cost-conscious" insurance plans have well over half of the under-65 market.

In an environment like this, competition takes on quite a different role than in the "traditional" health market. Insurance plans will, in effect, do the price shopping for consumers, by striking agreements with "preferred" providers for lower prices. In turn, if successful, these plans will be able to attract more enrollees, which gives them more market power to negotiate with (and channel patients to) providers who agree to their terms.

In order for this to work well, the insurance plans must have some reasonable alternatives to which they can turn if one provider will not bargain over prices. This emphasizes the importance of competition among providers (doctors, hospitals, etc.) in order to make a second type of competition flourish—competition among insurance plans.

Each of the various types of insurance plan now available in the market—standard health insurance, PPOs, IPAs, HMOs, second opinion programs, and so on—offers advantages and disadvantages to the consumer-patient. In general, the plans trade off financial risk for "convenience" of the patient or cost to the patient in terms of higher premiums. For example, the HMO plans provide lower-cost insurance than comparable full-coverage "standard" insurance, but with less convenience (the patient must get all care from the HMO staff only). IPAs and PPOs fall somewhere between HMOs and fee-for-service plans. Thus, the health insurance market now offers an array of choices not previously available, along an entire range of "cost-consciousness." Since people will inevitably have different tastes in such matters, we can expect that no single form of such plans will dominate the market, and a pluralistic system will persist.

In such a world, competition between doctors or between hospitals will typically not manifest itself very much to the patient, but rather to the insurance plan. The competition most important to the patient will be that among competing insurance plans, or—if you will—among competing styles or organizational structures offering different ways to reduce the financial risk of illness, while controlling the incentives for too much spending on medical care that traditional insurance puts in place.

PUBLIC POLICY TOWARD PERSONS WITHOUT INSURANCE

No discussion of the health insurance market would be complete without considering the question of those who do not participate in the market—people without health insurance. Obviously, this question is similar to the question about desirability of a comprehensive national health insurance plan, an idea that has been debated and rejected several times through the past decade at the federal level. Proposed solutions for both the uninsured population (a smaller-scale problem) and a more comprehensive national health insurance (NHI) plan have included (1) mandating of insurance coverage for those who are employed, and (2) a federal insurance program (perhaps a federalized Medicaid) to provide insurance (perhaps at subsidized prices) to those who do not have insurance otherwise.

Who Does Not Have Health Insurance Currently?

The group of persons without insurance has created persistent and considerable concern in public policy circles. Several reasons for this concern always appear in the discussions. First, the uninsured seem to be heavily (but not exclusively) concentrated among the young, lower-education and lower-income segment of society.[8]

---

[8] The data on this and other demographic characteristics of the uninsured population are drawn from Swartz, 1988.

Three-quarters of the uninsured are under age 35; nearly a third of all young adults (ages 18–24) have no health insurance.

Many uninsured persons work, and hence are not eligible for Medicaid, but do not have work-group insurance offered through their place of employment. Many workers receiving the minimum wage have no insurance coverage, for example. According to Swartz's studies, of the 25 million uninsured adults (under age 65) in the United States, 15 million (60 percent) are employed, and 70 percent of those are employed full time. However, the weekly earnings of those without insurance are quite low—mostly below $250 per week (i.e., below $13,000 per year). They tend to be employed either in the service sector (housekeepers, small businesses like shoe repair), as office help in professional offices like those of lawyers, doctors, and dentists, or in agriculture.

This group of the uninsured-worker population represents the most inviting target of opportunity for lawmakers to address, and that invitation has not gone unheeded. In almost every proposal for national health insurance from the early 1970s forward, the keystone of the proposal would mandate that every employer provide some minimum-standard insurance to employees and their families.[9] Beginning in 1989, the State of Massachusetts had planned to require such coverage for all employees, and the State of Washington has a similar (but less far-reaching) bill enacted. Implementation of the Massachusetts plan, however, has been delayed repeatedly and at this writing, it appears that the idea may be scrapped.

One problem with the mandated coverage approach is highlighted in the description of those without coverage provided by Swartz (1988); if the financial cost of such plans is eventually shifted back to workers (as would be expected in most labor markets—see Mitchell and Phelps, 1976), the incomes of an already low-income group will fall even further. The other alternative, perhaps even less desirable, is that they will simply lose their jobs. This is particularly likely to occur for people working at or near the legally mandated minimum wage. For such people, the employers' costs of hiring them (inclusive of their insurance coverage) may be so high that firms find other ways to organize their production. Hiring fewer people and having them work more overtime hours offers one alternative, since the insurance policy costs

---

[9] This concept was featured in national health insurance proposals by the Nixon administration, the Carter administration, and in one proposal by Senator Edward Kennedy, whose earlier proposals centered on a uniform federal plan.

the same whether somebody works 20 or 60 hours a week. An earlier study of mandated national health insurance (Mitchell and Phelps, 1976) estimated that the unemployment problem could be considerable, particularly in some industries with low wages and little insurance coverage currently. The Massachusetts law—if eventually implemented—will provide important information on the actual consequences of this approach to dealing with the uninsured working population.

What Do the Uninsured Do When They Get Sick?

The second persistent theme in the discussion of the uninsured population is their apparent mechanism for receiving medical care when they get sick. Commonly, these people appear either at a hospital clinic or a hospital emergency room, often leading to hospitalization. Then, the hospital either transfers the patient to a public hospital (if available) or pays for the care as charity care or writes off the expenses as a bad debt. (Hospitals seem to use either method interchangeably, so most people studying the problem lump together bad debt and charity care.) Thus, the argument goes, we are paying for this care anyway, either through taxes (to support local government hospitals) or through higher hospital bills for those with insurance, so why not provide the insurance directly? Providing the insurance would reduce the financial risk of those currently uninsured, and would probably not add considerably to the total bill, since hospital use is relatively insensitive to insurance coverage (see Chapter 5).

The ability to rely on hospitals' charitable intentions to supply care to this group appears tenuous. Particularly as competition increases in the hospital sector (due in large part to the actions of cost-conscious insurance plans), the ability of hospitals to cross-subsidize the care of the uninsured patients will decay. Uncompensated care (an amalgam of charity and bad-debt care) represents only about 5 percent of all hospital bills (Sloan, Valvona, and Mullner, 1986), but the distribution of that 5 percent falls unevenly on public hospitals. The solution to this problem is not transparent; at a recent conference on the topic (Sloan, Blumstein, and Perrin, 1986), while several speakers offered various preferred strategies, little consensus emerged. Each apparent solution (e.g., mandated worker coverage, expanding Medicaid, etc.) has its own drawbacks and political opponents. For the present, we can only be sure that the problem will continue to perplex policy analysts and legislators, but that will not make it go away. For better or worse, we will continue to confront the questions posed by uninsured persons and by uncompensated hospital care for some time into the future. Only a universal health insurance plan will make the issue "go away," and our country does not seem prepared to adopt such a strategy currently.

SUMMARY

Health insurance provides a way to diversify financial risks associated with illness, but the markets for such insurance plans have grown increasingly complex. Firms selling such insurance have important distinctions: Some sell group insurance, while others sell to individuals. Nongroup plans run the risk of considerable adverse selection, which group plans (especially those organized around the work group) avoid.

Another important distinction centers on the tax status of insurers. Some plans (Blue Cross and Blue Shield for hospital and doctor services, and Delta Dental plans for dental services) are organized as not-for-profit firms, and enjoy various tax benefits that their for-profit commercial competitors do not. The importance of these and other benefits of the "Blues" remains uncertain in the light of various empirical studies of the issue.

Self-selection seems a potentially important problem in insurance markets, particularly for the over-65 population. The only randomized controlled trial relevant to the problem found no significant self-selection in favor of HMOs, but several studies of persons enrolled in Medicare HMOs found relatively low prior-year use by new enrollees, suggesting favorable self-selection for HMOs.

Group insurance, coupled with favorable tax treatment for employer-paid premiums, has made health insurance a popular but not universal fringe benefit. A growing pool of persons remains uninsured through either conventional insurance or various government programs. Many of these people are employed with wages at or near the minimum wage. Various proposals to create universal insurance in the United States typically rely on mandated coverage for workers and their families as a cornerstone of these universal insurance plans.

PROBLEMS

1. Health insurance plans have various options for controlling cost. Describe those that affect prices paid and those that affect quantities of care delivered. For each, indicate reasons (if any) that consumers of health insurance might find such an intervention (by the insurer) *undesirable*, and hence might lead an insurance company not to adopt such a cost-controlling mechanism.

2. How are "adverse selection" (in insurance markets) and "demand inducement" (in physician-service markets) related, if at all?

3. (Question relating to the Appendix for more advanced students): Think of two otherwise-identical hypothetical countries, one of which has widely eliminated contagious disease and another country where contagious disease remains a major cause of death. Where would an active market for health insurance be less likely to emerge, and why?

4. Not-for-profit health insurers have several market advantages over for-profit insurers, including the ability to avoid premium taxes (taxes that function as a sales tax does, e.g., 3 percent of a premium paid). Would you expect that this would automatically result in a lower premium paid by people buying insurance? (Hint: What else might the insurance plan do with the money? Think of the model of hospital control in Chapter 8, and extrapolate to a firm providing health insurance.)

5. Describe the "typical" person in the United States without insurance, and discuss how that person's uninsured state is quite predictable.

APPENDIX

The Statistics of an Insurance Pool

An insurance pool reduces an individual's risks by trading in that person's risky gamble for an insurance premium with much lower variance. To see this, consider a very simple insurance policy against a random risk X_i for individuals $i = 1 \ldots N$, each with mean μ and variance σ^2. (The conditions are similar when means and variances differ by person.) The total

person premium is

$$\frac{P}{N} = \sum_{i=1}^{N} \frac{1}{N} \mu = \mu$$

The average per-person premium P/N is a weighted sum of the means (μ) where the weights are $w_i = 1/N$ for each variable. The variance of the average premium is

$$\sigma\frac{2}{T} = \sum_{i=1}^{N} w_i^2 \sigma_i^2$$

if the risks are uncorrelated. When each σ_i is the same σ and each w_i equals $1/N$, this becomes

$$\sigma\frac{2}{T} = \sigma^2/N$$

since the summation occurs N times and each has a weight of $(1/N)^2$. Thus, the variance confronting the "average" person in the pool falls proportionately as N rises.

If the risks are correlated, the average variance is $\sigma^2(1 + (N - 1)\rho)/N$, where ρ is the typical correlation between any two people's risks. If $\rho = 0$ (uncorrelated), then the average variance collapses to σ^2/N. If illness were perfectly correlated across individuals, so that when one person got sick, everybody got sick, then $\rho = 1$ and the ability to spread risks is completely nullified. In that extreme case, the average variance is $\sigma^2(1 + (N - 1))/N = \sigma^2$. In real data, risks are correlated across persons only a very small amount for medical expenses.

Chapter
12

Government Provision of Health Insurance: Medicare and Its Revisions

*I*n 1965, capping decades of political and legislative turmoil, the U.S. Congress passed and President Lyndon Johnson signed a law adding Titles XVIII and XIX to the Social Security Act. Title XVIII created a universal and mandatory health insurance program for the elderly called Medicare, subsequently expanded to include persons permanently disabled for at least two years, and persons with otherwise-fatal kidney disease (so called "End-Stage Renal Disease, or ESRD) who need kidney dialysis treatments or transplants to remain alive. Title XIX created federal-state partnerships for states to establish health insurance plans for low-income people, broadly called Medicaid. These amendments to the Social Security Act created several important changes in our society, both politically and economically. As a political step, they changed the role of the U.S. government in the provision and control of health care, increasing its scope and presence considerably.

The economic consequences of Medicare and Medicaid were only poorly understood initially, in part because no clear picture existed about the effects of insurance on demand for medical care at the time the law was under discussion (1964 and 1965). Thus, when Medicare and Medicaid actually took effect in July 1966, nobody had a really clear ability to predict what would happen. In retrospect, almost nobody understood the effects on demand in the short run or the long-run effects on demand for technology.

As time passed, the growth of costs in Medicare and Medicaid became *the* dominant issue in federal health policy discussions. Before Medicare and Medicaid had reached their twentieth birthdays, their structures had changed importantly to limit dollar out-flows from the Federal treasury. These changes have, in turn, spawned changes in the structure of private insurance.

The first changes (instituted during the Nixon administration's price controls) attempted to control costs by changing the fees doctors and hospitals received for each procedure. Later changes were more radical, shifting the payment base for hospital care from the smallest of procedures the hospital billed for a single payment for each admission. As we shall see, this shift in payment precipi-tated large structural changes in the health care delivery system in the United States. Some states even shifted their Medicaid sys-tems to a capitation system, in which providers received a single payment for a year's worth of care for a Medicaid recipient.

To understand the effects of Medicare and Medicaid and the importance of some of the changes in their structures, we first need to develop a clear picture of how these plans were originally struc-tured, and then we can portray the changes in their structures through time. We turn in this chapter to an analysis of Medicare, and then in the next for Medicaid.[1]

THE MEDICARE PROGRAM

Initial Structure

When Medicare was originally put in place, its designers had two options to which they could turn for comparison—the private U.S. health insurance market, and foreign health care systems. Most foreign systems involve a degree of nationalization (e.g., govern-ment ownership of hospitals) that probably would have proven politically unacceptable in the United States.[2] Instead, Medicare's designers turned unabashedly to the private health insurance in force during that period for a design template.

---

[1] The Health Care Financing Administration, the agency that operates Medicare and the federal activity in Medicaid, issues periodic reports on the Medicare and Medicaid systems, known in its fifth and most recent version as *Health Care Financing Program Statistics: Medicare and Medicaid Data Book, 1988*, published in 1990. This document offers an excellent description of the programs, and provides a wealth of data on the programs' history and current operation.

[2] Since the administration of Franklin Roosevelt in the 1930s, some people had actively sought a comprehensive national health insurance (NHI) plan. The closest the United States had come to this probably occurred during the Truman administration in 1948–1952, but NHI proponents were unsuccessful then, and their political advantage waned during the postwar years of prosperity during the Eisenhower administration (1953–1960).

The template had a fairly standard appearance: Hospital care would receive highest priority, with an emphasis on first-dollar coverage. Physician care would be paid under a "major medical"concept, with an initial deductible to accrue annually, a coinsurance provision to help control costs, and a "usual, customary, and reasonable" type of fee schedule to set provider fees.

The Medicare program had two parts: Part A paid for services provided by hospitals. Enrollment in Part A was mandatory for every person receiving Social Security benefits (including those over age 65, plus "permanently disabled" persons). The entire cost of Part A is paid by the Medicare Trust Fund, a separate government account funded by additions to each worker's current-year Social Security tax.

Part B, called "Supplemental Medical Insurance," was made voluntary, although the premiums paid by enrollees were so low that enrollment has been very nearly 100 percent from the beginning of the program. At the beginning, the program's planners had little idea how much the program would actually cost, but they intended the Part B premium charged to enrollees to cover about half of the Part B program costs, in order to get the largest possible enrollment. At present, the premium to enroll in Part B is set at one-quarter of the anticipated cost. This amounted to $17.90 per month in 1987, for example.

Hospital Coverage   At the time of Medicare's design, the predominant insurance for hospital care was through Blue Cross, which had historically offered 30, 60, or 90 days of coverage for hospital care, and nothing thereafter. Insurance sold by for-profit "commercial" insurance companies had also grown considerably, so by 1965, the group insurance market was nearly evenly split between the Blues and commercial insurance. The "model" set by the Blues still held considerable sway, and Medicare adopted the Blue approach directly.

In its initial design, Medicare paid for hospital care in the following way *for each episode of illness* of the patient:[3]

First day: Patient pays U.S. average cost of hospital care

---

[3] Episodes of illness begin when the patient enters the hospital, and ends when the patient is discharged. If the patient is readmitted within a week of discharge, Medicare rules consider it the same episode of illness. This prevents doctors and patients from "gaming" the Medicare system to eliminate copayments, e.g., by discharging a patient after 59 days, and readmitting him the next day..

Days 2–60: Medicare pays 100 percent of hospital charges

Days 61–90: Patient pays 25 percent of average U.S. cost per day, Medicare pays the remainder

Days 91–150: Patient pays 50 percent of average U.S. cost per day, drawing down a "lifetime reserve" of 60 days

Days 151 and beyond: Medicare pays nothing

The first-day "deductible" cost for hospitalization tracked directly with average hospital costs until 1987, at which point the Congress set it at $520, and established a new formula to update it.

From the viewpoint of standard economic models of insurance demand, this structure of hospital insurance is completely "upside down," since it provides nearly "first-dollar" coverage, but offers no serious protection against "catastrophic" (very large but unusual) financial risks. (Review the discussions on risk aversion in Chapter 10 if the logic behind this statement is not apparent to you.) We will discuss the two responses to this risk—private supplemental insurance and (later) the provision of catastrophic coverage by Medicare—later in this chapter.

Physician Care

The Part B coverage of Medicare actually includes a much broader range of services than physician care, but the great majority of Part B benefit payments occur for physician services. Medicare planners had two prototypes available within the private health insurance market in the United States after which they could model Part B. One model would have used the traditional Blue Shield approach, providing nearly full coverage for physician services, but only for hospitalized patients. The alternative model came from the "major medical" insurance sold mostly by commercial carriers. Under this approach, essentially all physician services would be covered (both in and out of the hospital), but the participants would have an annual deductible to meet before coverage began, and a coinsurance rate for all services. The Medicare Part B program closely followed the "major medical" model.

Under the structure of the initial Part B plan, Medicare paid for doctor services (as well as some prescription drugs, medical appliances such as wheelchairs and crutches, etc.) with a very traditional insurance model, encompassing a $50 annual deductible, a 20 percent coinsurance rate, and a fee schedule by which providers would be paid. Congress increased the annual deductible later to $60, and then to $75 in 1982.[4]

ECONOMIC CONSEQUENCES OF MEDICARE'S STRUCTURE

Risk to Medicare Enrollees

The old-style Medicare program created substantial financial risk for Medicare enrollees, particularly through some of the features of the Part A hospital coverage. We next discuss the nature of this risk, the private response to it (supplemental insurance), and finally, the policy change in Medicare itself (in 1988) to eliminate almost all of the risk of the old program's structure.

Nature of Risk

Recall from Chapter 10 that a key element driving the demand for insurance is the variability of financial risk confronting an individual. The old structure of Medicare, it turns out, left remaining both some moderately large and some very large financial risks, instead of removing them through the mechanisms of insurance.

We can summarize the sources of this risk by looking at the variance in spending confronting individuals within the Medicare program (Phelps and Reisinger, 1988). Table 12.1 shows these variances by characteristics of the program. Not surprisingly, the biggest sources of risk came from the failure to cover infrequent but highly expensive financial events—very long hospital stays.

Ironically, one feature of Medicare's original design that was possibly intended to *protect* patients from financial risk actually created the greatest possible risk—the "spell-of-illness" concept. Under the spell-of-illness concept, if a patient was discharged from the hospital, and then readmitted within seven days, the readmission was considered as part of the same hospitalization. This meant that the patient did not need to pay the first-day deductible for the new hospitalization. It had an important side effect, however. Patients with multiple hospitalizations were much more likely to

---

[4] The inflation-adjusted deductible that would equate to the original 1965 level of $50 per year would now be near $200. Thus, the "real" deductible has fallen substantially through time. Review the RAND HIS results concerning the effects of a deductible on health care use from Chapter 5.

Table 12.1   SUMMARY OF RISK CONFRONTING MEDICARE PARTICIPANTS
(STANDARDIZED TO 1987)

Medicare feature	Mean	Variance	Risk premium
Deductible Part A	$168	$183,500	$18
Coinsurance days[a] Part A	$19	$53,000	$5
Lifetime reserve[b] Part A	$18	Unknown	Unknown
Beyond coverage hospital days	Unknown	Unknown	Unknown
Part B (overall)	$200	$130,000	$13
Uncovered services	—	Unknown	—

[a] This risk represents the 1983 distribution of copay days, evaluated at 1987 prices. The risk of incurring a copay day fell considerably subsequent to 1983, but no data exist yet on the distribution of the risk, and hence we cannot calculate the variance.
[b] This is the dollar value of an average of 0.063 LR days per enrollee in 1983, at $286 per day. In 1984 the average LR days fell to 0.037 per enrollee, or about $11 in average risk.
Source: Reprinted from Phelps and Reisinger, 1988, Table 5.8.

spill into the realm where Medicare paid poorly, or not at all, for hospital care—that is, hospital stays in excess of 60 or 90 days. Simulations of patients' risk shown in the study by Phelps and Reisinger demonstrate well that the net effect of the spell-of-illness concept was a financial catastrophe for some people, because it greatly increased the chances of "long" hospitalizations.

As summarized by the risk premiums people would be willing to pay to remove this risk (according to the model that people seek to maximize expected utility), providing catastrophic insurance to remove this source of risk would represent an excellent change in the structure of Medicare. As we shall see shortly, a change in the Medicare law in 1988 did just that, if only temporarily.

Private Supplemental Insurance as a Response to Risk

Perhaps at least in part in response to the risk left uninsured by the structure of Medicare, a large fraction of the over-65 population purchased private health insurance to augment the Medicare coverage. This insurance covered Part A deductibles, Part B deductibles and coinsurance, and (less frequently) covered the high-risk "tails" in the distribution of hospital care that create the biggest amounts of uninsured risk under Medicare.

Table 12.2 shows the key data, from a household survey in 1977. The unusual thing about the insurance coverage selected by these people is that it seems "upside down" from the point of view

Table 12.2 PRIVATE INSURANCE COVERAGE OF THE MEDICAL POPULATION: PERCENT OF MEDICARE BENEFICIARIES COVERED BY AGE, EDUCATION, INCOME, HEALTH STATUS, 1977

	Any insurance	Hospital care	Inpatient physician care	Outpatient diagnosis	Physician office visit	Mental health care	Outpatient drugs	SNF care	Dental care
Total	14,518	97.6	88.4	64.5	61.6	58.0	40.6	60.8	4.1
Age									
65–69	5,244	97.9	88.6	68.4	65.7	63.1	48.0	58.3	5.2
70–74	4,131	97.9	90.8	65.0	62.3	57.2	41.0	62.0	5.5
75+	5,142	97.1	86.2	60.1	56.8	53.5	32.8	62.4	1.8[a]
Education									
<9 years	5,355	96.5	85.0	61.9	56.8	51.2	33.5	61.0	4.7
9–12 years	5,621	97.8	91.0	66.4	63.2	60.5	43.3	60.2	3.9[a]
13+ years	3,051	98.9	89.5	65.1	66.8	65.6	47.5	61.2	3.4[a]
Income									
Low	5,749	99.1	87.3	60.5	57.1	49.3	30.2	59.8	2.9[a]
Middle	5,197	95.7	88.7	68.0	63.1	63.5	46.6	62.0	4.8
High	3,572	97.9	89.7	65.8	66.5	64.1	48.6	60.8	5.0[a]
Health status									
Excellent	4,602	97.2	87.1	60.9	58.9	58.6	39.2	59.7	3.8[a]
Good	5,808	97.8	89.4	66.8	63.2	57.0	41.7	62.0	4.6
Fair	3,734	97.6	88.4	65.2	61.0	59.1	39.8	59.7	3.4

[a] Relative standard error > 30 percent.
Note: Income adjusted for family size.
Source: Reprinted from Cafferata (1984), Table 3.

of the expected utility theory. As Arrow first proved in 1963, a risk-averse person seeking to maximize expected utility, when offered the opportunity to purchase insurance with a positive loading fee, will always insure the biggest risks (in actual magnitude of the event), but will be willing to have a deductible so that the insurance does not cover "small"events. Exactly the opposite happens in the actual purchases of supplemental insurance.

Why the insurance plans people selected look like this remains somewhat of a mystery.[5] Either the standard model does a poor job of describing how people make choices about insurance, or the elderly are poorly informed about the nature of the risks they confront, or both. The implications of this for public policy seem considerable: The premise that catastrophic protection is "a good idea" rests primarily on the expected utility model. If people really want something different, then any attempts to impose catastrophic coverage in Medicare may be the wrong policy. Alternatively, if people would make this sort of choice when fully informed about the risks, then public policy should emphasize provision of appropriate information to Medicare enrollees and their families. Restructuring or eliminating the provisions of the Baucus Amendment would also follow, since that amendment to the Medicare law in many ways forces supplemental insurance companies to offer the "wrong" coverage, according to the expected utility model. Thus, resolution of this apparently arcane intellectual issue in the economics of uncertainty has some important real-world implications that follow.

CHANGES IN MEDICARE

Since its inception, Medicare has undergone numerous changes, almost all of which have the obvious intention of reducing Medicare outlays. (The since-repealed catastrophic insurance act of 1988 offers an important counterexample.) The changes in Medicare through time included the following (with the year these changes appeared):

---

[5] Regulations now require this sort of structure, but these plans primarily had this structure even before regulations. Section 1882 of Title 18 of the Social Security Act, put in place by the so-called Baucus Amendment (named after the congressman who sponsored the legislation) required minimum coverage standards for a policy to use the name "medigap." Conformity with these standards would force the same sort of "upside down" structure as the data in Table 12.2 show. However, the Baucus Amendment became law after the 1977 survey upon which Table 12.2 is based.

- elimination of most hospital copayments and some Part B copayments (1988), restored (1989)
- fee limitations for doctors (1972)
- per diem limits set on hospital reimbursements (1972)
- changes in the ways doctors can charge higher than the "allowed fees" (1983)
- shifts from per-service to per-admission payment for hospitals (October 1983)
- allowing Medicare participants to enroll in HMOs as an alternative to standard Medicare coverage (1972) and on a risk-sharing basis (1985)

In addition, several other changes in physician payment are pending for 1992 implementation, including a dramatic shift in the relative rates of payment across various medical and surgical specialties, and perhaps even a shift to per-year payment to physicians (capitation) rather than per-service payments.

Catastrophic Protection Act of 1988

Although the "Catastrophic Protection Act" was among the most recent changes in Medicare, it makes good sense to discuss it at this point, just after we have completed the discussion of financial risk and private supplemental insurance that the old-style Medicare program created. This change is all the more fascinating as a study of the political economy of health care financing, since the change (which made eminent sense from a standard economic perspective) was repealed after major citizen protest in 1989.

This major change occurred when the Congress passed the Medicare Catastrophic Coverage Act (MCCA). This changed several parts of the Medicare program importantly: (1) Hospital care was to become fully paid after 1990, except for the first-day deductible, which would be paid only once per year, even if the patient had multiple hospitalizations. This also eliminated the spell-of-illness concept and the copayment risk associated with that feature of the Medicare program. (2) Out-of-pocket cost sharing under Part B would have caps installed, beginning with a $1370 cap in 1990. This cap was designed to benefit the 7 percent of enrollees with the highest out-of-pocket costs, and had scheduled increases up to $1900 in the year 1993. Persons with expenses above the cap would have full-coverage protection. (3) Medicare would provide catastrophic drug expense protection. In 1991, when this plan was to become fully operational, persons with expenses above $600 annually would have had half their expenses paid. By 1993, the deductible would rise to $710, but the coinsurance would fall to 20 percent for the year 1993 and beyond.

The qualitative effects of this change in Medicare's structure are not difficult to predict, but the magnitude of those effects is more difficult to measure. Elimination of the risk under Part A to enrollees would have had considerable economic value, and the increase in hospital use due to this change probably would be small, because of the supply-side pressures to limit long-term hospital use as embedded in the PPS (DRG) program.

The catastrophic cap provisions of the Part B program were designed to affect only the 7 percent of recipients with greatest Part B expenditures. However, because of the skewness in medical expenses in general, a much greater fraction of all medical activity within Part B would have had the copayments removed. We should expect that Part B spending would have increased notably in response.

To estimate the effects of this change in Part B insurance coverage, we would need to know the proportion of all Part B spending that now sits above the cap's level, because this spending would have become subject to full coverage, rather than the current Medicare copayments and fee schedule limits. Recent data show that this upper 7 percent group accounted for over 40 percent of Part B reimbursements, and their out-of-pocket share (average copayment) averaged 30 percent (Ruther and Helbing, 1985, Table 2). For that group, the coinsurance would fall to zero. If their response is similar to the RAND HIS results, we could have expected their spending to increase by perhaps a third or more (see Chapter 5) once they exceed the catastrophic cap. Their hospital use may also rise. Thus, we could expect the overall use of medical services under Part B to have increased by over 10 percent, and Medicare's outlays under Part B to have increased by at least 20 percent, or approximately $12 billion annually.[6] This estimate

---

[6] The prediction is somewhat hazardous, because we don't know much about the episodes of illness of the people who will exceed the catastrophic limits of part B. The "worst case" for predicting new Medicare outlays assumes that all of this expense comes in a single episode, and that people anticipate the total correctly. Then, spending could rise by as much as 20 percent. The calculation works like this: The outpatient expense for full-coverage enrollees on the HIS was 32 percent higher than for the 25 percent copayment group. If this same differential applies to the 41 percent of Medicare spending that the catastrophic cap will now cover, then overall spending should increase by $0.32 \times 0.41 = 13$ percent. If this spending comes in multiple episodes, then the spending on earlier (below-cap) episodes will be unaffected, and the increased use will not be as large. (*Footnote continues.*)

contrasts starkly with Congressional Budget Office official estimates that the incremental effect of the catastrophic insurance
would cost Part B only $4 billion annually in increased outlays.

A Political Battle The MCCA generated a remarkable degree of
political hostility once it was passed, even before the provisions
went into effect. A key element, it seems, was the nature of the
financing of the change in benefits. As noted previously, almost all
of the financing of Medicare's Part A and Part B came either from
general tax funds (for half to three-quarters of the Part B funds) or
from a special payroll tax levied on working Americans (for Part A).
The MCCA changed all of that: The Congress voted in the insurance provisions of the Catastrophic Care Act, and they also imposed on the over-65 population almost the entire financial cost of
this program. Three-eighths of the new program's cost came from
direct premium payments by beneficiaries, equal for every Medicare enrollee. The remaining five-eighths of the program's cost
came from a special income-tax supplement for Medicare enrollees who have at least $150 in annual federal tax liability. This extra
tax amounted to a 15 percent surtax in 1989 that would have increased to a 28 percent surtax in 1993 as the benefits were phased
into place.

Some very high-stakes political pressure began almost immediately after the act's passage in 1988, causing the Congress to
rethink its previous action. The dispute centered on two ideas:
(1) Is the increase in insurance coverage a good idea? (2) Should
Medicare beneficiaries themselves pay for the program's entire
cost, or should a broader financing base be used, as has been
historically true for all other parts of the Medicare program? The
former question is one economists would characterize as an "efficiency" question, and the latter as a "redistributional" issue. Most
of the political pressure has centered on the latter question. The
new financing approach would have represented a substantial
change in the political philosophy of Medicare, since it would, for

---

In addition, Medicare would pay the 30 percent copayments that users had previously
paid. Although firm numbers do not exist, it appears that Medicare would cover at least
two-thirds of the total Part B expenses for this group using catastrophic coverage. Thus,
outlays by Medicare would rise by $0.3 \times 0.41 \times (2/3) = 8$ percent due to the shift of payment from patients to Medicare.

Combining these two, we can predict Medicare's Part B expense will rise by at least 20
percent (13 percent + 8 percent = 21 percent).

the first time, create a benefit that did not transfer wealth from the currently working population to the retired population.

The new financing mechanism, we can now see in retrospect, doomed the Catastrophic Protection Act. A massive political response, mostly from the elderly citizens who would have paid the highest new taxes, forced the Congress to reconsider, and finally repeal the MCCA in 1989.[7]

Limitations on Fees

We turn now to a completely separate change in the Medicare law. Remember from Chapter 7 that physicians commonly charge different prices for the same service, so that there is a generally broad dispersion of fees prevailing in a single geographic area. This dispersion creates a problem for insurance carriers—what is the "right" price to pay for (say) a doctor office visit under the insurance plan. The problem becomes more acute when one realizes that the dispersion in prices *might* be related to quality, and it also *might* be related directly to monopolistic or monopolistically competitive pricing practices.

At its inception in 1965, Medicare paid for doctor services under Part B using a technique following directly out of private insurance ideas—a "usual, customary and reasonable" concept of private insurance carriers. Medicare's terminology uses the terms *prevailing* and *customary*. *Customary* means the fee at the 50th percentile of the distribution of fees *each* doctor charges, so that half of each doctor's fees lies above and half below that doctor's customary fee for a given procedure. *Prevailing* is a concept that locates each doctor's fee within the overall fee distribution in the community. Medicare uses the 75th percentile in the distribution of fees within the community to define "prevailing." If the doctor's customary fee exceeds the prevailing fee for the same service, Medicare pays only the prevailing fee. This fee screen, which by definition affects 25 percent of all doctors' bills, constituted essentially the only limit imposed by Medicare on payments to physicians in its initial structure.

---

[7] One subtle issue bears mentioning. The new tax imposed only on the over-65 segment of the Medicare population, while the benefit was extended to all participants. The exclusion from obligation to pay premiums of about 1 million people covered on Medicare through permanent disability and the End Stage Renal Disease program created a special political uproar. Some of the political advertisements centered on a special population potentially eligible for the permanent disability program—those with AIDS.

All of this hands-off policy changed abruptly in 1971, when President Nixon instituted some economywide price controls, beginning with a 90-day freeze on all prices and wages on August 15 of that year, followed by a series of controls that limited the rates at which prices could increase. While these controls were soon lifted for the remainder of the economy, two sectors—petroleum and health care—remained controlled, with ongoing limits on the rate at which fees could increase.

The Nixon era price controls generally allowed physicians' fees to increase from their historic level (which varied, remember, from physician to physician) at a rate determined by a "cost index" designed (if not perfectly) to reflect the costs of a physician-firm's business. More specifically, changes in the "prevailing fee" screen in each region of the country were limited beginning in 1972 to changes in the Medicare Economic Index, a weighted average of overall inflation, changes in the costs of physicians' practices, and changes in the overall earnings level of the country.

A second change occurred in 1984 as a part of the Deficit Reduction Act. Physicians' fees were frozen again until the beginning of 1987, at which time a complicated series of rules allowed some doctors to increase their fees somewhat, while placing very strict limits on others. The important distinction was whether the doctor had chosen to become a "Medicare participating physician," a topic we discuss in the next section.

The fee schedule arrangements under Medicare produce additional financial risk for patients, but they also increase the incentives for patients to undertake comparison shopping. How much they actually do shop still remains unknown, but probably not much, given the dispersion in prices that Medicare's data reveal.

Medicare Assignment of Benefits

The economic consequences of fee limitations depend in part on another apparently arcane part of Medicare law, know as "assignment of benefits." Medicare pays for 80 percent of the doctor's fee *up to* a schedule of "allowable" fees determined for each region of the country. If any doctor's fee is higher than that allowed fee, the effective insurance coverage can fall. For example, if the allowed fee is $80, Medicare Part B would pay $64. If the doctor's actual fee is $100, Part B will still only pay $64. (If the doctor's fee is $50, Part B will pay $40.)

Physicians have had the option of billing the patient for any balance above the Medicare payment (called balance-billing). However, if they do this, they must collect the *entire* fee from the patient, and the patient submits the charge to Medicare for reimbursement. Alas for the doctor, some patients do not pay their bills. The physician-firm can circumvent much of this risk by taking "assignment of benefits," whereby the patient signs over the payment of the Medicare fee to the physician-firm. Thus, the doctor is virtually guaranteed the payment of the 80 percent share of the Medicare-allowed fee, and must try to collect only the remaining 20 percent from the patient.

Physicians must balance two economic forces in making this choice. If they accept assignment, they may be forced to accept a lower total fee, but they collect it with higher probability. If they do not accept assignment of benefits, they can charge higher fees but may not be able to collect all of them.

Physicians should be more willing to accept assignment the closer their own fees are to the allowed fee from Medicare. Indeed, if the doctor's fee is lower than the Medicare-allowed fee, the doctor can lose nothing by accepting assignment. In addition, the doctor should be more eager to accept assignment of benefits if the chances of default by the patient are higher. One study of physician acceptance of assignment of benefits found all of these phenomena actually occurring. Physician-firms accepted assignment of benefits less often when patients were deemed less likely to default (higher income *for the patients in each firm's practice*, less unemployment in the area), and less likely to accept assignment, the greater the amount by which the fee they would normally charge exceeded the Medicare allowed fee (Rodgers and Muscaccio, 1983). The proportion of claims for which doctors accepted assignment varied over time and space, but on average, it would not be misleading to say that about half of the claims were "assignment of benefits" and half were direct billing to patients.

Beginning in late 1983, Medicare introduced the concept of "participating physician," which altered the way assignment of benefits takes place. Before this change, physicians had to decide on a case-by-case basis whether to accept assignment or not. If they designate themselves as participating physicians, however, they *automatically* accept assignment of benefits for all patients. Of course, if they don't become participating physicians, they still have the option of accepting assignment on a case-by-case basis.

A substantial majority of physicians chose to become participating physicians, because of an important carrot-stick arrangement built into the program. The government had earlier frozen the fee structure of physicians (i.e., locked into place their profile of allowed fees). For physicians who did elect to become participating physicians, Medicare increased the allowed fees at a rate faster than that for nonparticipating physicians. The incentive worked: The assignment rate (including those from participating physicians), which had hovered in the 50 to 55 percent range for years, jumped to 69 percent of all claims submitted by doctors the year the rule was changed.

The Prospective Payment System (PPS)

Beginning in October 1983, Medicare radically changed the way it paid hospitals for care delivered to patients.[8] The "old" system had paid hospitals on the basis of activities performed for each patient: An hour of operating room time, five physical therapy visits, 20 doses of an antibiotic, a disposable enema kit, or a day in an intensive care unit each had a price that appeared on the patient's bill.[9] Additional days in the hospital each received an additional room and board charge, plus all of the associated physician billings for those days. While average hospital length of stay declined systematically after the inception of Medicare (from about 13 days in 1965 to under 10 days in 1982), almost certainly due to technical progress in medical science, the change in Medicare's structure initiated in FY 1984 abruptly altered the incentives confronting U.S. hospitals.

The new approach, phased in over a four-year span, began to pay hospitals on a per-case basis, rather than on a per-item or per-service basis.[10] For the first time, doctors and hospitals treat-

---

[8] The federal government operates on a fiscal year that begins in October of the previous year. Thus, FY 1984 was the first year of the PPS system, although it began in the autumn of 1983. This introduces some possibility for confusion when interpreting time-series data, because some will show calendar years (for which the last quarter in 1983 has PPS in effect) and some will show fiscal years, which are "pure" for purposes of evaluating PPS.

[9] Medicare paid according to a formula known as "Ratio of Costs to Charges Applied to Charges" (RCCAC) that formed the actual basis of payment. However, the key idea was that each item or service used led to an additional billing by the hospital.

[10] During FY 1984, 25 percent of the hospital's payment was based on the national (PPS) norm, and the remaining on hospital-specific prior-year cost rates, adjusted as allowed by government rules. In FY 1985, the mix went to 50 percent hospital and 50 percent PPS, and in the next year, the mix shifted to 75/25. By November, 1987, seven weeks into FY 1987, the system was 100 percent on the PPS system.

ing individual patients were confronted with a fixed budget for the care of each patient.

We can put this type of payment in perspective by returning to the models used in Chapters 8 and 9 about hospital behavior. Think for a moment about a hospital treating only Medicare patients. In the old-style insurance, Medicare patients would exhibit a demand curve that was nearly price insensitive, and hence quite "steep" in appearance in diagrams like D_2 in Figure 10.4. The PPS essentially offered a fixed price per hospital episode to the hospital, which we would draw in a figure like Figure 9.2 as a flat line, cutting across the price-quantity diagram just as we would normally draw a market price confronting a price-taking firm. In other words, the Medicare PPS system makes the demand curves confronting the hospital very price-elastic, in the sense that the hospital can get "all the business it wants" at the price offered by Medicare, but no business at a higher price. The hospital, as the model in Chapter 9 describes, would pick the quality such that the demand curve was tangent to an AC curve, and this would define the quality of the hospital's offering. Of course, if the demand curve is completely price-elastic (i.e., a flat line), then that tangency would only occur at the point of minimum average cost. By setting prices in this way, the PPS system at least potentially forces hospitals to operate efficiently (at minimum AC), and it determines the quality of care by the level of price offered.

The prices actually offered by Medicare come from a system known as Diagnosis-Related Groups (DRGs), wherein each patient admitted to the hospital is assigned to one of a fixed number of groups (the current list has 470 DRGs). The payment for a patient within any DRG varies somewhat by region and hospital type (e.g., teaching or nonteaching), but from the hospital's point of view, the price per *admission* is fixed. The hospital receives the same revenue from Medicare no matter what is done to the patient during the hospital admission, or how long the patient stays in the hospital.[11]

The physician who admits the patient to the hospital is now put in a quite different position than under the previous payment scheme. Before, the doctor could undertake any procedure or pre-

---

[11] This is not literally correct. A patient who stays in the hospital an extremely long time, relative to the average for patients within the DRG, becomes an "outlier," and the hospital begins to get paid on a basis quite similar to the old per-service system for continued care after that point. While this outlier system is an important component of the DRG system from the hospital's point of view, it represents a complication that we need not focus on here.

scribe any service or device that might benefit the patient. The hospital would get paid (by Part A), the doctor would get paid (by Part B), and the patient would feel that "everything possible" was done to effect a cure. Under the new arrangement, the hospital stands to lose money, perhaps a large amount of money, if the doctor continues treating patients in the old style, since each resource used costs the hospital money, but the added revenue under the new plan is zero.

Length-of-Stay Effects Hospital medical staffs in this situation face a problem similar to that of the "group practice" first described by Newhouse (1973)—each doctor's attempt to save money brings only a small proportional reward. However, the medical staff of a hospital somehow *must* respond to the incentives generated by Medicare's new DRG system, or the hospital runs the risk of going broke, particularly if it has been a relatively expensive hospital to operate.[12]

The most obvious dimension of adjustment for the hospital and its doctors is to shorten length of stay for patients. Length of stay is easy to monitor, providing something that committees of doctors can review readily. Guidelines on length of stay are easy to develop and interpret. Doctors who systematically deviate from such guidelines could come under pressure from other members of the medical staff to respond.

All such efforts would be fruitless, of course, if doctors had no discretion in length of stay. However, we know from overall hospital use data that length of stay differs substantially by region. (See the map in Box 3.2 showing the regional variability in LOS.) Studies also show how the length-of-stay decision differs by individual doctor, even after controlling for case mix and (at least partly) patients' illness severity (see discussion of study by Phelps, Weber, and Parente in Chapter 3). Thus, we have good reason to believe that *if the hospital's medical staff can find a way to coordinate doctors' behavior*, length of stay could fall in response to the DRG incentive.

Apparently, hospitals were successful in achieving such coordinated effort. Data from the Health Care Financing Administration (HCFA, pronounced "hick-fa" by almost everybody) show the

---

[12] Actually, even if the hospital has been "doing well" under the DRG system, it can still make more money by reducing LOS and ancillary expenses. These "profits" would allow the hospital to advance its other goals.

remarkable declines in LOS among Medicare recipients. Without doubt, the DRG system under PPS was extremely successful in its goal of reducing hospital lengths of stay, as the data in Table 12.3 demonstrate.

To put these changes in proper context, recall that the PPS system came in with a phased plan, whereby in the first year, only 25 percent of a hospital's costs were determined prospectively, and 75 percent retrospectively (i.e., based on the hospital's own historic costs). This phase-in period essentially ended at the end of FY 1986, so FY 1987 represents a year when the incentives no longer were growing tighter.

One study of Medicare spending in both Part A (hospital care) and Part B (physician services) found that government spending had fallen in 1990 in Part A by $18 billion, about a 20 percent savings over what would have otherwise been anticipated, with no offset in Part B spending (Russell and Manning, 1989).

The changes in LOS are all the more remarkable when one understands that a considerable volume of medical and (particularly) surgical activity that had previously been done within the hospital shifted to the outpatient setting, with a fast-rising trend in

Table 12.3   TRENDS IN LOS FOR MEDICARE PATIENTS IN SHORT-STAY HOSPITALS

	LOS	Percent change from previous year
Calendar Years		
1967	13.8	—
1970	13.0	−3.8%
1975	11.2	−2.6%
1980	10.6	−0.9%
Fiscal Years		
1981	10.5	−0.9%
1982	10.3	−1.9%
1983[a]	10.0	−2.9%
1984[b]	9.1	−9.0%
1985	8.6	−5.5%
1986	8.3	−3.5%
1987[c]	8.5	2.4%
1988	8.5	0

[a] Tax Equity and Fiscal Reform Act (TEFRA) rules in place.
[b] Phase-in of Prospective Payment System (PPS) begins.
[c] Phase-in of PPS complete.

"dry-cleaner" surgery.[13] Hospital admissions fell at an annual rate of 2.5 percent for the over-65, so by 1988, a mere five years after the DRG system was installed, overall Medicare admissions to the hospital had declined by 13 percent. A large fraction (perhaps over half) of these reduced admissions simply meant that a surgical procedure had been shifted to an ambulatory surgical center (ASC), a new technology experiencing phenomenal growth in the 1980s.[14] The ASCs obviously pull out of the hospital patients who would otherwise have had a very short length of stay. This means that the average LOS of those remaining patients has fallen considerably more than the Medicare aggregate data show.[15] In fact, this quite likely accounts for the growth in LOS in FY 1987 and 1988 shown in Table 12.3.

Use of Other Services   One reason why doctors were able to adjust the LOS for their patients is because the process of producing "health" allows for substitution. In terms of hospital length of stay, for example, one can substitute more intense treatment within the hospital, and one can also use nonhospital facilities as an alternative site of care. This clearly mattered considerably in reduced LOS in Medicare patients. In the first three years of the Medicare PPS system, use of skilled nursing facilities (SNFs, pronounced "sniffs" among the *cognoscenti*), a standard Medicare benefit, increased by 5.2 percent, 12.8 percent, and 5 percent, respectively, for a combined increase of 25 percent. Use of home health care, another standard benefit, increased by even more during the same period.

Finally, Medicare added a temporary benefit for hospice care in 1983, made permanent in 1986. This provides a low-technology "caring" environment for dying patients as an alternative to dying in the intensive care unit of a hospital (at the other extreme). Hospices emphasize supportive services such as home care, pain

---

[13] "In by 9:00, out by 5:00."

[14] In 1983, such centers performed about 375,000 surgical procedures. In 1988, they performed over 1.75 million, and projections for 1990 exceed 2.5 million procedures. These data include both Medicare and non-Medicare patients, but a considerable bulk of them are Medicare patients, because many of the most popular procedures to shift into the ASC include cataract surgery, hernia repair, and knee operations, all of which are much more common in the Medicare population than in the under-65 population.

[15] Some simple calculations suggest that the decline in LOS has been as much as 20 to 25 percent for the remaining Medicare patients. Problem 1 at the end of this chapter explores the mathematics of this problem.

control, and psychological, social, and spiritual services, rather than attempts to cure the patient. The annual number of patients using hospices under Medicare funding has increased to over 90,000 per year, and now appears to represent about 1 in every 16 Medicare persons who die.[16]

Sicker and Quicker?   An obvious issue in this program is the potential effect on patients' health. Numerous opponents of the PPS system implemented by Medicare made dire predictions about deleterious effects on patients' health outcomes. We have substantial reason to believe in advance that such concerns were not valid—the wide variations in length of stay across different regions of the country did not create any obvious differences in longevity or other measures of health, for example. Nevertheless, the government-mandated commission set up to monitor the effects of the PPS system, known as the Prospective Payment Assessment Commission (or in short form, ProPAC), had spent considerable effort documenting what effects, if any, the shorter stays generated by PPS might have on patients' health.

One indicator of poor patient outcomes is the rate at which patients get readmitted to the hospital.[17] The rate of readmission to the hospital within 30 days, a standard indicator of "poor outcomes," is about the same under PPS as in the previous five years. Readmission rates had been increasing systematically through time, in part due to the increasing average age of the Medicare population. However, the rate of increase slowed during PPS years, indicating that the incentives to reduce LOS had not contributed notably to any behavior that would cause readmissions to increase.

Another indicator of poor patient outcomes is mortality. Particularly because of the incentives to discharge patients early, a common measure of mortality is the death rate within 30 days of a hospital discharge. This comparison is not so obvious as it might seem, because two conflicting forces have altered mortality rates

---

[16] About 1.5 million people in Medicare's population die each year, about half in the hospital and half elsewhere (including private homes, nursing homes, and hospices). Thus, the 90,000 hospice users constitute about 6 percent, or one out of 16 dying persons.

[17] Recall that the hospital cannot benefit financially by discharging patients and then readmitting them to collect a new DRG payment: If the readmission takes place within 7 days, it is counted as the same spell of illness, and hence the hospital receives no new DRG payment.

through time. Improved technology has reduced the overall mortality among the Medicare (elderly) population from 6.6 percent (at the beginning of the program in the late 1960s) to 5.1 percent by the time PPS was introduced. This takes place, of course, in a period when the average age of the Medicare population was increasing, both through sociodemographic effects and because of the increases in life expectancy due to technological change. Thus, trying to predict effects of the PPS on mortality requires holding both of those things constant. The mortality outcomes during the first years of the PPS are completely consistent with the belief that no change occurred in mortality due to the PPS.

A separate analysis of mortality rates within selected disease categories shows a quite similar picture; in some of the categories, mortality improved, and in others, it got worse. In fact, in 35 disease groupings studied, there was no significant change in 20 groups, an improvement in 8 groups, and a decline in 7 categories. It seems safe to assume for now that the changes in LOS precipitated by the PPS have not created any systematic degradation in patients' health.

Consequences for Hospitals   The PPS system creates a dramatic change in the financial environment of hospitals in the United States, and one might expect that some of them will fare poorly, others well, in such a system. The whole idea of PPS was to put a financial squeeze on "expensive" hospitals and make them become more efficient, and to put a stronger sense of market control into an industry that had, in many ways, lost contact with normal economic forces.[18] What can we now say about the results, having seen PPS in place for over seven years?

Without a doubt, the downward pressure on hospital use has affected hospitals' financial conditions adversely. Hospitals' net revenue margins (as a percent of total costs) had climbed steadily from 1970 through 1983, with a brief two-year dip in 1971–1972 as a result of the Nixon-era price controls. That trend reversed itself in 1984, the first year of the PPS system, with a continued decline

---

[18] Many observers characterize the PPS system as an intense regulation, likening it to a price control. Others view it just as any other "contract" between a large buyer and its suppliers. Obviously, the force of government makes PPS different from private contracts, but the arrangement still has much in common with a normal contractual arrangement, perhaps more than it has in common with a "regulatory" regime. Perhaps most important, the PPS system does not directly alter the relationships that hospitals have with other insurance carriers or patients.

through 1989 (the most recently available data at this writing). Hospital occupancy rates have fallen throughout the country, from pre-PPS levels of about 72 percent (national average) to below 65 percent. Tables 12.4 and 12.5 show the basic data.

The concurrent shift to ambulatory surgery in the under-65 population has accelerated both of these trends. Table 12.5 shows the declines in admissions by size category. The strong relationship between size and loss of patients emphasizes the nature of patients for whom this type of substitution is possible. Small hospitals typically carry out procedures that are relatively uncomplicated, and therefore these hospitals are much more susceptible to competition from ASC-like organizations. While the Medicare PPS system may have contributed somewhat to the growth of the ASCs, it seems clear that at least some of the economic hardship facing small hospitals comes from other directions than the PPS itself.

With much of the hospital's costs fixed in the short run (see Chapter 8), these downward shifts in demand are bound to require response from the hospital. As one response, hospitals have tried to reduce costs. As Table 12.4 shows, hospital employment began to fall in 1984, the first PPS year, and has remained very stable through 1988 for inpatient full-time equivalent (FTE) personnel.

Table 12.4   ECONOMIC CONSEQUENCES OF PPS

	Hospital employment annual rates of change		Hospital occupancy	
	Total FTEs	Inpatient FTEs	Occupancy level	Rate of change
Year				
1980	4.7%	4.5%	75.9%	1.9%
1981	5.4%	5.1%	75.8%	−0.1%
1982	3.7%	3.4%	74.6%	−1.6%
1983[a]	1.4%	0.8%	72.2%	−3.2%
1984[b]	−2.3%	3.5%	66.6%	−7.8%
1985	−2.3%	4.3%	63.6%	−4.5%
1986[c]	0.3%	1.4%	63.4%	−0.3%
1987	0.7%	−0.7%	64.1%	1.1%
1988	1.1%	0.7%	64.5%	0.6%

[a] TEFRA rules in place.
[b] PPS phase-in begins.
[c] PPS phase-in complete.
Source: ProPAC, 1989.

Table 12.5   CHANGES IN ADMISSIONS, 1983–1985

Bed size	Percent change
<50 beds	−22.%
50–99	−17.1%
100–199	−11.4%
200–299	−8.4%
300–399	−5.1%
400–499	−5.8%
500+	−2.7%

Source: HCFA, 1989, Table 3.3.

This is quite consistent with studies of hospitals in New York State that confronted the strongest regulatory pressure on revenues (Thorpe and Phelps, 1990).

The external financial market has responded as well. Since hospitals in general cannot issue equity financing (since not-for-profit hospitals have no shareholders), bond financing remains by far the most important way of raising capital for these hospitals. Since 1983, bond ratings—the captial market's measure of the financial stability of the lending hospitals—have worsened dramatically. Over 300 bond issues have had their ratings downgraded, compared with only 60 whose ratings have improved during this period (ProPAC, 1989, Figure 4.5)

Physician Prospective Payment

The success of prospective payment in reducing hospital costs has prompted the inevitable question—why only hospitals? Indeed, the Congress anticipated this issue when it initiated the PPS system for Medicare hospital payment and at the same time authorized a series of studies of the potential benefits from prospective payment for other types of medical care, most notably, physician payment. It authorized the formation of the Physician Payment Review Commission (PPRC) in 1986, to provide recommendations about altering the way Medicare pays physicians.

Resource-Based Relative Values   The first large study in this area came from Harvard University's School of Public Health in late 1988, which developed a series of relative values of various procedures performed by physicians and surgeons, called the Resource-

Based Relative Value System (RBRVS).[19] We should not under-state the complexity of this task; the standard coding system for physicians' activities, the Current Procedural Terminology, edition 4 (CPT-4), describes some 7000 different activities that physicians might perform. Their study actually gathered data on 372 of these procedures, and extrapolated these findings to the remaining procedures.

This system has some similarities to a system originally developed by the California State Medical Association (called the California Relative Value System, or California RVS) that had been widely used for years until banned by the Federal Trade Commission.[20] The California RVS described the relative worth of procedures within single specialties, but deliberately avoided making cross-specialty comparisons. The RBRVS study deliberately made cross-specialty comparisons. The underlying logic of their methods was to produce what might be called "equal pay for equal work," or to pay physicians according to the time and complexity of their effort equally, whether the task at hand was neurosurgery, psychiatric consultation, or removal of warts.

The study drew the distinction between "invasive procedures" and "evaluation and management" (E/M), which corresponds (imperfectly) to the distinction between practice styles of surgeons and internists (respectively). The study's authors can apply their relative values to actual dollar charges, and simulate the effects of using their payment plan, holding total Medicare outlays unchanged, *assuming that the mix of services provided to patients in the long run would remain unchanged.* The results are quite startling: Fees for many surgical procedures would fall by one-third to one-half. Fees for "cognitive" (E/M) services would rise by comparable percentages. Overall, Medicare outlays for E/M services would rise by 56 percent, and for invasive procedures would fall by 42 percent, according to their simulations. Laboratory payments

---

[19] The study's result appeared in summary in the *New England Journal of Medicine* on September 29, 1988, and in considerable detail in the October 28, 1988, issue of the *Journal of the American Medical Association*, consuming the entire issue. The project's final report consumes four reams of paper per copy.

[20] The FTC banned the use of the RVS system because they thought it facilitated doctors' collusion on pricing. The argument was that doctors had to agree on only a single number—the dollar value of each RVS unit—and they then had a complete set of prices upon which they had agreed. The logic of this argument is incomplete; use of an RVS also simplifies the consumer's search problem, because it allows a complete understanding of any doctor's pricing structure simply by knowing the conversion factor from RVS units to dollars.

would fall by 5 percent, and fees paid for imaging (X-rays, CT scans, MRI scans, etc.) would fall by 30 percent (Hsiao, Braun, Kelly, and Becker, 1988).

This study leaves unanswered many of the most interesting and important economic responses to such a system, including (1) changes in the mix of medical interventions in the short and long run, and perhaps more importantly (2) changes in the patterns of specialty training in the long run. In part, of course, the answers to such questions would depend considerably on what private-sector insurance plans would do in parallel. If they adopted a similar or identical payment scheme as Medicare's RBRVS, this would strongly reinforce the effects of Medicare's actions, particularly on long-run specialty choice.[21]

Episode-of-Illness-Based Payment Plans (Physician DRGs)

The RBRVS really does not enter the realm of "prospective" payment, since it still essentially pays on a per-service basis. One could alternatively think of paying for physician services on a DRG-like system. The complexities of such a system seem considerable, however.

DRG payments "work" for hospital care because of three key things: (1) the activity is well defined, with a beginning (admission to the hospital) and an end (discharge); (2) only one economic agent (the hospital) delivers care that the DRG pays for; and (3) the DRG system is able to account for an important part of the overall variation in hospital costs. Without these factors, the PPS would be much more difficult to operate, and perhaps it would not be feasible at all. We can consider the role of each of these in turn.

Clearly Defined Episodes The clearly defined beginning and ending of the hospital episode seem crucial to operating a DRG-like system. In ambulatory care, particularly for patients with chronic illnesses, no such definition of "an episode" can exist. Thus, an episode-based PPS seems infeasible. It might still be

---

[21] For pure experimentalists, this change would offer a dramatic test of the demand-inducement hypothesis (see Chapter 8) and associated "target income" models. If doctors have the ability to generate demand willy-nilly and have a target income they seek, then dropping surgeons' fees in this fashion should lead to *attempts* to achieve large increases in the surgery rate. Thus, installing such a system would provide a crucial test of the demand-inducement theory.

possible to use such a system for inpatient care provided by physicians, but ambulatory DRGs seem completely impractical for this reason.

Multiple Economic Agents   A second complexity arises when more than one physician takes care of a patient within an episode of illness, as commonly occurs both within and outside of the hospital. The only meaningful way to operate a DRG system is to have one physician payment for each illness, no matter how many doctors participate in the care. Otherwise, one has made no meaningful change from a per-service payment system. This raises a second set of concerns, the most important being "Which doctor gets the DRG payment?" This matters considerably because, in effect, the doctor would become the "prime contractor" for the treatment and other doctors would become "subcontractors." This choice would likely affect, for example, whether "conservative" or surgical therapy would be recommended to the patient for some illnesses. Allowing the patient to designate who the "prime contracting" doctor is provides the greatest patient autonomy in these choices, and may also be the only legally admissible choice, since the Medicare benefits come to the patient initially, not to any doctor.

Time-Based Payments (Capitation)

An alternative payment plan would move to physician payments that were wholly prospective, no matter what the condition of the patient—a per-year "capitation" system. Capitation plans are not new, of course. Prepaid group practice plans have used these arrangements for decades. A big gap still remains between those plans and a physician-based capitation system for Medicare. Although capitation provides, in many ways, the strongest incentives for controlling costs, it imposes a large financial risk on individual providers (at least in its pure form). Prepaid group practices get around that problem by enrolling large numbers of individuals and families—tens to hundreds of thousands of enrollees per plan. With enrollments per plan on that order, the law of larger numbers begins to protect the plan with regard to overall variance of expenses, since the variance in expense per member falls with the square root of the number of enrollees (see the Appendix to Chapter 11 for details). However, with a single physician's practice, enrollees number in the hundreds and the financial risk that a pure capitation plan imposes on the physician is considerable. (A "pure" capitation plan would put the primary care physician at risk for all health care expenses of the patient, or at least all nonhospital expenses.) Some sort of sharing of the risk seems inevitable, either through insurance arrangements, formation of large groups of physicians, or (more likely within Medicare) some sort of stop-loss provisions that would limit the per-patient or overall risk of doctors involved in a capitation type of plan.

We should recall here that the Part A DRG plan, as it applies to hospitals, already includes a "stop-loss" provision for the hospitals for each hospital admission. If any patient's length of stay exceeds certain bounds (different for each DRG), the patient is declared an "outlier" and the hospital begins to get paid similarly to the old Medicare methods, on a per-unit basis. Some sort of protection like that seems inevitable for a physician capitation system as well. Balancing that risk bearing with the appropriate incentives for the doctor is a step in the formulation of Medicare policy that has not yet been taken. Indeed, proposals to use capitation within Medicare for physician payments typically speak of capitation payments to insurance plans, rather than to individual physicians.

Global Payment Caps

A completely separate system for controlling doctors' fees has been proposed in recent years, and has on occasion reached at least the level of committee approval in the U.S. House of Representatives.[22] The proposed system of global payment caps has (again) some roots in private insurance plans. Some IPAs have such overall limits on physician payments to their overall panel of physicians. In those IPAs' contracts with physicians, they typically "hold back" a portion of fees (say, 20 percent) that each doctor would normally get. If overall physician spending within the IPA exceeds targeted amounts, the deficit is taken out before "hold-back" funds are returned to the physician.

As a financial device to control costs, global caps have both advantages and disadvantages. On the advantage side, they allow the paying entity (insurance plan) a high level of guarantee about total costs for physician services. Unfortunately, the mechanisms to achieve this may almost guarantee that the overall limit is ex-

---

[22] The proposal was put forth in an overall deficit-reduction bill passing the Ways and Means Committee in June 1989.

ceeded, possibly in a way that penalizes those doctors most who have made the biggest efforts to make the system "work."

To see the problem in its clearest detail requires the use of mathematical models known as "game theory," but the ideas are relatively simple, and correspond closely to the standard "prisoner's dilemma." In the prisoner's dilemma, the usual story begins with two people who have both promised to each other in advance that if caught they won't "squeal" on the other. The dilemma arises when jailers take each prisoner aside and promise him or her harsh treatment for not cooperating, but lenient treatment for confessing in a way that implicates both prisoners. Since the prisoners have no ability to communicate (collude) with one another, the "best" private response for both prisoners is to defect from the original agreement, or to "squeal" on their partner, and—a key idea—this is the best strategy no matter whether the partner squeals or keeps quiet.

The physician payment problem differs somewhat, but not greatly. Suppose all physicians agree in advance to cooperate with the insurance carrier to keep costs down. Then each single doctor figures out his or her own optimal strategy: It involves charging the plan as rapidly as possible, no matter what other doctors do. If other doctors continue to "cooperate" to keep costs down, then the "fast-charging" doctor gets to keep all of the money billed, clearly coming out ahead. If other doctors all adopt the same strategy, then (collectively) they will trigger the holdback of the insurance carrier, and every physician's payments will be reduced (say) by 20 percent. Any doctor who has not "overbilled" by 20 percent will end up losing money under those circumstances. For this reason, all doctors' incentives lead them to charge fees as high as possible, and to bill as many procedures as possible, when such global spending caps are put in place.

The ability to use "group persuasion" to control such proclivities is quite real in closed physician groups, which is one reason why physician group practices can sometimes use "holdback" devices effectively. In large open-panel IPA plans, almost no successful persuasion is possible, because the doctors are so dispersed geographically and organizationally. At the national level, any possible use of group persuasion to control total spending seems wholly doomed. Any total-fee cap within Medicare can only lead to a "race to the cash register" that will surely trigger the mechanism. Ironically, if any physician-firms "cooperate" and try to keep total spending down, they will only suffer financial losses.

Box 12.1 GAME THEORY REPRESENTATION OF THE GLOBAL SPENDING CAP

	Other doctors' behavior	
	Control costs ("save")	Intense billing ("spend")
Save	10,000	8,000
Spend	12,500	10,000

(Individual doctor's strategy: Save / Spend)

Suppose the insurance plan sets up a target of $10,000 in fees billed by each doctor, and promises to withhold enough money so that if the total billings exceed $10,000 × N (where there are N doctors in the pool), each doctor will get paid proportional to their total billings, with the total paid to all doctors equaling $10,000 × N.

Each doctor must select a strategy—either to "save" costs or to "spend" rapidly—that is, bill as many procedures as possible at the highest fees allowed. If, under normal behavior, each doctor would bill the plan $10,000, so that if all doctors "save" to control costs, each bills and gets paid $10,000. Thus the doctor's payoff if he "saves" and all other doctors do similarly is $10,000.

Now, suppose that this single doctor "saves" costs, but all other doctors bill intensively. If all doctors bill (for example) $12,500 under such circumstances, then total billings relative to target billings are ($12,500 × N) ÷ (10,000 × N) = 5/4. According to the plan rules, each doctor will then get paid 4/5 of their billings. Thus, the doctor who "saves" gets paid $8,000, and all other doctors get paid $10,000.* Cooperation costs the doctor $2,000 if all other doctors don't cooperate.

Finally, if all doctors "spend" resources at the $12,500 rate, each gets paid $10,000 in revenue. Of course, if actual resources have been used to produce care to the patients that leads to the extra $2,500 in billings (above the usual $10,000 in care provided) then the doctors are all worse off. If the only thing that happens is that all doctors raise their fees by 25 percent, then the economic consequences of imposing the global revenue cap are nil; everybody raises prices, everybody gets paid a proportional share of them, and the total spending corresponds to what would have been there without the cap.

* This isn't exactly accurate, because the total billings are really $12,500 × (N − 1) + $10,000 if one doctor "cooperates" with the cost-saving venture. This approaches $12,500 × N very rapidly as N gets larger, so for expositional purposes, we can use $12,500 × N as the total billings.

In such economic games as this, Leo Durocher is always correct in saying that "nice guys finish last."[23]

SUMMARY

Medicare's structure has evolved through a series of important changes since its inception in 1965. Most of the changes have sought to control outlays by the Health Care Financing Administration (for Medicare) and by both HCFA and state governments (for Medicaid).

Initial steps to control costs centered on limits to increases in physicians' fees under Medicare Part B, originally instituted as part of the overall wage-price freeze by the Nixon administration in 1971, but persisting since then in many manifestations.

In 1983, hospitals also became the target of Medicare cost controls, with the introduction of the Prospective Payment System (PPS) using flat-rate payments based on Diagnosis-Related Groups (DRGs). This system of payments dramatically reduced hospital length of stay (predictably, given the financial incentives) but admission rates also plummeted, possibly because of the parallel growth of a new technology (ambulatory surgery) throughout the country. Declines in admissions hit hardest on small hospitals that tended to undertake relatively simple operations, the types that ambulatory surgery centers could most easily undertake.

A short-lived change in the structure of patients' insurance also was instituted in 1988 (catastrophic protection), but widespread protest caused the Congress to repeal these changes in 1989.

The most recent change in Medicare again centers on physician payments. Scheduled for implementation in 1992, Medicare will move to a new fee-schedule type of payment system based on a study of "work effort," formally called the Resource Based Relative Value System. Under this system, the remuneration for so-called "cognitive" services (thinking and talking with patients) would increase markedly, while fees for procedures (operations, specific diagnostic tests) would fall markedly. While the final implementation of this system is not yet complete at this writing, this system of paying physicians, especially if followed by private insurance carriers, could markedly alter the appearance of the U.S. health care system.

---

[23] Durocher was a combative baseball player and manager.

PROBLEMS

1. This problem considers the meaning of a decline in average LOS for Medicare patients when some short-stay patients have shifted out of the hospital at the same time the DRG system came into being.

   Suppose in 1982 that for Medicare patients the ALOS was 10 days overall, composed of two groups of patients of equal size. The short-stay group (cataracts, etc.) had an ALOS of 2 days.

   (a) What was the ALOS of the remaining group?

   Now suppose that the overall Medicare average had fallen to 8.5 days five years later, and that all of the short-stay patients from previous years now received their surgery in ambulatory surgical centers (ASCs).

   (b) What was the relative decline in LOS for the "long-stay" group? To find this, compare your answer to part (a) of this question with the 8.5 day LOS.

   (c) How would your answer change if only half of the "short-stay" group were getting their surgery in ASCs in the later year?

2. What effects on hospital use would you most expect from the Prospective Payment System (PPS) as developed using Diagnosis-Related Groups (DRGs) as the method of hospital payment? What effects might this have on patients' health outcomes? (Hint: The relevant catch-phrase is "sicker and quicker.")

3. What single feature of the original Medicare program is most likely to cause an economist to say that "Medicare stinks as insurance"? (Hint: Think about aspects of the Medicare coverage that create high financial risk.)

4. When people buy private insurance to supplement Medicare insurance, it most commonly covers up-front deductibles and least commonly covers "high-end" risks (see Table 12.2). Does this pattern of coverage accord well or poorly with the model of demand for insurance set forth in Chapter 10? What should we conclude from this?

5. Describe "balance-billing" in the Medicare program, and discuss what it does to affect the financial risk confronting Medicare enrollees.

6. The new method of paying for doctor-services under the Medicare program will increase payments to "cognitive services" (thinking and counseling) and decrease the payments to "procedures" (surgery, invasive diagnostic tests).

   (a) What do you think will happen to the demand for residency training in orthopedic surgery, geriatric medicine, and pediatrics?

   (b) What effect, if any, would you expect for hospital use?

Chapter
13

Other Government Health Care Programs

*F*ederal and state governments have a broad array of health care programs in addition to Medicare. While Medicare is the largest—and in many ways the most interesting for economic analysis because of the recent innovations in payment methods—these other programs have important issues associated with them, and deserve study on their own account. Other governmental health care programs include:

- the Medicaid program for low-income persons
- military health care systems
- Veterans Administration (VA) system
- mental health programs
- maternal and child health programs

The scale and purposes of these programs differ considerably, and we cannot analyze each here. However, we can study some of these programs in more detail, and at least briefly describe the purpose of each of the others, as well as indicating some of the important economic issues associated with each.

THE MEDICAID PROGRAM

Medicaid provides for low-income people the same services that Medicare provides for persons over 65 years old in this country—protection against the financial risk associated with illness, and a direct transfer of income. Medicaid accomplishes this in a considerably different way for low-income persons than does Medicare. These two programs differ in five important ways:

- The class of people eligible for Medicaid depends upon in-

363

come, where the class eligible for Medicare depends only on age or illness status (ESRD, disabled).

· Individual states design and operate their own Medicaid programs, under guidelines and with financial assistance from the federal government. The Medicare program provides uniform benefits and eligibility across the country.

· The scope of benefits differs; most importantly, Medicaid covers long-term (nursing home) care while Medicare does not. Thus, some Medicare recipients receive Medicaid assistance for long-term care expense.

· Medicaid plans, by federal law, had no copayments for recipients, and may contain "nominal" copayments only under current law (e.g., $1 per office visit).

· Providers who participate in the Medicaid program must do so on an "all or nothing" basis—that is, they must accept all patients seeking care under the program if they treat any eligible patients.

The Medicaid Program Structure

Medicaid began as a federal-state partnership, for which Title XIX of the Social Security Act spelled out the overall structure of "eligible" Medicaid programs, and provided for federal support of state programs. Its predecessor, the so-called Kerr-Mills program, had provided similar care for the elderly; Medicaid extended the program to welfare recipients (mostly families receiving Aid to Families with Dependent Children—AFDC). The federal law required that state programs include hospital care, physician care, diagnostic tests, family planning, and (importantly) nursing home care in skilled nursing facilities (SNFs). For these services, states could not require any copayments by enrollees. States could also optionally include nursing home care in intermediate-care facilities (ICFs), dental care, drugs, eyeglasses, and other miscellaneous services. Initially, these services also were without copayment, but later revisions in the federal law allowed states to use copayments for these "second tier" services.

Medicaid offers a cost-sharing basis to the states that depends intensively on the per capita income (PCI) in the state. States with lower PCI received a larger subsidy from the federal government, ranging from more than 83 percent at the highest rate (for lowest PCI states) in the program's early years to 50 percent (for highest PCI states). Overall matching rates have fallen slightly since changes in the federal law in 1981, and now the highest rate is about 78 percent. We will see momentarily the effects of this subsidy rate on states' choices of program structure.

The Scope and Scale of Medicaid Programs

Unlike Medicare, Medicaid does not offer recipients either guaranteed or permanent "health insurance coverage." States must offer enrollment to any person receiving an AFDC or Supplemental Security Income (SSI) grant, but they may also offer the health coverage to persons who otherwise receive no grants, a group commonly labeled as "medically needy." Eligibility depends upon income and (in some cases) assets held by the family. The program mostly enrolls people in two categories—persons in AFDC families (mothers, children, and, in some states, fathers who are present in the home) and elderly persons (many of whom reside in long-term care institutions).

While the composition changes from year to year (for example, through business cycles), the combined total of state Medicaid programs covers about 11 million children and 11 million adults, or about 9 percent of the U.S. population. The program, for two separate reasons, provides much more coverage to female adults than to male adults, although children covered by the program (for strictly biological reasons) are split evenly by gender. In total, of the 11 million adults covered, 8.5 million are female, and 2.5 million are male.

Why does this program's coverage diverge so much by gender? In the AFDC population, most states provide coverage only for families where no father is present, so the eligible adult population by definition is almost exclusively female.

To understand why the medically needy portion of the program so heavily favors women, we must look to life-expectancy tables and habitual marriage patterns in our society. The SSI program provides income (and health care coverage) for two groups of people, the permanently disabled and the aged. The permanently disabled tend to include both genders about equally, but the aged SSI recipients include far more females than males. This occurs because women live longer in our society—the average life expectancy of females exceeds that of males by almost 5 years[1]—and

---

[1] Review Chapter 1 concerning the effects of life-style on health. Some life-style events that reduce male health include cigarette smoking and alcohol consumption, activities disproportionately undertaken (pardon the expression) by males in our society.

because marriages in our society commonly have the male as the older person. Thus, widows far outnumber widowers, and this group has the greatest chances of becoming eligible for SSI coverage or "medically needy" coverage.

The combination of these two factors—AFDC program structure and the heightened survival of elderly females—makes the Medicaid program very much a program, borrowing from the old sea chantey, to "save the women and children first." While the language of Title XIX makes no reference to gender of recipients, the structure of the program and its income eligibility criteria tend to draw far more women than men into its coverage. (Alternatively, one might equally well say that in our society, there are substantially more poor women than men.)

Patterns of Expenditure The Medicaid program's spending patterns quite naturally depend upon the medical conditions of this eligible population. All eligible groups tend to use hospital and physician services more than the "average" person. The AFDC enrollees obviously have a nontrivial demand for maternity care (both prenatal and delivery), plus well-child and acute pediatric care. The elderly also use more physician and hospital care than the average person (see Table 2.6 in Chapter 2), and the SSI recipients who are permanently disabled also have unusually large medical use.

Table 13.1 sets out Medicaid expenditures by category of recipient. The elderly and the permanently disabled, while accounting for only a quarter of the program's enrollees, generate two-thirds of the medical spending, with per-recipient payments (in 1985) of about $4500.

Another portrait of this spending pattern appears in Table 13.2: These high-cost groups of recipients are most likely to use institutional care, which creates by far the largest cost per recipient using the service. Because of this, nearly three-quarters of the total spending in the Medicaid program occurs on behalf of institutionalized patients—those in hospitals and long-term care facilities. One reason for this is that Medicaid usually pays hospitals at the same rate as Medicare does (essentially, average cost), while the program can (and does) pay doctors at rates far below "market" rates. (We study below the effect of such strategies on physicians' willingness to participate in the Medicaid program and the effect on access to medical care.)

Table 13.1   MEDICAID RECIPIENTS AND PAYMENTS BY
ELIGIBILITY CATEGORY

	Eligible persons— 1985		Payments—1985		
	Number (millions)	Percent of total	Dollars (billions)	Percent of total	Per recipient
Age 65+	3.1	14 %	$14.0	37 %	$4500
Blind	0.08	0.4%	$ .25	0.7%	$3100
Permanent and total disability	2.9	13 %	$13.2	35 %	$4550
Dependent child under 21	9.8	45 %	$ 4.4	12 %	$ 450
Adults in families with dependent children	5.5	25 %	$ 4.7	13 %	$ 850
Other	1.2	6 %	$ .8	0.2%	$ 650
Total	21.81	100 %	$37.5	100 %	$1700

Source: Health Care Financing Program Statistics: Medicare and Medicaid Data Book, 1988, Tables 4.4 and 4.14.

Table 13.2   COMPOSITION OF MEDICAID EXPENDITURE BY TYPE OF
CARE—1985

	Amount (billions)	Percent of total
Hospital		
General	$ 9.45	25.2%
Outpatient	$ 1.80	4.8%
Mental	$ 1.20	3.2%
Skilled nursing	$ 5.06	13.5%
Intermediate Care		
Mentally retarded	$ 4.72	12.6%
All other	$ 6.52	17.4%
Physician	$ 2.36	6.3%
Clinic	$ .71	1.9%
Dental	$ .45	1.2%
Prescription drugs	$ 2.33	6.2%
Home health	$ 1.12	3.0%
All other	$ 1.79	4.8%
Total	$37.51	100.0%

Source: Health Care Financing Program Statistics: Medicare and Medicaid Data Book, 1988, Table 4.17.

ECONOMIC POLICY ISSUES IN MEDICAID

Political Economy—Demand for Medicaid by States

The Medicaid program, because of its diversity across states, allows the study of several issues that Medicare does not allow. How states actually select the extent of their Medicaid coverage represents one such issue.

If we thought for a moment about a "state" as having an omnipotent governor with a single utility function (much as we characterized the "director" of a hospital as a single person with a utility function initially in Chapter 9), then we could think about that "state's" demand for providing medical care to its elderly and poor as two "commodities" that provide utility. Obviously, if the "omnipotent governor" is a heartless despot, caring nothing for the health of the state's citizens, then the "state" will select no medical care coverage for its low-income and elderly citizens, but if those people's well-being create utility for the "governor," then we can expect that the state will purchase some Medicaid to help them. It does that by allocating some of its own income (tax revenue), and can buy Medicaid at prices that vary from state to state, according to the Title XIX sharing formulas. We can think of each state as "demanding" a level of Medicaid, and the federal government "supplying" Medicaid, where the price of Medicaid is the sharing rate between the federal and state governments set by the federal government.

Figure 13.1 shows the indifference curves for such an "omnipotent governor," allowing Medicaid payments and "other goods"

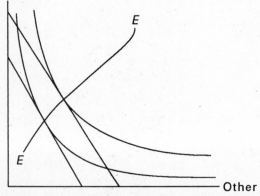

Figure 13.1 Indifference curves for Medicaid versus other goods.

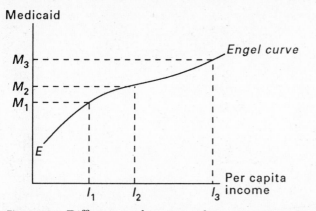

Figure 13.2 Different combinations of state income and Medicaid spending.

(e.g., roads, parks) to affect utility. We can and will break apart the bundle of Medicaid payments into finer categories in a moment. As the state's income expands, if the relative costs to the state of Medicaid and other goods remained constant, we should expect to see Medicaid spending expand along the EE path. We could, in turn, map each possible combination of income and Medicaid spending onto another figure, showing the amount of M that the state would choose at each income level I. Figure 13.2 does this, and has a direct analog to the "Engel curve" in consumer demand theory. At any point along the Engel curve, we can calculate the income elasticity of demand—that is, the percentage change in quantity demanded for a 1 percent change in income.

   In the "real" Medicaid program, however, life is not so simple. The federal government doesn't sell Medicaid to the states at a constant price, but rather, it "price discriminates," or sells at different prices to different states, depending on the state's income. As noted before, the federal government pays as much as 75 percent of low-PCI states' Medicaid costs and as little as half of high-PCI states' costs.[2] Thus, as any state's PCI changes, the price of Medicaid also changes. Figure 13.3 shows the appropriate budget constraints, along with the original indifference curves from Figure 13.1. Suppose the state originally has income level corresponding to the II line in Figure 13.3, but that income grows to the line $I'I'$. If the story ended there, consumption would shift from the combi-

---

[2] This works just as "coinsurance" works in an individual's health insurance plan. See Chapter 4. The only difference here is that the Medicaid rules link the "coinsurance rate" to income. In almost all private health insurance plans, no such link exists.

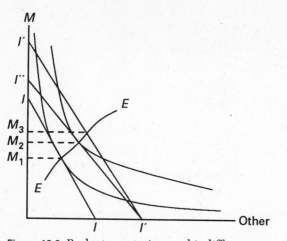

Figure 13.3 Budget constraints and indifference curves.

nation M_1 to M_3 and we could stop. However, because of the rules of Medicaid, the price of Medicaid goes up, so the relevant budget line is not $I'I'$ but rather $I'I''$, showing that Medicaid has become relatively more expensive. Final consumption by the state will be M_2. The growth from M_1 to M_3 represents the "income effect" in demand, and the shrinkage from M_3 back to M_2 represents the "price effect."

Since each state has different PCIs, and the federal Medicaid rules make the price differ for each state along with its income, we can estimate separately the effects of income and price on states' demand for Medicaid.[3] Granneman (1980) has done precisely that, using the techniques of multiple regression analysis. He estimated the "demand" for Medicaid by states, separating out the effects of income and price. Table 13.3 shows the results. For the moment, concentrate on the "total Medicaid benefits" result; the price elasticity of -0.78 and income elasticity of 1.23 indicate that states' demand is quite sensitive both to the federal matching rate (price) and to income. Suppose, for example, the federal government decided to reduce its assistance to states by 20 percent (say, from a peak matching rate of 75 percent to a peak matching rate of 60 percent, with corresponding changes from the 50 percent match

---

[3] To do this, one needs to assume that the "governor" of each state has the same utility function. We also need to have Medicaid use the same price for some states with different incomes; otherwise, we'd never be able to untangle statistically the separate effects of income and price. Fortunately, Medicaid does this, by providing a specific subsidy rate over a range of incomes. Thus, one really can estimate the effects of income and price separately.

Table 13.3   INCOME AND PRICE ELASTICITIES FOR STATES' DEMAND FOR SCOPE AND GENEROSITY OF MEDICAID PROGRAM

Component	Price elasticity	Income elasticity
AFDC child recipients	−0.30	2.17
AFDC adult recipients	−0.25	2.32
AFDC child benefit level	−0.26	0.26
AFDC adult benefit level	−0.39	0.61
Total Medicaid Benefits	−0.78	1.23

Source: Grannemann, 1980.

rate for high-income states to a match rate of 40 percent). The price elasticity of −0.78 tells us that the overall scale of the Medicaid program would change by $(-0.78 \times 0.2) \approx -0.15$, or about a 15 percent decline.

The other parts of Table 13.3 tell us how the states would achieve this change—they would scale back about equally on both the per-person benefits and on the number of people eligible for the program, and these cutbacks would take place about proportionally for both children and adults. How do we know this? The answer is that the price elasticity for the total program (−0.78) must represent the added-up total of the elasticities of the component parts, weighted appropriately for their relative size, and in this case, Granneman has estimated that the price elasticities are all about the same. (Box 13.1 shows the algebra for this, and problem 1 at the end of this chapter uses the same concepts.) Since the component-part price elasticities are all about −0.25 to −0.3, each part would be scaled back by about the same proportion in response to a "price increase."

Changes in income do something quite different. As the income elasticities in Table 13.3 show, if a state's income increases (or, if we look at states with higher incomes), we can predict from these results that the number of recipients will expand rapidly (the income elasticities exceed 2.0), but the per-person benefits will not expand so rapidly (the elasticities are 0.26 and 0.61 for per-person benefits).

Thinking about the question of how a state governor's "utility function" gets established, these results make decent sense. If a governor is really elected, rather than being an all-powerful dictator, then it makes sense that if the state planned to spend more money on its Medicaid program, it would do so in a way that

---

Box 13.1 THE (RESPONSIVENESS OF THE) WHOLE IS EQUAL TO THE (RESPONSIVENESS OF THE) SUM OF ITS PARTS

Medicaid spending is composed of numerous subparts, each of which can in concept respond differently to the subsidy provided by the federal government. The algebra of this problem shows how the responsiveness of each part adds up to create the responsiveness of the whole.

Suppose that there are two groups that Medicaid might serve (A and B). Both the number of such persons served (N_A and N_B) and the generosity of coverage (benefits) per person served (G_A and G_B) can vary independently. Suppose that each of these components responds differently to the federal subsidy F.

Define total Medicaid spending as

$$M = N_A G_A + N_B G_B \tag{*}$$

and the rates that each change with respect to F as dN_A/dF, dG_A/dF, and similarly for group B. Then

$$dM/dF = G_A(dN_A/dF) + N_A(dG_A/dF) + G_B(dN_B/dF) + N_B(dG_B/dF). \tag{**}$$

Now define the elasticities associated with each of these derivatives, so that η_{MF} = the elasticity of M with respect to F, η_{NF}^{A} = the elasticity of N_A with respect to F, and similarly for G_A, N_B, and G_B. A little algebraic manipulation (try it yourself!) shows that equation (**) becomes

$$\eta_{MF} = [N_A G_A/M] \{\eta_{NF}^{A} + \eta_{GF}^{A}\} + [N_B G_B/M] \{\eta_{NF}^{B} + \eta_{GF}^{B}\} \tag{***}$$

Note that $N_A G_A/M$ is just group A's share of total spending on M, and a similar term exists for group B. Note also that, for example for group A, the elasticity of their spending total is the *sum* of the elasticities for the number of persons served (N_A) and the per person benefits (G_A). Thus, the elasticity of the total (M) is a weighted average for groups A and B of the *sum* of their respective elasticities for enrollment and per person spending, which we might also call the elasticities on the extensive margin (for N) and the intensive margins (for G).

---

directly benefited more voters, rather than increasing the per-beneficiary "goodies" for a small group who already receive some benefit. Thus, we would expect to see income elasticities for "numbers" that exceeded the income elasticity for "generosity" of the program.

Physician Participation

Physicians who treat any Medicaid patients must agree to treat all such patients who come to their offices. By agreeing to treat Medicaid patients, the physician-firm agrees to accept the Medicaid fee (determined differently by the government bureau administering Medicaid in each state). The choices made variously by the different state government agencies create a natural experiment, allowing the study of the response of physician-firms to different fee levels.

Several studies have measured the rate of physician participation across states as a function of the per-visit price paid by Medicaid. Not unexpectedly, participation increases with the generosity of Medicaid payment. Studies using alternative data sources have estimated that a 10 percent increase in the Medicaid payment level will produce anywhere from a 2 percent increase in Medicaid participation by doctors (Sloan, Cromwell, and Mitchell, 1978) to a 17 percent increase (Hadley, 1979).

Physician Participation and Patient Access   The rate at which Medicaid programs pay for treatment varies considerably across states, and as we just found, the payment rate affects the willingness of doctors to participate in the program. Does this affect the ease of access to medical care by Medicaid program beneficiaries? To analyze this, we need a standard of comparison of a program's generosity. One useful standard of comparison is the Medicare fee schedule, which is fairly similar (but not perfectly so) across all states (with minor variation for differences in costs of living and other differences for some procedures). Ten states pay the full Medicare rate in their Medicaid programs, at one extreme. At the other extreme, some states pay less than half of the established Medicare fee schedule. Even with these large differences in payment rate, which *do* affect the participation of physician-firms, the overall access to care by Medicaid patients does not seem to differ systematically (Long, Settle, and Stuart, 1986). In states with stringent payment schedules, patients are somewhat more likely to receive their care from a hospital outpatient department than from a private doctor's office, but the overall physician visits do not differ significantly by Medicaid patients.

Long-Term Care in Medicaid

The Medicaid program offers the only significant government-provided long-term care coverage for individuals in the United States, and this type of payment accounts for a large fraction of overall Medicaid expenditures. Thus, it is worth exploring the economic and policy issues of the long-term care program.

Before we can make complete sense out of the Medicaid long-term care program, we should first go back to the Medicare program to see what type of long-term care coverage it does—and does not—provide, including the provisions of the new catastrophic care coverage in Medicare. The previous discussion of Medicare concentrated mostly on the coverage for inpatient services within Part A and physician care in Part B. Recall that Part A does contain coverage for skilled nursing facilities (SNFs) and intermediate care facilities (ICFs). SNFs offer limited coverage for rehabilitation that *must* follow a hospitalization of at least three days, and then only under fairly strict rules of eligibility. (These rules vary from region to region according to the interpretation of the agencies administering the Medicare program in different places.) Medicare coverage provides for 30 days of SNF coverage following a hospitalization, and then (like the hospital care), added days at decreased rates of coverage. To remain eligible for SNF coverage, individuals must also participate in an active program of rehabilitation, designed to return them to their homes.

Many elderly people and almost all young people requiring long-term care services[4] find either that they are ineligible for Medicare SNF coverage or that the coverage rapidly runs out, while the nursing home stays continue indefinitely. Elderly persons who gradually become unable to function for themselves, for example, may have no "crisis" event that leads to hospitalization, yet they may become unable to function in their homes. Nursing homes constitute one way to provide for the care of such people. (Home visitation services and other programs may help as well.) Obviously, such people meet none of the requirements for SNF coverage under Medicare, because they neither could enter the SNF through a hospital nor would they be actively participating in a rehabilitation program designed to get them back on their feet. In simple terms, Medicare will not pay for custodial care, and this type of care constitutes the requirement for many elderly, institutionalized people.

Medicaid provides long-term care coverage for some persons in this type of situation. However, the Medicaid program has its

---

[4] Young people can be eligible for Medicare through permanent disability. Permanently disabled persons, say, from automobile crashes, may well receive Medicare, and may also have disabilities leading to permanent institutional care, e.g., in a nursing home or rehabilitation hospital.

own eligibility criteria, which generate a complicated set of economic, legal, and public policy issues. These issues focus attention on the primary purposes of the Medicaid program—provision of care for "the poor"—and the legal and economic difficulties arising when attempting to define what "poor" means in this population.

In normal conversation, "poor" probably refers to somebody with little income. Income, however, reflects the *flow* of money coming into the household per unit of time (e.g., a year). Official federal definitions of poverty focus exclusively on a family's income, relative to some predefined standard ("the poverty line").

By these criteria, a large majority of elderly families and individuals are "poor," because they have retired. Their incomes depend upon the social security program, returns from investments they have made previously in their lives, and state welfare programs. However, this picture can become deceptively complicated. For example, many elderly persons own their homes with no remaining mortgage payments or rental payments. As economists normally think about such matters, these families have an "imputed" income equal to the rental value of their homes. (One can think about them both receiving and paying the rent on the home.) Another way to think about it would say that the "needed" income to escape "official poverty" should be lower for such persons than for those who had rental payments to make. Ownership of cars, a large closet full of clothes, and a freezer full of meat all provide the same problems, although on a smaller scale. The problem, obviously, is that "poverty" definitions ignore assets, a measure of the *stock* of wealth, and focus exclusively on income, the measure of the *flow* of wealth.

Medicaid programs have had to confront the question of the stock assets as they decide on eligibility rules. (Remember that these programs differ from state to state, so the discussion that follows does not describe perfectly the program of any single state.) Would a person become eligible for Medicaid on the basis of low *income* alone, or would a combination of low income and low assets be required? Most states have come to the conclusion that Medicaid eligibility requires both low income and low assets. Realistically, for elderly persons in long-term care institutions (nursing homes and custodial care homes), this means that they must *spend down* their assets, paying for the care on a month-by-month basis, until their assets fall so low that they become eligible for Medicaid.

The so-called spend-down rules created in turn a whole raft of economic and legal issues. Because our society allows the transfer of wealth between generations, the spend-down provision serves in many ways to take wealth away from the children of the elderly, rather than the elderly themselves. Families realized this soon after the spend-down rules emerged, and took steps to counteract this effect. Most notably, parents would give large gifts to their children, thus reducing their assets so much that they would become eligible for Medicaid.

Medicaid programs responded in turn. They defined any gifts made (say) in the last two years before a person applied for Medicaid as gifts with the intent of avoiding the Medicaid spend-down rules, and thus counted those assets as belonging to the applicant (rather than the recipient of the gifts), even if the gift were in truth made without such intent.[5]

A broader social issue emerges from this debate—do families have a responsibility to care for elderly (or other) family members who may require institutional care? The Medicaid rules officially say "no," but the *de facto* result of the anticipatory gifts rules say "yes."

The Medicaid spend-down rules have other odd implications. For example, the effect of the rules differs greatly depending on whether an elderly household has one or two members. As noted previously, a large majority of the oldest households contain only a female, because of the differential survival of females over males in our society. If one elderly member of a household attempts to enter a nursing home with Medicaid coverage, the spend-down provisions can heavily reduce the assets available for the spouse. The effects of such rules will differ from state to state, and can even depend on whether the state has "common property" rules. (In such states, wives and husbands are presumed to own equal shares of the family's assets, no matter who earned them.) Obviously, even if the Medicaid programs wished to continue the spend-down program, they could protect surviving spouses by "splitting" the assets of any couple, requiring the spend-down of only half the assets before one spouse became eligible.

---

[5] Previous tax laws had the same sorts of rules regarding gifts "in anticipation of death." Recent revisions in the tax law have combined the overall treatment of gifts and bequests in a lifetime perspective, eliminating many of the complicated gift rules that previously existed.

We can also consider the Medicaid long-term care coverage from the point of view of an optimal insurance program (as implied by standard economic models of demand for insurance—see Chapter 10). Using those concepts, optimal insurance programs contain a deductible and a coinsurance rate for some spending, and eventually offer a catastrophic cap that provides full coverage for really large expenses.[6] We can cast the Medicaid long-term care program as an insurance policy in the same way: It has a deductible equal to the assets held by the family, no coinsurance rate, and full coverage beyond the deductible.

From an insurance point of view, requiring the consumption of all (or almost all) of the assets of a family as the "deductible" in an insurance policy seems strange, to say the least. Most people buying voluntary insurance would select far smaller deductibles, and would probably be willing to pay some coinsurance for some of the long-term care in exchange for the lower deductible. Medicaid programs could do this, for example, by requiring families to pay for (say) the first 10 days of nursing home care themselves, and then to pay for (say) 30 percent of all care until (say) half of the family's assets had been depleted; then Medicaid would pay full coverage.

MILITARY HEALTH CARE SYSTEMS

While many people never come in contact with it, the largest single provider of health care in the United States is none other than the Department of Defense. The scope alone might make it of interest to the student of health economics, but the military health care systems offer several important examples of the role of incentives in the provision of health care that make it worth some analysis. In particular, we can see

- how incentives for doctors alter the rate at which they hospitalize patients, and how they change patients' visit rates to clinics
- how multiple objectives of an organization sometimes make their behavior look very strange from the point of view of a single-objective organization

---

[6] Arrow (1963) proved many of these ideas. Keeler et al., 1988, explore the expected utility arising from various insurance plans using the RAND HIS data. They find that a plan with a moderate deductible (say, several hundred dollars), a coinsurance rate of about 25 percent, and a catastrophic cap provides the highest expected utility.

The military health care "system" actually contains three separate hospital and clinic systems operated by the Army, the Navy, and the Air Force. Each of these operates within its own military service, under the general guidance of an office within the Department of Defense (the Office of Health Affairs). Each system contains multiple hospital-clinic facilities, located at military bases scattered throughout the country and in parts of the rest of the world. The most famous of these—the Walter Reed Army Hospital and the Bethesda Naval Hospital in the Washington, D.C., area—often treat high government officials, including the president.

These hospital-clinic facilities have a triple role to play—and these roles often conflict with one another. The primary role is to provide health care treatment (for all care, not just hospital care) for active duty military personnel. In this role, the hospitals and clinics correspond (say) to health clinics at many large corporations. (Indeed, the Kaiser Permanente prepaid practice plans in California, Oregon, and elsewhere initially were formed during World War II to provide Kaiser employees with health care. They became open for enrollment to the general public after the end of the war.)

A second role is to provide standby health care capability in case of war. This role leads to the construction of facilities that are much larger than the "peacetime" requirements of the system, often with a considerable portion of the capacity (e.g., hospital beds) remaining empty or seldom used.

In part because of the presence of the available capacity, a third goal has emerged for the military health care systems—provision of care for the families of active duty personnel, and for retirees and the families of retirees from the military.[7] This system—and this is one of the highly unusual parts of the military health care system—operates in tandem with a fairly standard health insurance package provided to families of active duty personnel, retirees, and their spouses. This insurance plan, called CHAMPUS (Civilian Health and Medical Program for the Uniformed Services) has provisions closely matching any standard "major medical" insurance plan, with a modest deductible, copayments, and so on. People covered by CHAMPUS can use this insurance plan just like

---

[7] A person can serve for 20 years in the military and then "retire" into a civilian job. Thus, for example, an 18-year-old enlistee can serve until age 39, retire from the military, and then work for another 25 years or so in the civilian sector before again "retiring." During this latter period, the retiree is eligible for health care from the military. After age 65, Medicare acquires the primary liability for providing health insurance coverage for such persons.

any other insurance plan, and can purchase care from any private doctor, hospital, or other provider.

The one important "wrinkle" in CHAMPUS is the requirement that any hospital care be received from a military hospital, if a nearby hospital can provide the appropriate care. In many small military hospitals,[8] routine care is available, but nothing very complicated. Using transport aircraft, the military services sometimes fly patients to larger hospitals for more complicated treatment, but more commonly, the local base hospital commander (always a doctor) will just certify that the care is not available, and the patient can use CHAMPUS to obtain care privately.

Ambulatory care for this CHAMPUS-eligible population presents a more complicated matter, because no comparable provision exists about seeking care at the military facility first (before using CHAMPUS). Patients can use CHAMPUS first with a private doctor, or can go to any military hospital-clinic to receive care. In some areas with more than one military hospital in the same vicinity, patients can choose on a day-by-day basis which facility they might visit, for example, depending on which has a shorter waiting period for patients. As a result, quite commonly, patients will begin treatment at a military hospital, and end up being treated by a private doctor using CHAMPUS, or vice versa. This approach to providing ambulatory care also makes it impossible to manage the military health care systems effectively, because nobody can determine the effective population served by any military hospital.

The management problem arises through a classic "shirking" problem, as described in Chapter 7. Since the military services cannot meaningfully determine how many patients each facility cares for, they provide resources in next year's budget based on the *inputs* used in this year's budget. Thus, hospitals have strong incentives to keep the beds full, creating unusually high rates of hospitalization for the military and their dependents, and long lengths of stay for those patients hospitalized.

Base commanders have little way to measure a doctor's efficiency or productivity, except for setting targets of how many visits a doctor must produce each day in the clinic. The responses to incentives appear clearly here as well. Given a target (say) of 30 visits per day in a clinic, any doctor realizes that 30 new patients

---

[8] Many hospitals in these systems have under 50 or 100 beds, and are much like small rural U.S. private hospitals.

present a lot more work to be done than 30 visits from established patients. The doctors have a strong incentive to generate repeat visits from established patients, because these fill up their schedules with easy cases. Some doctors in this system report, for example, that they routinely prescribe enough medication for a diabetic to last only a month. The patients must then come to see them monthly, for the simple task of getting a new prescription for insulin. By contrast, a diabetic under routine care by a private doctor would commonly see that doctor perhaps three or four times a year. Patients cooperate with this system because the visits cost them no money, although the time cost is obviously considerable.[9]

The outcome of this type of demand generation is quite predictable. One study (Phelps, Hosek, Buchanan, et al., 1984) showed that military personnel—a group originally selected partly on the basis of excellent health—has nearly three times as many physician visits per year, on average, as a comparable civilian population (9.6 versus 3.4 visits). Some of this may reflect military requirements, but the families of active duty personnel treated within the same system receive a considerably larger number of visits than their civilian counterparts (7.8 versus 5.0 visits), and these people bear considerably larger time costs than the active duty personnel. The incentives built into the military health care system, coupled with the inability of higher officials to monitor individual physician activity closely, apparently create a standard problem of shirking that every large organization faces to some extent.

A final consequence of this approach to providing patient care is also inevitable—the clinic systems clog up with scheduled visits far into the future. Patients attempting to schedule routine visits with doctors commonly find waits of months before they can get an appointment. Not uncommonly, a patient will begin treatment within a military clinic, and then find that follow-up care is not available for many months. An obvious solution is for the patient to go to a private doctor using CHAMPUS. This doctor is likely to repeat any diagnostic tests previously conducted within the base hospital, obviously adding to costs of the overall system. Continuity of patient care also suffers in this environment, with possible degradation of quality of care.

---

[9] For personnel themselves, the visit to a doctor provides an excuse to leave their standard job, so they don't bear any financial cost from this time cost. The military pays this cost as well.

OTHER GOVERNMENT PROGRAMS

Governments operate other health care programs for a wide array of persons, sometimes offering a specific set and sometimes a wide spectrum of services. Governments offering such services range from the federal government (maternal and child health, Veterans Administration), to state governments (mental health services, participation in the federal Medicaid program), local general governments (county hospitals, health department clinics), and sometimes, special "districts" created specifically for the purpose of providing health care (hospital districts, sanitation districts, and— if one wishes to broaden the definition of preventive health— mosquito abatement districts).

Many of these programs serve special and limited populations, defined either on the basis of some particular illness (e.g., mental illness), or limited to a particular population (the Veterans Administration hospitals serve only persons who have worked in the U.S. military, and Maternal and Child Health programs obviously serve populations defined by specific age and gender criteria). County and city hospitals (and local hospital district hospitals) and clinics associated with those hospitals, as well as free-standing clinics operated by county health departments, provide services that are nominally available for every citizen of a geographic region, but in many cases, these programs end up serving low-income persons who have neither private health insurance nor private doctors whom they typically would visit when ill.

Provision of treatment for mental illness in special government hospitals occurred historically for a variety of reasons. Probably one important reason is that while the treatment of mental illness was nominally a part of traditional medicine (and a specialty of some physicians), in actual fact, mental illness has often been set aside as something different, and was not a part of the treatment portfolio of many U.S. hospitals for a long time. Indeed, until quite recently, many serious mental illnesses were considered incurable, and persons with such illnesses were simply "committed" (often with associated legal proceedings) to permanent care in state mental hospitals.[10] Recent innovations in pharmacologic

---

[10] Several movies have focused on these legal proceedings and life in mental hospitals. Perhaps the most famous was *One Flew Over the Cuckoos' Nest*, starring Jack Nicholson. A movie simply entitled *Nuts*, starring Barbra Streisand, portrays vividly the issues associated with commitment of persons branded as mentally ill. The recent *Rain Man*, starring Dustin Hoffman, shows another facet of the issues associated with mental illness and the quasi-legal processes involved in commitment.

treatment of these illnesses has led to widespread reductions in long-term and permanent hospitalization of many of these persons, although substantial concerns now exist that these moves have importantly contributed to the increasing presence of homeless persons living on the streets of our cities.

Mental illness and incarceration of persons described as dangerous either to themselves or to others has been a controversial area of law, and in some societies (perhaps most prominently in the Soviet Union), incarceration under the guise of treating mental illness has provided a common way to get rid of political or social troublemakers.

While we cannot analyze each of these programs in detail, it would be reasonable to generalize as follows: These programs commonly have arisen to serve patient populations that fall through the cracks of the "mainstream" U.S. health care system. The VA system, for example, has always provided a substantial proportion of its services to low-income veterans, commonly persons with mental illness, chronic alcohol or drug dependency problems, or persons with other illnesses related to long-term care. The VA patient population is predominantly male (because of the veteran-status requirement) and, even historically, quite low in income, although any veteran was previously eligible for care. In the late 1980s, the Congress formally limited access to the VA to those veterans with income below $18,000 per year (Hollingworth and Bondy, 1990). County hospitals and clinics, while nominally intended to provide care for any resident of a community, mostly end up serving low-income persons. Medicaid often becomes the primary third-party payer for county hospitals, and Medicaid budget cuts end up mostly shifting financial responsibility for caring for low-income persons from state and federal governments to local governments.

Maternal and child health programs have a special situation, because in this area more than any other, evidence exists that provision of primary and preventive care (prenatal care for mothers, vaccinations and other well-care services for children) actually saves money as well as improving health. Maternal and child health programs in the United States are relatively limited compared with those in other countries, and many analysts of the U.S. health care system predict that universal insurance for this subset of the population will become politically acceptable in the United States sooner than universal insurance, and perhaps even if universal insurance is never enacted in the United States. The potential effects of maternal and child health programs appear more vivid when international comparisons of infant mortality are made; Chapter 17 explores these issues in further detail.

SUMMARY

A number of governmental programs exist to provide health care for special populations, defined sometimes on the basis of income (Medicaid and its predecessors), sometimes on the basis of geography (county and city hospitals, but these too are basically designed to serve low-income persons), sometimes for specific subpopulations defined by age and gender (maternal and child health), and sometimes on the basis of illness category (mental illness). In each case, these programs seem to be societal responses to the lack of care provided for these populations in the mainstream health care system. Analysis of the extent to which society is willing to pay for care for these populations becomes an exercise in the analysis of political behavior more than economics of individuals, firms, and markets.

In some cases, one can use the standard tools of economics to help understand what the political system will decide to do. Studies of the generosity of Medicaid programs, which vary considerably from state to state despite a commonality imposed by federal government rules for participation, offer one opportunity to study such political behavior. One study, for example, showed a very high income elasticity of "demand" for generosity of enrollment eligibility, but a much smaller income elasticity with respect to the level of benefits for persons enrolled.

Governmental programs also offer interesting opportunities to study nontraditional ways of providing care, and to help understand the roles of incentives in different environments for providing and using medical care. The health care system operated by the U.S. military, for example, creates numerous interesting opportunities to study nontraditional modes of providing care (some of which were discussed in Chapter 7 regarding substitution of production). The direct provision of care within the U.S. military health care system and interactions between that system and the CHAMPUS insurance system offer another opportunity for studying the role of incentives in the provision of care.

PROBLEMS

1. Participation in the Medicaid program is like a foreign-trade problem, with a downward-sloping demand curve (for the individual physician) composed of "regular" patients and a source of additional patients that Sloan et al. describe as infinitely elastic, with a price ("willingness to pay") determined by Medicaid.
    (a) Show a picture of this market, and describe what the Medicaid demand curve means.
    (b) What happens if the Medicaid market has a finite number of patients in it, so that (after some quantity) it becomes completely inelastic? What would be the effect on the private-market price of a decrease in the Medicaid compensation rate?

2. The method of financing Medicaid requires low-income states to pay a low share of program costs (say, 25 percent), while for higher-income states, the states' share is higher (say, 50 percent). What effect do you expect this to have, other things being equal, on per capita Medicaid spending in low- and high-income states? (Hint: Use an indifference curve graph to show the effective price to states of Medicaid and other services they might provide.)

3. What effect would you anticipate on private fees charged by a nursing home if Medicaid *lowered* its payment for Medicaid patients? How does this compare (or contrast) with the common assertion by nursing home operators that they must *raise* their private fees when Medicaid doesn't increase payments sufficiently to keep up with cost increases? (Hint: Think about a monopoly pricing model, where the marginal revenue curve from private patients begins above, and eventually descends below the price paid by Medicaid.)

4. Medicaid recipients include 11 million adults, 8.5 million of whom are female and 2.5 million of whom are male. One feature of the program and one biological fact together explain most of this differential. Discuss the program feature and the biological fact, and show how they lead to the preponderance of females among Medicaid beneficiaries.

5. Physician participation in the Medicaid program comes on an all-or-nothing basis, whereas in Medicare, physicians can choose to treat some patients but not others. Particularly given the low fees paid by Medicaid for physician services, would you expect the delivery of Medicaid physician care to be spread evenly across most physican-firms, or to concentrate in a few? Can you make any predictions about amenities and quality of firms providing Medicaid care as a major portion of their business?

6. The military health care system commonly places a "quota" on each doctor, requiring that she or he see a certain number of patients each day. What strategy might doctors take in such a system to decrease their daily work load? (Hint: Think of selective demand inducement for patients of varying degrees of illness.)

Chapter
14

Medical Malpractice

*E*arlier portions of this book have alluded to the important role of the medical malpractice system. In this chapter, we will explore the structure of this system, and learn what is known about the ways in which malpractice law affects providers and patients. These roles and effects are not clearly known at present, and the subject generates considerable controversy in many circles. No discussion, including this one, can avoid offending some parties involved in malpractice law and/or the delivery of health care, because most participants in the health care system and the medical-legal system hold strong views on the subject, often in conflict. Perhaps only the issues of reforming medical malpractice law stand as importantly in the public eye as those of cost control that pervade much of the U.S. health policy debate. As we shall see, even these issues are not wholly separate, since many people blame the increasing costs of our health care system on the increases in medical-legal risk and expenses associated with the legal system.

BACKGROUND ON THE LEGAL SYSTEM IN THE UNITED STATES

"Medical malpractice" is a legal concept, related not so much to the practice of medicine as to the law of personal injury, the law of contracts, and in a very few cases, to criminal law. Thus, before we can study medical malpractice events, malpractice insurance and the consequences of both, we must review the legal system in the United States.

The first issue of importance arises from the U.S. Constitution, which does not withhold for the federal government that part of the law relevant to medical malpractice. Therefore, by default, this area of the law becomes the domain of each state. This means that we do not have "a" malpractice law to consider, but rather, 50 of them. Each state specifies its own laws regarding everything about medical malpractice. The states also control the way doctors, nurses, hospitals, and all other providers of care get licensed.[1] While the laws have evolved in similar ways across most states, important differences remain, and these allow some analysis of the effects of these laws on the rates at which malpractice occurs, the costs of defensive medicine, and the premiums paid for medical malpractice insurance, all issues we will discuss in detail below.

Figure 14.1 shows the important parts of the legal system in every state that affect medical malpractice. This figure looks something like the standard biological classifications, but the "kingdoms" represent "Criminal" and "Civil" law, not "Plant" and "Animal." With very few exceptions, the relevant law for medical malpractice is part of the civil law. Within the civil law, three major branches pertain to medical malpractice—most prominently tort law, contract law, and, finally, the laws regulating the insurance industry.

Tort law provides the basic apparatus to define medical malpractice.[2] This law provides the basis by which one person may bring a lawsuit against another to recover damages for personal injury, one of the many forms of a tort.[3] In tort law, the *plaintiff* makes a claim with the court that the *defendant* has harmed him or her. The lawsuit specifies the acts done by the defendant and the damage those acts imposed on the plaintiff, and asks for *relief*. In medical malpractice suits, the relief most commonly means the

---

[1] The federal government directly licenses use only of drugs controlled by federal narcotic laws.

[2] As will become apparent, while many words in medicine have their origins from the Greek, most legal terms arise from Latin. The word *tort* comes from the Latin word *tortum*, meaning "twisted or distorted." The context here probably better translates as "made wrong." We get the word *tortuous* from the same origin, and those who have encountered the legal system may feel that this sense of the word, as in a twisted maze, better represents what happens in a legal context such as medical malpractice.

[3] The verb *to litigate* means to bring a dispute to a court of law. This also comes from the Latin word *litigare*, which means "to dispute, quarrel."

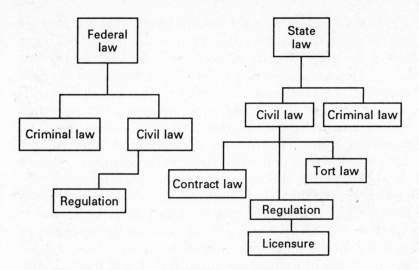

Figure 14.1 The legal system as it affects medical malpractice.

payment of financial damages, the amount of which the lawsuit commonly specifies. Sometimes the relief will include an order from the court to the defendant—for example, that the defendant stop doing the sort of thing that caused the damage (a "cease and desist" order).

The key issue in medical malpractice cases is one of *negligence*.[4] Many medical events have bad outcomes—the patient does not improve, gets worse, or dies. The law recognizes that many of these events would occur no matter what the doctors, hospitals, nurses, and other medical professionals might have done. In the most simple terms, upon which we will elaborate shortly, the law says that a plaintiff has been harmed by the defendant if and only if the injury to the plaintiff was preventable, and that it was "reasonable" to undertake the activity that would have prevented the injury.

Some medical malpractice lawsuits stem not so much from whether an injury occurred (or was preventable) but rather from whether the doctor (or other provider) appropriately warned the plaintiff about the possible risks of a treatment that the plaintiff received. These cases have their roots in the law of contracts, and the broad set of activities that these encompass includes issues of

---

[4] This comes from the Latin word *negligere*, meaning "to neglect." We get the word *negligee* from the same base, via the French.

"informed consent." In the most simple terms, a doctor may not undertake a procedure on the patient unless the patient consents. (Otherwise, the doctor would, in the eyes of the law, be attacking the patient, a form of assault and battery.) For the consent to have legal meaning, the law requires that the patient have reasonably full information about the possible risks involved. Failure to provide or to document that they did provide information sufficient for an informed consent has led to many a lawsuit for doctors. Particularly those doctors who undertake specific "procedures" (such as surgical operations) spend considerable time documenting how they have described the risks to the patient, and almost invariably have patients sign a statement that describes the risks involved.[5] Box 14.1 describes an apocryphal legal event involving informed consent that contains other legal lessons as well.

Lawsuits brought by patients suing for alleged medical malpractice may or may not end up in a trial, and that may or may not include a jury. Either the defendant or the plaintiff may insist on a jury trial, but if both agree to waive that right, the trial can proceed with only a judge hearing the case and rendering the entire verdict. Both sides hire attorneys-at-law (lawyers) to present their cases to the court and to provide legal counsel. Both sides may hire experts to testify about the case, including, in medical malpractice cases, issues of negligence, medical causality, and the extent of damages incurred by the plaintiff.[6] Other "witnesses of fact" may testify about events that transpired, but only experts may render opinions about those events. As we will see below, this process consumes large amounts of resources, so that even plaintiffs who win their litigation often recover far less than half of the total resources devoted to the case.

Most lawsuits in this country, including medical malpractice cases, do not culminate with a trial. Rather, the case is settled in advance of trial, quite often with no payment from the defendant to the plaintiff. (In these cases, we usually say that the plaintiff just drops the case.) One study of claims closed by medical malpractice insurers found that over half the cases were settled with no payment at all to plaintiffs (Danzon and Lillard, 1982).

---

[5] Some doctors tape-record the session during which they describe the risks to the patient, and keep the recording in the patient's record.

[6] A cynic once said that the world consisted of three kinds of persons: liars, damned liars, and expert witnesses, presumably ranked in order of their tendencies toward mendacious behavior.

Box 14.1 **AN APOCRYPHAL STORY ABOUT INFORMED CONSENT**

Issues of informed consent loom large in some medical malpractice litigation, and as noted in the main body of the text, some doctors take considerable effort to document the information they give to patients to "inform" them before receiving their consent to operate. One case reveals how informed consent issues can affect the strategy involved in a specific trial. This case involved a doctor who did repeated back surgery on patients with apparently unfavorable outcomes being quite common. This doctor had been successfully sued by numerous patients, and indeed, one plaintiff's attorney in town apparently made a substantial living from his fees in lawsuits involving Doctor X.

Each medical malpractice case stands on its own merits, so even if a doctor has lost 20 malpractice cases in recent years for the same procedure, information from those previous cases cannot be admitted as evidence into the twenty-first case. In one of Dr. X's cases, his own testimony opened up the floodgates that allowed his past history of lost litigation (and poor outcomes for the patients) into evidence. This specific case involved the question of informed consent. The dialogue in the trial apparently went something like this (this is *not* from the trial transcript):

> PLAINTIFF'S ATTORNEY: Now, Dr. X, did you advise the plaintiff of the risks associated with having this type of surgery?
>
> DR. X: Yes, *just as I always do.*

This response contains five too many words, from the point of view of legal strategy. (It also demonstrates why lawyers advise their clients to answer unfriendly questions as briefly as possible, and not to volunteer anything beyond that.) The last phrase opens up the question about Dr. X's common practices. It admits into evidence other patients who have had the same procedure.

The case proceeded next by having a long series of Dr. X's former patients appear on the witness stand as witnesses for the plaintiff, all in various stages of pain, permanent injury, and in some cases inability to walk. The questions from the plaintiff's attorney proceeded along the following lines:

> PLAINTIFF'S ATTORNEY: Did Dr. X operate on your back?
>
> WITNESS: Yes.
>
> PLAINTIFF'S ATTORNEY: Before that operation, did Dr. X inform you of the possible risks of that surgery?

The answer, of course, was immaterial. The plaintiff's attorney had succeeded in parading a long sequence of Dr. X's former patients into the courtroom. Their mere presence demonstrated that Dr. X had commonly achieved very poor outcomes with this operation. Without the issue of informed consent arising in this case, the attorney would never have been able to bring those previous patients into this case. But of course, that option also required an overly wordy answer on the part of Dr. X to the key question.

The logic for early settlements was first explored by Gould (1973). He showed that, because of the costs of carrying out a trial, settlement was very likely to occur in tort cases whenever the parties agreed on the probability that the plaintiff would win the case. If they agree that the plaintiff's case is very strong, the defendant will settle for an amount quite near the amount requested in the lawsuit. If both believe that the plaintiff's case is weak but has some chance of success, they may well settle for a much smaller amount, making the plaintiff happy to get something out of a "thin" case, and the defendant happy to get out from under a potentially expensive risk. The logic for settling such cases closely resembles the logic for purchasing insurance that we discussed in Chapter 10. (A further discussion appears two sections below regarding the economics of settling and the value of establishing a reputation in the legal arena.)

Danzon and Lillard's study (1982) found that settlements averaged three-quarters of the size of their best prediction of what a trial award would have been had the case proceeded to trial. They also found that settled cases tended to have much smaller awards than those going through trial; the larger stakes in cases with most severe injuries led both plaintiffs and defendants to show reluctance to settle. Danzon (1983) has concluded that "self-serving rationalism [i.e., pure economic incentives] largely explains the outcome of malpractice claims on average." Of course, this still leaves considerable latitude for court error, different motives (including malice and spite), and suboptimal strategy pursued by plaintiffs and/or defendants in individual cases.

Each state's tort law defines the general terms by which plaintiffs may file medical malpractice suits against providers of medical care. These laws specify the nature of what constitutes negligent behavior, specify the statute of limitations (the time that may pass after an injury during which the patient may still bring suit), the nature of allowable evidence, and in many states now, the size of damages that plaintiffs can recover for some forms of injury.[7] Naturally, these laws also affect the propensity to settle a

---

[7] Many states now set limits, for example, on the size of awards for "pain and suffering" associated with an injury. Because of the highly subjective nature of pain and suffering, and because of the occasional extremely large awards granted by courts for pain and suffering, these types of award became an obvious point of contention during the medical malpractice crises of the 1960s, 1970s, and 1980s, and many state legislatures chose to set limits for this area of damages in particular.

case in advance of trial to some degree, as does the backlog of cases within the court system.

Another relevant body of the law is that providing regulation of insurance within each state. Hospitals, doctors, dentists, podiatrists, nurses, psychologists, and other providers of care all face financial risk associated with their attempts to heal people's injuries and illnesses. The tort laws within which they must operate create financial risk for those providers. These providers of care come equipped—at no extra charge—with utility functions. If those utility functions make the individuals risk averse (as we would normally presume in economic analysis), then the risk created by the law leads them to seek insurance against those risks. The risks arise in several forms. First, all providers know that they have some chance of engaging in negligent behavior, either through mistaken knowledge, sloppiness, or fatigue. The imperfection of the court system creates additional risks; sometimes the patient may suffer an adverse outcome, but not through negligent behavior of the provider. Even so, patients cannot always distinguish an injury caused by negligence from one caused by something else, and they may sue because of the injury itself. An imperfect court system will allow awards to some of those patients, even though negligence did not occur in some "pure" sense. This aspect of risk—the whimsical court system—bears heavily in the minds of many providers of care. Every reader of this book has probably read about or personally encountered some medical malpractice lawsuit that appeared to be "without merit" by many standards. Providers fear this risk considerably—indeed, perhaps pathologically—and it forms a strong incentive for them to purchase medical malpractice insurance.

As with tort law, the federal government has no control over the provision of insurance, so each of the 50 states defines the rules by which medical malpractice insurance can be provided. Some states actually enter the business of providing such insurance to doctors and hospitals, and others regulate the premiums that insurers can charge for such insurance. The various rules adopted by the different states have provided another "laboratory" to study the effects of insurance regulation on insurance rates, malpractice itself, and the effects of insurance costs on providers' participation in the market.

Finally, we can note in passing the (small) role played by the criminal law in medical malpractice. Criminal cases differ from civil cases in two important respects. First, guilty defendants can be sentenced to prison, in addition to monetary fines; civil cases cannot lead to this outcome. Second, only the government can bring criminal charges against an individual, a role served by "district attorneys" and their counterparts at various levels of government. In medical events, the most common source of a criminal charge involves financial matters (fraud), commonly perpetrated against a government provider of insurance such as Medicare or Medicaid. For true medical issues, a doctor's behavior sometimes becomes so bad that the state files a charge of *criminal negligence* against the doctor. Cases sporadically reach the courts under charges of homicide when doctors terminate life support for individuals, or assist patients in committing suicide.

THE ECONOMIC LOGIC OF NEGLIGENCE LAW

> We have done those things we ought not to have done, and we have left undone those things we ought to have done. [Prayer of Confession]

Why do we have a law of negligence? Laws defining negligence (and responsibility to pay for damage) have two obvious purposes: to compensate victims ("fairness") and in an efficient way, to deter people from harming others. Note the important condition—in an efficient way. One could think about trying to deter *all* damage that one person might inflict upon another, but (as we shall see) the costs of doing that would become prohibitive. Because of the costs of preventing every possible harm, a zero-damage world is not optimal. However, in any overall analysis of the negligence law, these goals stand out. If we cannot accomplish either one of them well with the negligence system, then consideration of alternative legal systems seems desirable.[8]

We can now turn to an analysis of negligence law and its application in medical malpractice. Negligence occurs when a provider of care harms a patient, and in addition, could reasonably have prevented that harm. In a classic negligence case, in order to gain compensation from the defendant, the plaintiff must prove not only that he or she was harmed, but also that the actions of some provider of care ("doctor" hereafter, for simplicity) was responsi-

---

[8] The alternatives include a "no fault" insurance system, as some states have adopted for auto insurance, or a "strict liability" system wherein the provider of care pays for any harm, whether negligent or not. Workers' compensation for on-job injuries offers an example of the latter system.

ble for the harm. Further, the plaintiff must show that the doctor's behavior did not meet "reasonable" standards of care.

In many cases, the relevant standard of care derives from "local custom," so in many malpractice cases, the testimony of other doctors is required to establish the relevant local custom of care. Put simply, if a doctor's behavior deviated importantly from local custom, that allows the inference of negligence. This standard of care has fallen by the wayside, giving way to a more "national" standard, reflecting the widespread dissemination of medical information through journals, continuing medical education seminars, national and regional medical meetings, and even a specialized cable television channel devoted in part to providing doctors with "current" information about modern methods of diagnosis and treatment. Interestingly, the old "local" standard of practice accurately captured the reality of widespread differences in "culture" about the correct ways to use various medical interventions, as documented well in the past several decades in the "medical variations" literature. (See the discussion in Chapter 3 about variations in medical practice to refresh your memory about this phenomenon.) Given the widespread variations in practice that we can readily observe, it would appear relatively easy to find a doctor from *some* community who would testify (correctly) that the custom of care in her own community differed from the choices made by a doctor in some other community. Reliance on "local" custom eliminates that as a possible strategy by plaintiffs' attorneys. The converse also appears to be true: Requiring that experts come only from the local community inhibits the ability of plaintiffs to establish an act of malpractice, because it would require doctors to testify, in effect, against their personal acquaintances and friends.[9]

The particular ways in which medical malpractice law has defined negligence (e.g., comparison with local custom) have been only a proxy for the more general approach. Negligence law in its most general structure offers a standard that has compelling economic logic. This standard, first set forth by Judge Learned Hand,[10] forms the "classic" basis for defining negligence. In the

---

[9] Kessel (1958) argued that the unwillingness of doctors to testify against one another was part of a widespread strategy of the medical profession to feather their own economic nests at the expense of patients.

[10] It may be hard to imagine a person with a name more fitting for a judge. One can just visualize the "learned" hand writing "correct" opinions of law.

Hand Rule, negligence occurs when doctors fail to take some action to prevent harm, *and* when it would on average cost less to prevent that harm than the costs of the harm itself. This definition closely matches the economist's prescriptions in cost-benefit analysis. More formally, suppose p = probability of some harm occurring without any intervention by the doctor, D = the amount of damages incurred if the harm takes place, and C = the cost of preventing the harm. Then the Learned Hand Rule says that negligence has occurred if

$$C < p \times D$$

This rule, written in this way, suggests that the doctor has yes-no choices (either do or don't do some preventive activity). However, the logic of the rule also holds up when the choice involves "how much" or "how often" to do something. One interprets the rule in such cases in terms of incremental costs and incremental effect on the probability of harm, the extent of damage, or both.[11] Negligence occurs whenever a doctor doesn't undertake "enough" activity to prevent harm, with "enough" defined specifically in terms of incremental costs versus incremental benefits.

This formulation of the law has a powerful economic logic to support it; it places the burden of behavior on the person with the greatest knowledge about the appropriate technology—the doctor. In effect, it tells the doctor, "If you behave in a way that would minimize social costs of harm, you will never harm a patient negligently." Rationally behaving doctors would always advise their patients to do things in a way that corresponds exactly to economists' prescriptions for efficient use of resources. Rational patients would also always accept such advice, for obvious reasons.

Notice also that *in concept,* the Learned Hand Rule deters doctors from doing "useless" procedures, and indeed from doing "too much" medical intervention. Broadly interpreted, the

---

[11] In terms of the calculus, consider a "social cost" function $SC = (p \times D) + C$, with these terms defined as above. The problem is how to minimize social cost. If both p and D fall as C rises (i.e., the "preventive" or "safety" activities actually have some effect), then we could find the point of minimum social cost by taking the derivative of SC with respect to C and setting the result equal to zero. This would occur if $Ddp/dC + pdD/dC + 1 = 0$, or equivalently, $-(Ddp/dC + pdD/dC) = 1$." (Remember that these derivatives are negative if prevention "works.") If one stopped doing preventive activities at some point before the optimum, then social costs would still fall as preventive effort increased, i.e., $dSC/dC < 0$. Under the logic of Judge Hand, stopping "too early" in the application of preventive activity would constitute negligence.

Learned Hand Rule would say that a procedure was unwarranted if the costs exceeded the benefits. However, in practice, the rule seldom if ever gets applied in this way. Patients almost never sue doctors for doing (for example) useless surgery. The reasons why such cases seldom appear are complex, but depend at least partly on the notion that the patient agreed to have the doctor do the procedure ("consent"). Thus, cases like this would usually rest on the question of whether the consent was really "informed" or not. In addition, the patient would have to show harm; courts have proven reluctant to award damages unless the procedure turned out badly. They will not award damages for the cost of the procedure itself, except in very rare cases.

JUDICIAL ERROR, DEFENSIVE MEDICINE, AND "TOUGH GUYS"

In a world with perfectly functioning courts, doctors behaving according to the Hand Rule would never lose a medical malpractice suit. Alas, the legal system does not perform perfectly.[12] Courts make errors of omission and commission, just as doctors do. Doctors often feel that courts award cases to plaintiffs just because a bad outcome has occurred, rather than only when negligence has occurred. (We will see some evidence on this specific issue shortly.) In order to prevent this sort of problem, many doctors assert that they undertake "defensive medicine" that is, the carrying out of medical procedures only for the purpose of preventing lawsuits, rather than something that they think is medically appropriate.

"Defensive medicine" is difficult to measure, perhaps impossible. Given the widespread variations in behavior of doctors' use of various interventions, it may well be that a considerable fraction of doctors believe that something is "medically inappropriate" when in fact, by a careful application of the Learned Hand Rule, it would be appropriate. In this case, negligence law would actually increase patients' well-being by "forcing" doctors to do things (such as carry out diagnostic tests) that they would not normally do. If one conducted a survey, those doctors would describe such activities as "defensive medicine," and bemoan the added costs to patients of such activities. However, under the logic of the Learned

---

[12] This may come as a shock to some attorneys.

Hand Rule, total costs will fall with the use of such interventions, so patients have actually been made better off. Thus, negligence law can *in concept* serve to offset untoward variations in the use of medical care.

Danzon (1990) has studied how patterns of medical care differ across regions with different rates of malpractice claims filings. (Recall that we have 50 distinct malpractice environments, creating somewhat of a natural experiment for studies like this.) She found that rates of use of standard "defensive medicine" activities like X-rays and laboratory tests were generally unrelated to measures of patients' propensity to sue.

Doctors commonly complain that medical malpractice trials represent an expensive form of Russian roulette, wherein plaintiffs make outrageous claims, seek a settlement for a relatively small fraction of the claim, and occasionally go to trial and win a large verdict by convincing a sympathetic jury to help them, even when no negligence has actually occurred. In the view of many doctors, the medical malpractice system is worse than an "imperfect" system, because it allows legal "blackmail" of doctors who have done no wrong.

The essence of the "blackmail" idea comes from a game-theoretic analysis of the problem. Individual doctors facing a medical malpractice suit confront the costs of defending the case, not only in financial outlay to lawyers, but also in their own time in preparation for the case and in court. Most doctors carry malpractice insurance that covers not only payments to plaintiffs but also the legal costs of a defense, but their own time and effort (and also the mental anguish associated with the event) are not insured. As a result, each individual doctor faces some incentives to settle a case before trial, if for nothing else than to avoid the legal costs of a complete defense.

In game theory, when an individual repeatedly plays the same "game," it often pays off to establish a reputation as a "tough guy" player. If a plaintiff's attorneys know that doctors (and hospitals) have such reputations, the attorneys will have less interest in filing frivolous suits. However, no single doctor has much incentive to spend the time and effort to establish such a reputation, because the chances of their being able to take advantage of the investment are small. In this sort of game, particularly when the courts even occasionally err in awarding verdicts to plaintiffs where negligence has not occurred, settlement often seems the wiser choice for each individual doctor. Thus, the effort spent in establishing a "tough guy" image has many aspects of a public good with regard to the medical community; every doctor would benefit from a "hard line" stance relative to frivolous lawsuits, but no individual doctor has the incentive to help establish such a reputation.

One important party in the "lawsuit game" does have an incentive to establish such a reputation, however—the insurance company that provides the medical malpractice insurance to doctors. These companies often guide (or at least advise) the defense of a medical malpractice claim, and they sometimes have the contractual right to determine when a settlement will take place. The ability of an insurer to establish a "tough guy" reputation may be one of the reasons why medical malpractice insurance sales are typically quite heavily concentrated (few companies in a single region). If any single insurer can establish a "tough guy" reputation in bargaining with plaintiffs' attorneys, it will have created a competitive advantage that will allow it to dominate a market.

MEDICAL MALPRACTICE INSURANCE

Almost all doctors and hospitals carry medical malpractice insurance. This insurance pays for the costs of defending medical malpractice cases and also for any awards against the provider. For individual doctors, the amount of insurance varies, but commonly doctors carry something like "$1 million, $3 million" in "basic" coverage, meaning that the coverage will pay up to $1 million for a single judgment, and $3 million total during the life of the contract. (Despite the occasional news story to the contrary, very few medical malpractice judgments exceed $1 million, and those that do often have multiple defendants—a hospital and several doctors— to share the cost.) A 1989 survey showed that nearly half of the doctors carried this type of insurance (Paxton, 1989). This same survey showed that about half the doctors carried a second "excess coverage" policy that most commonly added another $1–$2 million in coverage. Doctors carry such insurance for precisely the same reason that individual persons carry health insurance—to reduce the uncertainty associated with a financial risk—for precisely the same motivations. Hospitals also commonly carry liability insurance, since they too can be held accountable for harm to patients, even if caused by a doctor's negligence rather than that of any person employed by the hospital.[13]

EVIDENCE ON ACTUAL DETERRENCE

Does the medical malpractice system actually deter negligent behavior? This question sits centrally in the debate over medical malpractice reform. If we achieve no deterrence from this system, then reform seems more desirable, for we could certainly accomplish the goals of compensation of harmed persons more cheaply with alternative systems such as a no-fault system or a social insurance system.

In order for deterrence actually to occur, several things must take place. First, injured patients must bring suit against providers. If the patients do not sue doctors, then no deterrence can possibly occur. Second, the doctors must bear the financial brunt of their mistakes. Insurance may undo any incentive effects that the legal system generates, depending on how the insurance policies are priced to doctors.

Do Injured People Bring Lawsuits?

The question of whether injured people actually bring lawsuits turns out to be fairly difficult to answer in practice. To do that, one must first find a set of people who have been injured by negligent behavior and then determine whether they filed lawsuits, or accomplish some comparable method to estimate the numbers of negligently injured patients and the number of suits filed. Two separate studies have now undertaken that task, using similar methods, and reaching quite similar conclusions.

The first of these studies took place in California (a hotbed of medical malpractice cases and precedents), relating to hospital care in 1974 (Mills et al., 1977). The California Medical Association and the California Hospital Association hired experts to study

---

[13] Recall here the usual relationships doctors and hospitals have in general: Hospitals do not usually employ doctors, but rather grant them "attending privileges." However, courts of law have commonly held hospitals liable for harm created by doctors on their staffs, in part because the hospital's own staff (e.g., nurses, pharmacists) sometimes participates in the negligence, and if for no other reason, because the hospital's board of directors has the ultimate legal responsibility for assuring the quality of the doctors.

the medical records in 23 hospitals scattered throughout the state, sampling from their medical records. The medical-legal experts studied the records looking for *documented* evidence of injury due to negligence. They concluded that approximately 1 in every 125 patients was negligently injured during a hospital stay. From this sample, they estimated how many injuries occurred for the entire state in 1974, and then compared this estimate with the number of lawsuits actually filed relating to that care. (The lawsuits could have been filed immediately or any time in the next four years and they would still count.)

The results proved somewhat disheartening for those seeking important deterrence from the medical malpractice system. The California study estimated that less than one injured person in ten files a lawsuit. Of those (remember—these were people that the CMA and CHA experts thought had been injured), less than one-half actually received compensation. Using the experts' judgment as a "gold standard" this suggests substantial court error in favor of hospitals and doctors. Only 1 out of 25 injured patients received compensation! This probably overstates things as well, since there were almost certainly other injured patients for whom hospital personnel did not record the event in the medical record, thus escaping detection by the CMA/CHA experts. It also completely ignores malpractice events occurring in doctors' offices rather than in hospitals (although suits arising from such events would have been counted).

Since this study was conducted, the rate of malpractice suits has dramatically increased. By 1985, the rate of malpractice suits had doubled compared with 1978, the period of this CMA study (Danzon, 1985a). Even so, this says that at maximum only about one in five injured patients brings suit.

A study conducted in 1989 using 1984 hospital data from New York State produced quite similar results (Harvard Medical Malpractice Study, 1990). The researchers sampled over 30,000 medical records from 51 hospitals throughout New York State. They again found (as in the earlier California study) that only a small fraction of negligent injuries led to patients' filing claims.

Specifically, they estimated the rate of malpractice events per patient in one part of their study, and malpractice legal claims per patient in a separate analysis. Combining these two sources of evidence allows an estimate of the rate of claims per incident. Using a variety of estimates of the number of claims filed (a fact not readily determined, due to the dispersed nature of the legal process and the fact that only cases ending in trial get "counted" for sure), they concluded that about 8 times as many people were injured as filed suits, and that about 15 times as many people were injured as received compensation. (The step from 8:1 to 15:1 occurs because not all injured patients win their cases or receive settlements.)

The New York State study also attempted to learn what sorts of injuries did not end up in litigation. They took all cases with "strong" evidence of negligence, and classified the degree of injury from one (not serious) to five (disability resulting in at least 50 percent decrease in social function) and six (death). They concluded that those not filing claims tended to be those with less serious injury (usually self-healing in less than six months) or very old persons with limited life expectancy.

Why do patients bring so few lawsuits? The economics of the problem dictate the answer in part: If the injury had small consequences, any court award would presumably also be small. If the activity of bringing lawsuits to court has any important fixed costs (such as shopping around to find an attorney), then small claims will tend to get filtered out. Indeed, the California study classified 80 percent of the injuries as having only temporary consequences, perhaps a better testimony to the healing power of the body than to the efficacy of medical care. The propensity to sue seems to increase with severity of injury, consistent with this "fixed cost" idea. For minor injuries, roughly 1 out of 13 patients brought suit, while for permanent injuries, 1 out of 6 brought suit.

A common complaint by doctors and providers says that patients bring lawsuits whenever *any* poor outcome occurs during medical treatment, whether due to negligence or not. The CMA/CHA study refutes this strongly. For each negligent event they found in the 23-hospital sample, they found over five more nonnegligent "incidents" that should not count as malpractice. If any significant proportion of those nonnegligent injuries had filed suit, such suits would have swamped the suits filed for negligent injury.

Of course, due to the design in the California study, we have no way to determine how many of these nonnegligently injured patients also brought lawsuits (and possibly won them, through a different type of court error). The actual mix of lawsuits observed in California relating to the care delivered in 1974 surely contained some "false positive"suits—people who were not *negligently* injured but did bring suits.

As for the central question of the extent of deterrence, we still have no conclusive evidence on this matter. We do know (from the studies discussed above) that relatively small fractions of injured patients bring suit, and this logically tells us that any incentives for deterrence have been seriously blunted as a result. However, direct evidence on the extent of deterrence still remains elusive.

Malpractice Insurance and Deterrence

Malpractice insurance presents an interesting and important dilemma for those trying to assess the role of liability in deterring injury. The incentives for individual providers—particularly doctors, but also hospitals—to purchase malpractice insurance seem quite clear. Because of risk aversion, when the courts allow plaintiffs to bring lawsuits against defendants, the financial risk creates obvious incentives to seek insurance. However, just as health insurance creates incentives for patients to increase medical care use, malpractice insurance *may* dull physicians' incentives to treat patients with proper caution. The importance of this effect hinges on the way the insurance companies do business, including how they select their "customers"(i.e., select which doctors they will insure), the information that they use about doctors in setting insurance premiums, and finally, the amount of "risk reduction" activity they undertake.

To begin, suppose that only two types of doctors existed in terms of their propensity to injure patients negligently. Call the probabilities of injury p_L and p_H (L and H for low and high risk), and suppose for greatest simplicity that when doctors harm patients, they create the same level of damage (D). Thus, the expected damage they would create would be $p_L D$ and $p_H D$ respectively for low- and high-risk doctors. Finally, suppose that some effort to prevent harm, costing C, sufficed to make any high-risk doctor a low-risk doctor (this makes doctors equal except for their efforts at risk prevention) and that the Learned Hand Rule would make it negligent *not* to use this level of effort (see previous discussion defining negligence).

We can now see the effects of insurance company pricing practices. If insurers could identify high- and low-risk doctors in advance, and charge them insurance premiums reflecting their expected damage, then every doctor would decide to undertake the risk prevention activity, thereby both reducing malpractice insurance premiums and minimizing the amount of harm created. In such a case, malpractice insurance would not blur the incentives of the legal system at all.

Now suppose in contrast that the insurance company charges every doctor the same premium, no matter what their risk. If some share of the doctors (s) decides to spend the money and becomes low-risk doctors, the insurance premium each doctor paid would be $R = sp_L + (1 - sp_H)D$. It turns out that under a remarkably wide range of circumstances, this "community rating" form of insurance blurs and may nearly completely eliminate at least the most direct and obvious economic incentives to undertake any damage-preventing activity. (Box 14.2 shows why this happens.) Thus, although negligence law makes doctors liable for harm in the case we have constructed, community-rated insurance may well lead all doctors to choose not to prevent the risk, driving s to zero, and the insurance premium will rise from $p_L D$ to $p_H D$ for everybody.

Premiums do vary considerably by specialty, a characteristic of physicians that insurance companies can readily observe. Every insurer rates doctors by their type of practice, a specific form of experience rating. Box 14.3 shows typical premiums paid by specialty across the country. Another way to look at the effect of specialty examines the premiums within a specific region. This "holds constant" the legal structure, and provides a clearer picture of the specific effect of specialty on malpractice premiums and (by inference) costs of malpractice claims.

In one important study of medical malpractice events, Rolph (1981) showed that insurance companies have at their disposal—but seldom use—information equally as useful in predicting future medical malpractice costs as the information they do use (medical specialty). This study showed that, holding specialty constant, previous claims experience explained as much of the variability in awards against doctors as did the information about specialty. For example, the *average* claims experience against Class VII doctors (e.g., neurosurgeons) is about seven times larger than the average for a Class I doctor (e.g., family practice). Premiums paid by doctors in these specialties reflect these costs. However, it is also true that the average claims experience from those with the biggest costs in the past four years (holding specialty constant!) will be about seven times larger than those in the lowest-cost group over the past four years. Perhaps even more remarkable, insurance

Box 14.2   SHOULD DOCTORS BOTHER TO PREVENT RISKS? AN APPLICATION OF GAME THEORY

A simple game theory model shows that, when the pool of doctors gets large enough, and all doctors pay the same malpractice insurance premium ("community-rating") nobody would rationally take any costly steps to prevent injuries to patients. To do this, we can define the payoffs to a doctor taking steps to reduce negligent harm, and then show that for a sufficiently large pool of doctors (or sufficiently costly harm-prevention activity), the strategy of undertaking no prevention activity *dominates* the other choice, no matter what other doctors do. In such a case, rational behavior by doctors will lead to no harm-prevention being undertaken.

Suppose first that the insurance pool contains 20 otherwise identical doctors, each of whom will create $100 in damage if they take proper injury-prevention steps (costing C each). Without the injury prevention, each doctor creates $200 in damage. Thus, the Learned Hand Rule would require them to spend C each, so long as C were less than $100 per doctor.

Now consider the payoffs to spending money on prevention if the doctors all pay the same malpractice insurance premium. The following table shows these outcomes, where the first number shown in each pair shows Doctor A's outcome (total costs of malpractice insurance plus prevention costs), and the second number of each pair the outcome for all other doctors.

		All other doctors	
		Prevent	Do not prevent
Dr. A	Prevent	$100 + C$, $100 + C$	$195 + C$, 195
	Do not prevent	105, $105 + C$	200, 200

Think about the problem from Dr. A's perspective. If all other doctors undertake prevention, Dr. A's lowest-cost choice is to "not prevent" if $C > 5$. (Compare $100 + C$ with 105.) The same also holds true if all other doctors choose "Do not prevent" as their strategy (compare $195 + C$ with 200). Thus, no matter what other doctors do, Dr. A. will select "Prevention" if and only if $C < 5$. It's easy to prove that the same result holds if some intermediate share of other doctors choose to prevent.

Obviously, this presents a very different incentive than the no-insurance case. In effect, the incentive has been cut so that doctors will undertake prevention not at a cost of $100 or less, but only at $5 or less. The fact that $100 ÷ 20 = $5 is no accident: The 20 doctors in the pool blur each other's incentives by a factor of 20. It is also easy to see that if we make the pool contain 100 doctors, rather than 20, the incentive gets worse. In that case, doctors would individually choose to prevent harm only if $C < 1$. As the pool becomes very large, the incentive to undertake costly injury prevention activities vanishes to zero.

This problem is a classic "noncooperative game" problem; if everybody could collude (cooperate) they would all be better off, but if they can't collude, they individually take actions that make them all worse off. Here, "patients" and "society" are also made worse off by the uncooperative game solution, since doctors rationally choose (individually) not to undertake economically desirable prevention activities. In settings like this, regulation sometimes helps out.

companies typically do not make use of previous claims history to set premiums for doctors within their specialty groups. Within specialty groups, they practice something akin to "community rating," whereby everybody in "the community" pays the same premium, even if their costs will differ predictably in the future.

Why insurance companies typically choose not to use previous experience by doctors in setting malpractice insurance premiums remains something of a mystery. It certainly has the effect of reducing incentives for doctors to choose appropriately safe care (see Box 14.2).

Box 14.3 COSTS OF MALPRACTICE INSURANCE

How large are medical malpractice premiums? The answer is about as diverse as the profession of medicine itself, with premiums varying hugely by specialty, region, and year. Premiums have grown greatly over the past 15 years, explosively during periods of medical malpractice "crisis" as the data in Table A, a (sporadic) time series reveal.

Table A

	AVERAGE PREMIUM	AVERAGE ANNUAL GROWTH DURING INTERVAL
1974	$ 1,300	—
1976	3,000	52%
1981	3,650	4%
1983	4,170	7%
1986	8,040	25%
1987	9,630	20%
1988	10,950	14%

Premiums differ hugely across specialty. For example, in 1988, doctors reported the following median premiums (the amount such that half of the doctors pay more, and half less); the last column of Table B compares each specialty's median premium to that of general practice doctors.

Table B

SPECIALTY	PREMIUM	RATIO TO GP
Cardiovascular surgeons	$32,120	4.68
Family practitioners	6,580	0.96
General practitioners	6,860	1.00
General surgeons	22,500	3.28
Internists	6,090	0.89
Neurosurgeons	43,500	6.34
Obstetrics/gynecology	34,170	4.98
Orthopedic surgeons	30,630	4.47
Otolaryngologists	19,830	2.89
Pediatricians	5,680	0.83
Plastic surgeons	26,500	3.86
Urologists	14,680	2.14
Average—all surgical specialties	25,000	3.65
Average—all nonsurgical specialties	6,420	0.94

These median premium figures mask a large variability, some of which is explained by cross-state differences and urban-rural differences. Consider, for example, the distributions of claims for two extreme specialties in terms of median premiums costs—neurosurgery and pediatrics, in Table C. Obviously, neurosurgeons' premiums center at the high end of the distribution, while pediatricians' premiums at the low end. Both distributions are quite wide, but they barely overlap.

Table C   Percent of Doctors Paying Premium

PREMIUM	NEUROSURGERY	PEDIATRICS
Over $50,000	34%	1%
$40,000–$50,000	25%	—
$30,000–$40,000	20%	1%
$20,000–$30,000	9%	2%
$15,000–$20,000	4%	4%
$10,000–$15,000	5%	11%
$ 8,000–$10,000	—	13%
$ 6,000–$ 8,000	2%	13%
$ 4,000–$ 6,000	1%	29%
$ 2,000–$ 4,000	—	20%
Under $2,000	—	6%

Even if individual insurance companies do not use "experience rating," (i.e., charging according to past risk, on the presumption that this predicts future risk), another common industry practice could lead to a similar result. Some insurance companies specialize in "good-risk" doctors only, while others accept any applicant. (See Box 14.4, which illustrates this phenomenon.) However, some states negate this market-based experience rating by operating a state-owned malpractice insurance plan that sells insurance to any doctor. Obviously, if the lowest-risk doctors can all get malpractice insurance through private firms, the state-operated plan will have an unusually large concentration of poorer risks. Unless the state-operated plan prices the insurance appropriately, the overall distribution of insurance costs will not reflect risks appropriately, and the incentives for careful practice of medicine will get blurred or eliminated.

Box 14.4 INSURANCE COMPANIES SPECIALIZE IN VARIOUS TYPES OF RISK (MARKETS PRODUCE "EXPERIENCE RATING" EVEN WHEN FIRMS DON'T)

Even if individual insurance firms don't use experience rating to price their insurance, the market may produce an equivalent result. That is, every firm might charge each of their customers the same price, yet each firm may accept different classes of risk. This can readily lead to high-risk customers paying higher rates, and low-risk customers low rates, even if no single firm charges different rates to different risk classes. All that it takes is for the "good risk" customers to shop around and find the best rates. The "bad risk" customers will end up in a pool containing mostly bad risks, and the rates will necessarily reflect this sorting in the marketplace.

This phenomenon occurs commonly in automobile insurance. Look in the Yellow Pages of your telephone directory, and you will find good evidence of this. For example, in the Rochester, N.Y., directory, one advertisement appears with the following message:

> UNLIMITED INSURANCE SERVICES
> No one refused
> Drivers of any age
> We can help:
>   DWI/DWUI
>   Accidents
>   Cancelled Policy
>   Moving Violations

By contrast, consider this advertisement from the same Yellow Pages:

ARE YOU PAYING TOO MUCH?
Business and Personal Insurance
Preferred   Financial
Driver Rates   Planning

To which firm would you turn for insurance if you were 40 years old and had never had a moving violation or accident? If you were 18 years old and had a DWI accident last month? Which firm will sell insurance with the lowest total premium for (say) $150 deductible collision insurance for a 1984 Chevrolet sedan?

Some states even *require* that auto insurance companies charge the same premium to every customer. If a company wishes to evade that regulation, it establishes a portfolio of four or five companies with similar names but different rates. When a potential customer comes to one of their agents, the agent determines which risk class the customer belongs in, and then "assigns" him to the corresponding insurance company from the portfolio available. Of course, "independent" insurance agents (those selling the insurance of many companies, rather than working for a single insurer) have the same opportunity available even if no single company establishes a portfolio such as the one just described.

Why do states operate such plans? One reason they do this is try to keep doctors from leaving practice within their state, on the grounds that a "doctor shortage" exists, and they need to do everything they can to retain doctors. Ironically, getting a few doctors out of practice may be the best thing that could happen to malpractice insurance premiums, as we shall see momentarily.

MALPRACTICE AWARDS: "LIGHTNING" OR A "BROOM SWEEPING CLEAN"?

The belief held by many doctors, some patients, and probably even some attorneys says that malpractice claims are completely whimsical, striking providers at random, much like lightning bolts from the sky, unrelated to their quality of care. A senior official of the American Medical Association recently summarized this view, asserting:

> The physician who gets sued is not the incompetent one, but rather the physician at the height of his career who is using frontier medicine on very sick patients, where the uncertainty of outcome is very

high. All the doctor-owned insurance companies have documented this: they do not have large numbers of repeaters in terms of physicians being sued. [J. S. Todd, M.D., quoted in Holzman, 1988.]

The alternative view—indeed, the one embedded in the theory of deterrence—says that tort awards punish doctors for economically inappropriate behavior (as defined by the Hand Rule), and that if a doctor persists in such behavior, we should expect to see multiple awards against that doctor.

Several studies have now produced strong evidence on this issue: The awards do not appear at random, but rather concentrate heavily within a few doctors. One analysis of claims in Southern California showed that doctors who had previous claims stood a much larger chance of having successful claims against them subsequently. More precisely, of over 8000 insured physicians, 46 (0.6 percent of the total) accounted for 10 percent of all claims, and 30 percent of all awards. The chances of this occurring randomly if all doctors had equal probability of getting "struck by lightning" are vanishingly small, refuting the idea that the awards really are unrelated through time (Rolph, 1981). Note also that the assertion by Dr. Todd quoted above does not conflict with these results; indeed, it supports them in many ways—very few "repeat offenders" exist in this Southern California study—but they account for a disproportionately large fraction of total costs!

The awards may be related through time, but does this imply that the doctors with multiple awards against them have poorer quality of care? A more recent study (Sloan et al., 1989) studied insurance claims against doctors in Florida, and found few relationships with usual markers of quality. Board-certified doctors had worse, not better, claims experience than their uncertified counterparts. They found no consistent effect of medical school rankings, foreign versus American medical schools,[14] or between solo versus group practice. They *did* find that doctors with more frequent claims against them were more likely to have had complaints filed with the state medical examiners board.

The overall evidence is thus somewhat scattered, but leaves several clear impressions. First, if malpractice awards are like "lightning," then it clearly strikes a small handful of doctors with much higher than normal frequency, even after accounting for the

---

[14] Many people believe that foreign medical schools produce a lower-quality doctor than U.S. schools.

effects of their specialty choice. We can conclude either that they practice unusually poor medicine or that they have a manner of dealing with their patients that invites lawsuits, independently of the quality of medicine they practice. It would also appear that many of the usual markers for quality, such as board certification, do not help detect such doctors.

TORT REFORM

Beginning with the malpractice "crises" of the 1970s, state legislators have been under common if not constant pressure to reform medical malpractice law. Part of the pressure obviously comes from doctors footing the bills. (They have incentives to do this so long as they cannot pass along all of the costs to patients in the form of higher fees.) Tort reforms have steadily eroded the opportunity for extremely large plaintiff awards. A common change capped the award for "noneconomic" damages (e.g., the so-called "pain and suffering" awards) to $250,000 or some comparable amount.

Other changes aim more at the plaintiffs' lawyers. Common practices in the legal profession have plaintiff lawyers essentially becoming equity partners in a claim, taking the case on for a share of all awards, rather than for per-hour fees. From a purely economic standpoint, this "contingent fee" system has some positive features. First, it should prevent the filing of frivolous claims, since the attorney confronts substantial fixed costs even to file a claim, and would receive no compensation if no award materialized.

incentive to take on cases for lower-income patients, since many medical malpractice awards use lost wages as a basis for computing awards. For example, in a study of wrongful death awards, one study found a strong relationship between lost earnings and actual awards from the court (Perkins, Phelps, and Parente, 1990). Thus, some people have expressed concerns that the current malpractice system, coupled with a contingent fee system, "shuts out" low-income people from receiving compensation.

Highly visible awards to plaintiffs—shared commonly on a one-to-two basis with attorneys (one-third to attorneys, two-thirds to plaintiffs)—created a strong public sentiment against the common structure of plaintiff attorneys' awards. For example, the landmark California malpractice reform of 1975 limited attorney fees to 40 percent of the first $50,000 of award, descending down to 15 percent of incremental awards at higher levels. Thus, plaintiffs' attorneys in California and elsewhere now face a declining marginal revenue schedule with the size of the award, both reducing the incentive to file for large cases and the incentive to gather evidence that would increase the magnitude of awards.

Another approach has particular importance in a few select cases—namely, the requirement of "structured settlements" for persons permanently injured. Under previous approaches, awards (say) for a child brain-injured at birth might reflect the expected lifetime costs of care for the person. Under a structured settlement, the award would continue only so long as the child remained alive. Since such infants commonly die quite young, sometimes after a few years of life, the structured settlement ceases payments at a far smaller amount than would occur with many lump-sum awards. This approach has been mandated into law for some patients in recent tort reform in Virginia.

The effects of these various tort reforms on medical malpractice insurance premiums and awards have been documented only recently, in part because only by now has enough statistical evidence accumulated to measure the effects of these changes. However, the evidence seems to point toward substantial effects of at least some of these approaches. Table 14.1 displays the evidence on award sizes by year for six recent years, collected by a firm that specializes in gathering data on court awards. These data point to a clear peaking of the awards in the mid-1980s, with a precipitous drop in every measure of awards in 1987 and 1988, in particular. (Data are not yet available for later years.) Note that these data only

Table 14.1 MALPRACTICE AWARDS IN SINGLE-PLAINTIFF MEDICAL MALPRACTICE CASES

	Average award (thousands of dollars)	Median award (thousands of dollars)	Largest award (thousands of dollars)	Number of awards exceeding $1 million
1983	$ 888	$260	$25,000	69
1984	$ 640	$200	$27,000	71
1985	$1,179	$400	$12,700	79
1986	$1,478	$803	$15,800	92
1987	$ 924	$610	$13,000	62
1988	$ 732	$400	$ 8,100	54

Source: Jury Verdict Research, as quoted in The Wall Street Journal, April 28, 1989.

reflect cases going to trial, not those that settle at some earlier stage of the legal process. To assess the overall effects of these changes, one can look as well at the effects on medical malpractice premiums, since these should ultimately reflect changes in insurers' risks. One study (Zuckerman, Bovbjerg, and Sloan, 1990) used multiple regression analysis to study malpractice insurance premiums with data from 1974 through 1986. (Note that their data do not include the years of dramatic drop in awards shown in Table 14.1.) Their study uses data from all 50 states, thus taking advantage of the substantial differences in the phasing of tort reform across different states. Thus, in effect, they let the various state legislators carry out experiments for them. They analyzed effects not only of tort reform, but of other factors such as the number of trial lawyers in the region per capita. The most important changes in the tort law, they found, were those that capped physician liability (such as limitations on pain and suffering awards) and those that placed limits on how many years plaintiffs had to file a claim after an injury occurred (statute of limitations or limitations on the "discovery period"). The malpractice insurance premiums were *unrelated* to the number of attorneys per 1000 persons, deflating the idea that plaintiffs' lawyers are the engine driving the malpractice cost explosion. Interestingly, insurance premiums were strongly and negatively related to the number of physicians per capita. This may reflect specialization effects: We know in general that as physicians per 1000 population increase, specialization increases. (See discussion in Chapter 7 relating to physicians' location decisions.) If so, this provides indirect evidence that specialty training reduces injuries, although direct studies on this question have shown no effects on propensity of doctors to get sued (Sloan et al., 1989, using data from Florida).

These substantial reductions in premiums and awards seem to reduce the ability of "malpractice-bashers" to claim that the medical-legal system is responsible for "all of" the increase in medical care costs that has taken place over the last several decades. In many ways, such an assertion seems obviously incorrect, given the clearly important roles of increased insurance coverage, increased per capita income (see Chapter 17 for further details), and increased availability of medical technologies. However, these data provide an important direct test: In a time when malpractice awards and premiums fell considerably (see Table 14.1 in particular), we did not see a corresponding reduction in medical care costs. As the legal phrase might suggest, *res ipsa loquitur*.

TORT REFORM WRIT LARGE

The most radical type of tort reform considered sporadically would completely scrap the current tort system in favor of a "no-fault" insurance plan to compensate persons injured in medical misadventures. The common model—workers' compensation–has functioned for most of this century with apparently favorable response from both workers and industry. Under this approach, no determination of fault is made. Rather, "the system" makes a determination of the magnitude of injury, on which basis payments are made by insurers. While this system does not completely eliminate "duels of experts" (there is still considerable dispute about the magnitude of loss in some of these cases), issues of fault do not exist, and the "overhead" borne by the system to determine fault is much lower than would otherwise occur.

No-fault systems obviously completely eliminate any direct economic incentive to undertake loss-preventing activities. The incentives are the same as those under an insurance scheme where every doctor pays the same premium (see Box 14.2). The economic loss from such a plan, however, may be only very small if the current system produces little or no deterrence of negligent behavior.

Danzon (1985) has made some calculations to suggest that, even with all of its apparent problems, the medical malpractice system "pays its way" in a social cost framework if it can deter as few as 20 percent of the injuries that would occur in its absence. Alas, we have no good evidence yet (and it may never appear, given the research problems associated with the issue) on just how much injury reduction actually takes place due to the risk of malpractice claims.

SUMMARY

Negligence law is designed to serve two functions—to deter negligent behavior, and to compensate victims of negligence. Formal definitions of negligence closely approximate a cost-benefit analysis, defining a provider's behavior as negligent if a harm occurs to a patient and it would have cost less to prevent that harm than the expected amount of damage (in a statistical sense).

The current system of negligence law makes sense to operate if and only if an important amount of deterrence actually occurs, since the system compensates injured persons poorly, with estimates that only 1 in 15 to 1 in 25 negligently injured persons actually receive compensation.

Several strong forces work against successful deterrence within the current tort system: Few patients bring suit when injured, and medical malpractice insurance blunts the incentives of providers to exercise due caution, since insurance premiums do not closely reflect individuals' propensities to get sued. Nevertheless, no clear evidence exists yet on the amount of deterrence produced by the tort system, and the case for moving to an alternative system has not yet been persuasively made.

PROBLEMS

1. "There are not enough lawsuits in the current medical malpractice system." Discuss.

2. "The major purpose of medical malpractice insurance is to provide financial relief for people who are injured by medical care going awry, and it does this very well." Discuss.

3. Describe the Learned Hand Rule, and discuss the economic logic underlying it.

4. What is the major role of the federal court system in medical malpractice law?

5. What role does malpractice insurance have (if any) on the rate of malpractice injuries to patients? Does it matter how that malpractice insurance is priced?

6. Many doctors claim that the malpractice system is purely random, striking good doctors and bad doctors with equal propensity. What evidence do you know that supports or refutes this assertion?

Chapter
15

Externalities in Health and Medical Care

*T*his chapter discusses externalities in the medical care system, both positive and negative. We can define externalities as events when one person's actions impose costs on (or create benefits for) other persons. We encounter externalities every day of our lives: Traffic jams, cigar smoke drifting across the restaurant, and a boom-box blasting obnoxious music across the beach all create externalities. Some people may consider an "event" a cost, and others a benefit. The boom-box provides an obvious example: One person may greatly enjoy Beethoven's Ninth Symphony coming from a nearby boom-box, and another may hate it, but their positions may reverse if the boom-box owner switches to tapes of Madonna's "Material Girl." Most externalities don't have this sort of characteristic, however—they either uniformly provide benefits for or uniformly impose costs on others.

Much of the final portions of this book deals at least in part with externalities, and ways of dealing with them. Chapter 16 on regulation and Chapter 14 on the legal system discuss ways in which our society confronts and attempts to control some externalities. Chapter 16 also discusses the production and distribution of knowledge (biomedical research) and the production of knowledge about intelligent ways to use known medical interventions. This production of knowledge has at least some characteristics of externalities as well.

EXTERNALITIES, PROPERTY RIGHTS, AND THE CONTROL OF EXTERNALITIES

A fruitful discussion of externalities must begin with a discussion of property rights, because in many ways, externalities cannot occur when property rights are fully defined, and (conversely) when property rights are not fully defined or enforceable, externalities will occur commonly, if not inevitably.

Property rights, at least as considered in English-based legal systems such as ours, define the conditions under which a person may own, use, and transfer an "object." The object may be a parcel of land (the traditional topic of property law), a personal object (a hat, an automobile), or a series of ideas (such as the manuscript for this book) or musical notes (such as Beethoven's Ninth Symphony). Important distinctions arise when we deal with existing objects (such as an automobile) or the *invention* or creation of a new class of ideas or objects. The branches of law that deal with invention (patent law and copyright law) have considerable importance in health and medical care.

From an economic perspective, the most important parts of property law include those defining

- the ability of the owner to use the object
- ability of the owner to exclude others from using the object
- ability to transfer the object's ownership
- responsibilities of the owner toward others who use the object, and to third parties involved in its use

The simple example of an automobile shows the importance of these characteristics of property. If Henry and Marsha Ford own an automobile, it has little value to them if they cannot use it for their own transportation. It also has little value if they cannot legally exclude other people from using it; otherwise, it will never be available when they want to use it themselves. It has less value to them if they cannot sell it to another person. For example, the Fords may wish to move to Manhattan, where a personal car is an expensive millstone around the owner's neck.

Liability law also determines the responsibilities that Henry and Marsha face if they let somebody else drive their car. If the brakes fail while their friend is driving it, are they responsible for the friend's injury? If the friend drives it negligently and injures or kills somebody else, do the Fords have any liability? If so, they will exercise more caution and prudence in their decisions about letting other people use the car, and possibly even take more steps to prevent unauthorized use (if they would be liable for damage created by an unauthorized user).[1]

The same issues surround the invention of a new idea or device, and patent and copyright law protect the inventor's rights similarly. A patent or copyright provides the creator the right to use the idea, and most importantly, to exclude others from using the idea without permission. The inventor can also sell the patent, or "lease" it to others—that is, give them permission to use it in exchange for money. The legal system defines these rights, but the owners of property must often exert considerable effort to enforce them. An automobile parked on the streets of New York City may have the legal protection of exclusion, but in reality it stands a high chance of having an unauthorized user break and enter it, possibly to drive it, possibly to dismantle it and sell the parts. Alarm systems and garages provide some private enforcement of the property right. Police and the criminal justice system provide some public enforcement of those rights.

Particularly in the case of the rights surrounding inventions (patents and copyrights), the exclusion of illicit users presents a difficult problem. Industrial theft of an invention occurs commonly. Illicit copies of products ranging from Apple computers to Calvin Klein jeans to Chanel No. 5 perfume abound, and the manufacturers of such products spend considerable resources trying to detect and stop such illicit copies, often with private lawsuits to collect damages from the illicit user. Illicit copying of some copyrighted items seems impossible to stop. Probably few students and professors have not at some time violated copyright laws by illegally photocopying a journal article or book. Fewer still modern Americans have not illegally duplicated a recorded musical performance. In this case, while the law defines the property right, its enforcement is so costly to the copyright holder that illegal copying cannot be meaningfully stopped.

With this background of ideas about property, we can now begin our discussion of externalities in health and medical care with a bold assertion: Externalities occur if and only if the system

---

[1] This idea is not without substance. Swimming pool owners, for example, are sometimes held liable for drowning accidents, even if their pool is enclosed by a fence. Such pools have been held as an "attractive nuisance," and the owners of such property have special standards of protection required of them.

of property rights fails to define ownership and/or liability surrounding an event or object. As we shall see through repeated example, the failure of property rights and/or liability law seems an essential part of every type of externality we will find in the health care system and in other things that affect our health. This does not imply that "the solution" to such problems always rests in the parts of the legal system defining property and liability. Regulation often serves as a better tool to control externalities, as Chapter 16 discusses. Economists often consider tax and/or subsidy schemes to control externalities as well, and these concepts also help convey the nature of externalities, even if "the solution" to the problem does not involve a tax or subsidy. Nevertheless, focus on the property rights idea helps us examine externalities more fruitfully.

EXTERNALITIES OF CONTAGION

Contagious diseases and their control provide perhaps the "classic" example of externalities in health and medical care. Contagious diseases fit the classic example of an externality perfectly. Your action (sneezing) imposes costs on others (increased risk of getting a "cold") that you do not account for fully in your own decisions. Those decisions range from the most simple (carrying a handkerchief) to more costly (buying and using decongestant medicines that reduce your sneezing) and even more costly still (staying home from work and missing a day's pay).

The common cold provides a ludicrous but revealing example of the failure of property rights and liability law. If you sneeze into the air (which nobody owns), you create extra risks for others. The law also might make you liable for health damage you impose on others, so they could sue you for lost work time if they catch a cold from you.

If property rights and liability were perfectly defined, *and enforcement costs were trivially small*, then people's behavior regarding sneezing would change. For example, suppose that you owned the air space within 2 feet of you, by law (and similarly for all others). Then, not only would we all stay at least 4 feet apart unless given permission to trespass, but if you sneezed into someone else's air space, they could claim damages.

In a remarkable analysis, Coase (1960) has shown that the same behavior regarding sneezing would occur no matter whether we each owned "2 feet of air space," or someone else owned it all, *so long as transactions costs were trivially small.*[2] Alas, transactions costs in cases like this overwhelm all other considerations. If you had to sue everybody who sneezed in your vicinity, you would have no time remaining for any other activity.

In the case of sneezing, social customs and "manners" create society's best control mechanism. These work reasonably well, because most people whom you might infect with a cold see you repeatedly, and you select your behavior knowing that they can retaliate if you repeatedly impose costs on them. This sort of retaliation might include cutting you out socially, deliberately infecting you when they have a cold, or even reducing your pay at work (if you give your boss a cold too many times). The famous "Golden Rule" prescribes appropriate behavior in such settings. Amazingly, this sort of rule even seems to have good "survival" characteristics in a biological sense.[3]

Sometimes these sorts of social controls will fail. For example, you have less incentive to control your sneezing on a crowded public bus than you do in a classroom or office, even though you are more likely to inflict a cold on somebody else. The obvious reason is that you have almost no chance of seeing any of those

---

[2] Coase's landmark article "The Problem of Social Cost" revolutionized economic and legal thinking about externalities. The main idea, known as "the Coase theorem," says that if transactions costs are very small, then you get the same amount of "externality" costs in a society no matter how you assign property rights. He used the example of farmers' fields and a train throwing off sparks. If you assign the property rights to the farmers, their ability to sue the railroad will cause its owners to install the correct amount of spark arresters, i.e., the amount where the marginal expected damage equals the marginal cost of prevention. If you assign property rights to the railroad, then the farmers will "bribe" the railroad owner to install spark arresters, again, just to the point where marginal benefit equals marginal cost.

Many readers erroneously interpret Coase as saying that the assignment of property rights "doesn't matter" in efficiency considerations. The main thrusts of Coase's work are (1) that one should carefully attend to transactions costs when considering the allocation of property rights, and (2) when transactions costs are large, the problem won't go away, even with full property rights.

[3] More precisely, experiments using computers have shown that a strategy of cooperation, coupled with a tit-for-tat punishment strategy for defectors, dominates almost any other known strategy in games of repeated interaction. In this type of strategy, you "cooperate" until your partner "defects" from the cooperative strategy on one play of the game. You immediately punish your partner by defecting for one play, and then return to a cooperative strategy until your partner defects again. In cooperation games where players with various strategies compete against one another, this tit-for-tat strategy outperforms almost all others, including some that have been designed specifically to defeat a tit-for-tat strategy. For an excellent discussion of these ideas, see Axelrod (1984).

people again, so they have no chance to retaliate. In some societies, such as Japan, social custom then takes precedent. There, people wear surgical masks when they have a cold, to cut down on the contagious effects of sneezing. The emergence of such a custom in a crowded society like Japan probably makes more sense, say, than in Wyoming, which has an average population density of about 3 persons and 6 cows per square mile.

More Serious Contagious Diseases

Some diseases have more serious consequences than the common cold, and we take more expensive steps to respond to them. These types of disease also highlight the importance of property and enforcement costs.

Some diseases (such as dysentery) get transmitted readily through water systems. The famous case of "Aspenitis" involved the vacation town of Aspen, Colorado, where tourists and natives alike suffered common and serious intestinal illness. Research into the causes of the disease finally determined that the town's water lines and sewer lines, which ran in parallel through much of the town, had both broken and cross-contaminated each other. The contamination in one direction (from sewers to water system) had more serious consequences than in the other direction (from water system to sewers). However, the potential liability of the owners of the water system (the town, in this case) caused them to find and repair the break rapidly.[4]

The threat of poliomyelitis in the 1950s caused dramatic changes in behavior. Before the Salk and Sabin vaccines became available, and even before people understood fully that a virus caused polio, they did understand that person-to-person communication of the disease was possible, although the exact vector was not known. A common response during the height of the polio epidemic of the 1950s was to close public swimming pools. Some of this occurred by regulation (i.e., by order of county health department officials) but some as a "voluntary" action by owners of swimming pools who might have been held liable for transmission of the disease.

---

[4] Berton Roueché's book *Eleven Blue Men* describes a number of fascinating episodes of detection of contagious diseases.

Vaccines and Vaccination Policy

For some contagious diseases, scientists have discovered vaccines that make people much less susceptible to the disease, often totally eliminating the risk for the vaccinated person. (The discovery of such vaccines represents a separate problem of externalities, as Chapter 16 explores.) The nature of vaccines and contagious diseases offers a useful study of externalities.

Consider a society of, say, 1000 inhabitants on an island, confronting the risk of a contagious disease, perhaps carried back to the island from elsewhere by a vacationing citizen. Call this citizen Patient Zero.[5] Each Person j who comes in contact with Patient Zero has some probability π_{j0} of contracting the disease from Zero, depending on the virulence of the disease and the nature of their contact. Each Person j getting the disease also has a subsequent probability of transmitting it to some other Person i, whom we can call π_{ji}. Some people may be "naturally immune" to the disease (e.g., because of previous exposure, and hence a well-developed antibody system), so their probability of catching and spreading the disease is zero.[6] Suppose that we call the economic cost of getting the disease C (treatment costs, lost work, pain, etc.). Then Person i's expected cost is $C \sum_{j \neq i} \pi_{ij} = C\pi_i$, where π_i is Person i's probability of getting the disease—that is, $\pi_i = \sum_{j \neq i} \pi_{ji}$. This cost avoidance creates each person's private willingness to pay (WTP) for the vaccination. If people's private costs differ from each other, then a graph of the WTP for the vaccination will create what looks just like (and is) a downward-sloping demand curve at the societal level for the vaccine—that is, the demand curve one gets by adding up each individual's demand curves. (See the discussion with Figure 4.7 in Chapter 4 to remind yourself how this "horizonal aggregration" works.)

Each person will rationally get vaccinated if the expected costs of the illness exceed those of getting the vaccine (including time, travel, fees, the pain of the vaccination process, and the expected side effects of the vaccine), which we can call C_v.

---

[5] The transmission of AIDS in North America follows a clear path from a single person, whom public health authorities call "Patient Zero."

[6] Most viral diseases have this characteristic. Sometimes, we know with high likelihood whether we have had a specific disease. "Mumps" offers a good case in point. In other cases, we may not know at all whether we've had a disease. Many people became infected with the polio virus, for example, and recovered completely after a minor illness much like "the flu," but they did receive "natural" immunity thereafter.

Two things appear immediately in this problem. First is the concept of "herd immunity." If any other persons in the society are immune—for example, if they have already become vaccinated—then Person i's chances of getting the disease from any vaccinated Person k fall to $\pi_{ki} = 0$. At the extreme, if everybody else in society had already become vaccinated, then that last person would never bother, since Person i's chances of getting the disease (π_i) would fall to zero. In the language of externalities, herd immunity creates a positive externality for Person i.

We can turn the question around, and ask what private and social benefits occur if Person i gets vaccinated. Holding constant the number of other people who get vaccinated, Person i will decide to get vaccinated if $C_v < C\pi_i$. However, the social benefit extends past Person i, since, once vaccinated, Person i will contribute to the herd immunity for everybody else. To be precise, the net benefit for the entire society is $C\pi_i$ *plus* $C\sum_{j\neq i}\pi_{ij}$, since each Person j's chances of contracting the disease from Person i fall from π_{ij} to zero. Each person's *private* willingness to pay (*WTP*) for the vaccine is $C\pi_i$, and the *social WTP* is $C\pi_i$ plus the contribution to the herd immunity. The difference, $C\sum_{j\neq i}\pi_{ij}$, is the externality benefit. Figure 15.1 shows both the private and social *aggregate WTP* curves that vary with the proportion of the society vaccinated. The vaccine cost C_v appears as a flat line in this diagram, because

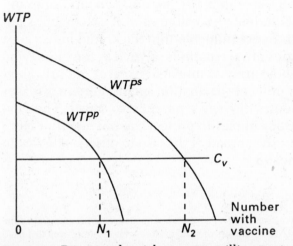

Figure 15.1 Private and social aggregate willingness to pay.

(by assumption) the costs per vaccination will not vary with the proportion of people vaccinated.[7]

Private decisions will lead to N_1 persons becoming vaccinated, the number where WTP^P equals C_v. If the number vaccinated exceeds N_2, then the cost C_v exceeds even the social benefit, WTP^S, for the vaccine, and getting Person i vaccinated makes no sense from either a private or social standpoint.

Between N_1 and N_2 of the population vaccinated, a conflict arises between the private and social decisions to get vaccinated. The private decision says, "Forget it!" The social decision says it is worthwhile. The public policy problem is how to induce enough people to get vaccinated in order to reach the proportion N_2, since private decisions will only lead to N_1 percent of the people getting vaccinated.

Economists usually think about such problems in terms of an optimal tax or subsidy. Looking at Figure 15.1, to reach N_2 percent vaccinated using a subsidy, one would have to extrapolate the WTP^P demand curve down below the N axis and to the right until it reached N_2, and then pay a subsidy amounting to the difference between that curve and C_v. As the figure is drawn, merely giving away the vaccine (so $C_v = 0$ to the individual) wouldn't suffice, since the WTP^P curve crosses the N axis at a smaller value than N_2.

While the idea of an "optimal subsidy" has clear meaning to economists, this almost never turns out to be the actual *vehicle* that public policy uses to increase vaccination rates. Sometimes vaccines are provided free and at convenient locations (the mass polio vaccinations of the 1950s and 1960s provide a good example), but commonly, compulsion also enters the picture. Schools, for example, often require that students have some set of vaccinations before they are admitted. Every inductee into the armed forces goes through a standard series of vaccinations. Overseas travelers (who may be subject to a whole array of unusual diseases not prevalent at home) must present proof of vaccination against certain diseases before their government will issue a passport to travel. The government takes these steps in part to protect the traveler, but more importantly, to prevent the spread of disease to others—that is, in the name of "public health."

---

[7] If economies of scale in vaccination exist, e.g., through school vaccine programs, then the C_v line would fall. Similarly, if diseconomies of scale existed—for example, due to a fixed supply of some input necessary to make the vaccine—then the C_v line would rise.

As with every compulsory act of government, some individuals may be harmed when such methods are used. Some people's religious beliefs, for example, prohibit their use of medicines. When the government imposes the requirement of vaccination on such people, it has inflicted a real cost. Some vaccines also have side effects that occasionally strike the vaccinated person. On occasion, these side effects prove fatal, as with the infamous "Swine Flu" vaccine, involved in an influenza vaccine program personally promoted by the president of the United States (he received his flu shot on national television), but where the vaccine turned out to have some bad side effects for some people.[8] One of the advantages of a "subsidy" program is that it induces those people to get vaccinated who bear the smallest costs, because only those people who derive a net positive benefit (including the subsidy) from getting the vaccine will respond to the subsidy.[9] However, subsidies increase the "on budget" costs of the government, whereas compulsion creates only "off budget" private costs. This distinction apparently drives many political decisions.

HEALTH-AFFECTING EXTERNALITIES: THE EXAMPLE OF ALCOHOL

An important class of externalities has nothing directly to do with the health care system, but nevertheless constitutes an important part of the study of the economics of health. This class of externalities includes events when one person's behavior affects another person's health, but the medical care system need not become involved (although it often will). Such events include the use of alcohol (drunk driving), tobacco (secondhand smoke, apartment building fires), illegal drugs (increased crime and increased spread of infectious diseases through shared needles), and similar "lifestyle" choices. As an example of these events, we will study the externality costs imposed by the use of alcohol in more detail.

Alcohol has many effects on people, many predictable and well-known to most people and some more hidden and dangerous.

---

[8] Because of the problems of this program, one wag asserted that the public health officials had sold a "Pig in a Poke" to the American people.

[9] The ideas behind an "all volunteer" military force versus a compulsory draft rest on the same logic.

Consumers of alcoholic beverages directly gain the pleasures and directly absorb many of the costs of alcohol use, thus involving no externalities. However, alcohol involves many changes in behavior that also produce external costs, not accounted for even by a "rational" drinker. The most obvious costs come from alcohol-involved driving. Other alcohol-related externalities include fetal-alcohol syndrome, spouse and child abuse and other crimes of violence, and (at least by some definitions of externalities) the costs of health care programs like liver transplantation, and government provision or subsidy for treatment of mental illness and other alcohol-related illnesses. We will analyze these costs here, and consider ways to confront them using public policy.

Alcohol and Driving

Everybody understands to some extent that "drunk driving" creates additional danger beyond "sober driving." Indeed, most people living in the United States will know at least one person during their lifetimes who died in an alcohol-related vehicle crash.[10] Many people have heard the statistic that half of all highway deaths are alcohol-related. The 20,000 to 25,000 annual highway deaths in the United States attributable to alcohol far exceed the peak rate of deaths of U.S. soldiers in the Vietnam War. Even with these grim statistics, few people understand fully the nature of the externalities involved with alcohol-involved driving.

One can readily obtain information about alcohol involvement in drivers who die in vehicle crashes, but each source of information has potential flaws.[11] The national Fatal Accident Reporting System (FARS) describes alcohol involvement on several bases. If officials investigating a crash obtain a blood alcohol concentration (BAC) test from a dead driver, they can tell specifically the role of alcohol in that driver. Reporting officers also can indicate their beliefs about alcohol involvement, even when no BAC tests are

---

[10] Each year, about 1 person in 10,000 in the United States dies in an alcohol-related vehicle crash. If the average person in the United States knows only 200 other people, the probability each year of *not* knowing one of those dead persons is 0.9999 raised to the 200th power, or 0.98. The odds of living 70 years without such an event happening are 0.98 raised to the 70th power, or about 0.24. Thus, anybody with 200 friends at any time has about 3:1 odds of knowing at some time at least one person killed in an alcohol-related crash.

[11] Note the use of wording here: "Accident" implies something beyond the control of the individual. "Drunk driving" accident implies that only a person substantially inebriated has any increased risk. "Alcohol-involved crash" avoids both of these linguistic mistakes.

Table 15.1   DISTRIBUTION OF BAC LEVEL AMONG DEAD DRIVERS IN 15
STATES WITH HIGH RATES OF BAC TESTING AMONG FATALLY
INJURED DRIVERS—1980
(Percent of all drivers in category)

	Age						
BAC level	16–19	20–24	25–34	35–44	45–54	55–64	65+
None	42%	32%	29%	36%	49%	57%	77%
0.01%–0.05%	5%	5%	5%	4%	3%	3%	3%
0.05%–0.09%	10%	9%	6%	5%	4%	4%	3%
0.1%+	43%	54%	60%	55%	44%	36%	17%
Total	100%	100%	100%	100%	100%	100%	100%

Source: NHTSA Technical Report DOT HS 806 269, May, 1982, "Alcohol Involvement in Traffic Accidents: Recent Estimates from the National Center for Statistics and Analysis," Accident Investigation Division, National Highway Traffic Safety Administration, U.S. Dept. of Transportation.

taken. The FARS system reports alcohol involvement if either of these sources indicates it. Unfortunately, both sources have potentially important biases. Many states do not automatically have a BAC test done on dead drivers. In addition, reporting officers often understate the role of alcohol, apparently considerably.[12] Because of these biases, it seems likely that the FARS system underreports the role of alcohol in fatal crashes.

Some states get autopsies from a large fraction of all drivers who die in fatal crashes. Using data from these states, a clearer picture emerges about the role of alcohol in these deaths. Table 15.1 shows the distribution of BAC levels among deceased drivers of various ages. Overall, more than 60 percent of all driver fatalities show positive BAC levels, with most of those exceeding 0.1 percent, the legal "DWI" limit in most states. By contrast, the standard reporting method in many states (relying partly on officer reports) shows only a 38 percent alcohol involvement for all driver deaths, demonstrating the extent of bias in those data.

---

[12] Two studies have reported the BAC levels taken from drivers hospitalized after vehicle crashes, and compared these rates with reports from officers at the crash site. In one study (Maull, Kinning, and Hickman, 1984) of 56 drivers with BAC > 0.15 percent, none received any legal citation for alcohol-involved driving at all (DUI or DWI). Reporting officers affirmatively declared that 30 percent of these drivers were not alcohol-involved, and reported that 18 percent of these drivers were "not drinking."

Knowing what fraction of highway deaths have "alcohol involvement" doesn't alone tell much about the additional risk of driving after drinking. For example, if two-thirds of all drivers had consumed alcohol before driving, then one would conclude that alcohol made driving safer, not more dangerous. In order to determine the extent of the externality, we need to know the proportion of all drivers with different levels of alcohol involvement, and then compare that with the proportions involved in fatal crashes.

Each of the methods that might help measure the mix of drinking levels of drivers has potential flaws, but two very different methods have produced quite similar results, lending credibility to the conclusions drawn. One method directly measures BAC levels of *all* drivers passing a given checkpoint at a given time. This approach requires considerable police cooperation and imposes large costs on drivers (traffic delays, etc.), and obviously the results depend on the time and place where the roadblock gets established. One Canadian study used this approach, and then compared the resulting distribution of BAC levels of all drivers with that of drivers involved in fatal crashes on the same roads (but over a longer period of time). The results appear in Figure 15.2.

These data show precisely the nature of the externality associated with drinking and driving. As the BAC levels rises, the risk of

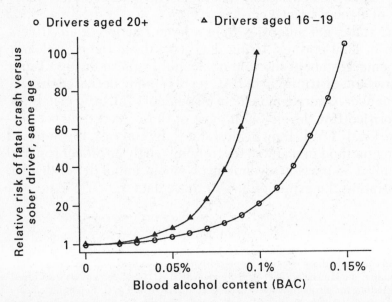

Figure 15.2 Relative rise of fatal accident related to BAC and age. *Source:* Mayhew and Simpson.

a fatal crash also accelerates. There also appear to be important differences between young drivers (age 16–19) and others (ages 20+).[13] An adult who has had 6 drinks and then drives (the amount putting an average-sized adult into the legal DWI category, with BAC > 0.1 percent) has about a 12-fold increase in risk, compared with the same type of driver without alcohol. By the time the person has had 9 or 10 drinks (BAC > 0.15 percent) the risks increase to about 100 times that of a sober driver. Each of these fatal crashes, of course, has some probability of killing one or more innocent victims; on average, each 10 fatally injured drivers have an additional 7 dead passengers, pedestrians, and so on involved.

For younger adults (ages 16–19), the risks of drinking and driving are far worse than for older adults. A BAC of 0.1 percent leads to a relative risk of teen-aged drivers 100 times greater than when sober. A BAC exceeding 0.15 percent creates a stupendous 400-fold increase in risk, according to the Canadian study.

A separate study (Phelps, 1987, 1988) reached almost the same estimate of relative risk using quite different data. That study used a national survey of drinking habits to estimate the proportion of young persons who had consumed a specified number of drinks on any randomly chosen day (e.g., no drinks, 3 drinks, etc.). With the assumption that the propensity to drive doesn't change any after drinking, this provides a separate estimate of the mix of drivers with different drinking levels, from which the same type of risk calculations can be made. Those calculations reached almost exactly the same conclusion as the Canadian study did; for young drivers, the relative risk of driving after 6 drinks was 100 times higher than the risk of driving sober. The calculated risks at lower levels of drinking also corresponded well with the Canadian study's conclusions. Thus, we can probably take the portrait of risks from these studies as reasonably reliable.

The externality associated with these alcohol-involved crashes at least includes the deaths and injuries of "innocent victims" in these crashes, the drivers of other vehicles, passengers, pedestrians, bicyclists, and so on. The value of their lost lives is a classic externality—a cost imposed on others by one person's behavior.

Controlling an externality requires more than just identifying it. Economists commonly talk about imposing a tax on activities

---

[13] The original study by Mayhew and Simpson looked at the risks for a larger number of age groups, but concluded that all of the over-20 groups had similar risk, and thus could be pooled together for purposes of analyzing risk.

that create externalities, to "internalize" the costs that had been imposed on others. The logic goes that by making users of the externality-creating good or service pay a tax equal to the costs imposed on others, they (the users) will reduce the amount demanded, and thus reduce the externalities. The optimal tax, under this way of thinking about externalities, constitutes the difference between the "private" cost paid by the user of the good or service and the "social cost," the latter reflecting not only the private cost to the user but also costs imposed on others.

Taxation of alcoholic beverages to control the externality of drunk driving shows how frustratingly complex the simple idea of an optimal tax becomes when one attempts to apply it to real-world settings. We *really* want to tax the combined activity of driving and drinking, because that combination (not either alone) creates the greatly heightened risk of a fatal crash. This turns out to be almost impossible to do, because it would literally require that "tax officials" find and tax each alcohol-involved driver according to his or her BAC. Of course, instead of using "tax officials" to do this, we use highway patrols, police, and other law enforcement officials, and we levy "fines" for alcohol involved driving, rather than "taxes."

Prevailing enforcement efforts appear not to apprehend most drunk drivers. One study (Filkins et al., 1970) of hospitalized alcoholics who drive commonly showed that only one in six of that group had been arrested for drunk driving within the past six years before hospitalization. Two studies of drivers injured enough to require hospitalization studied the arrest outcomes for those with BAC levels exceeding legal limits. In one city in Virginia (Maull et al., 1984), none of 56 postcrash hospitalized drivers with BAC > 0.15 percent was even arraigned for drunk driving.[14] In a similar study in Connecticut, of 84 hospitalized drivers with BAC > 0.1 percent, only three were arraigned, and none convicted (Colquitt et al., 1987). Thus, we can presume that existing enforcement efforts cannot appropriately "tax" alcohol-involved driving.

Another approach to the problem directly taxes either "drinking" or "driving" separately. Either of these approaches is obviously somewhat of a blunt instrument, because one either taxes a lot of driving that doesn't involve drinking, or one taxes a lot of

---

[14] In criminal law, "arraignment" is the step in which a person is charged with a violation. The trial then determines guilt or innocence.

drinking that doesn't involve driving. Can economic analysis tell us which (if either) offers a more sensible approach to controlling this externality?

One way to think about this problem returns to the discussion on demand curves from Chapter 4, recalling that the information contained in the demand curve for a "good" reveals not only the marginal willingness to pay (for an additional unit of consumption) but the total value received by consumers for each amount they consume. The area under the demand curve shows that total area and the difference between the area under the demand curve and the costs of production represent "consumer surplus," the value of the extra pleasure consumers get above and beyond the costs of production. If one taxed driving, but not alcohol, to reduce the problem, one would incur the welfare losses associated with the reduction in all driving. Since most *drivers* have not consumed alcohol, this would create large welfare losses. In parallel, one could tax all drinking, which would also impose taxes on many drinkers who do not drive, but this turns out to be the more socially efficient approach of the two. Taxing drinking, even at moderate levels, has the capability of reducing alcohol-involved fatalities considerably (Phelps, 1988).

By focusing on highway fatalities created by alcohol, one should not lose sight of the other important health-related consequences of alcohol consumption. The U.S. Public Health Service has estimated the mortality-increasing effects of alcohol through a variety of diseases and events, including (of course) highway crashes.[15] They estimate that a total of 70,000 male and 35,000 female deaths occur annually due to alcohol, including (of course) 20,000 total deaths in highway crashes. They add over 9,000 homicides annually attributable to alcohol, and a comparable number of suicides' deaths. Alcoholic and alcohol-related cirrhosis of the liver adds another 14,000 deaths annually, the other single largest alcohol-related cause of death. The Public Health Service estimates a total of 2.7 million potential life-years (taking the difference between each dead person's age and normal life expectancy) lost annually due to alcohol ingestion, about a quarter of which are inflicted on others (mostly highway crashes and homicides), with

---

[15] These data come from the *Morbidity and Mortality Weekly Report*, published by the Centers for Disease Control in Atlanta. These reports focus each week on a particular health-related issue, and provide a wealth of data for the interested health policy student. You can get on their mailing list by writing to the editor of MMWR, CDC, Atlanta, GA 30333.

the remaining three-quarters of the losses borne directly by the persons consuming alcohol (and their families).

EXTERNALITIES FROM TOBACCO

Another important externality affecting health arises from the widespread use of tobacco products, primarily cigarettes, throughout the world. Two types of externalities exist from tobacco use. For the first, the smoke itself is unpleasant to many people, possibly more than smokers themselves realize. For persons with hay fever and asthma, any irritant, including tobacco smoke, can set off allergic reactions that are at least unpleasant, and in the case of asthmatics, potentially fatal. However, for most persons "bothered" by cigarette, cigar, and pipe smoke, the unpleasant smell is the most obvious manifestation of an externality, and until recently was thought to be the only one of importance.

Response to this externality forms an interesting study in the use of regulation instead of relying on property rights and markets. Over the past decade, many cities and states have enacted regulations that limit the areas where smokers may smoke in "public" buildings (i.e., those open to common access by strangers), including office buildings, restaurants, airports, and so forth. Typically, these rules require that restaurants (for example) provide nonsmoking sections and offer every customer a choice of smoking versus nonsmoking seating. In areas with more strict rules, smoking cannot take place in office buildings, even in "private" offices, if such offices might be used by persons other than the primary occupant.[16] Some companies have privately gone beyond the requirements of the law, and not only totally banned smoking on their premises, but also offered to pay for smoking-cessation programs for their employees. Numerous hospitals have totally banned smoking by patients, staff, and visitors, not only within the hospital itself, but in surrounding areas.

The logic for a regulatory approach to control smoking in public buildings arises directly from the "random" nature of access to such buildings. Trying to use agreements (contracts) between peo-

---

[16] These rules have created an entirely new social network in some businesses—namely, employees who meet at office building entrances where they go outside to smoke. They also create interesting euphemisms such as people saying that they are "going outside to get a breath of fresh air" when they are in fact going outside to smoke a cigarette.

ple in a restaurant to determine whether smoking would take place would be the height of absurdity, and nobody would think seriously of a full "property rights" approach to such a problem. The transactions costs of reaching agreements would overwhelm the problem. Similarly, even if owners of (say) restaurants wish to establish smoking and nonsmoking sections in their restaurants, they may fear the loss of customers who wish to smoke. (This is another example of a noncooperative game problem such as was discussed in more detail in Chapter 14 on medical malpractice. You may wish to look at Box 14.2 discussing this problem.) Regulation offers an alternative—albeit Draconian from the point of view of smokers—to solve the problem. Notice that previous rules freely permitting smoking have an exactly parallel nature, except they are viewed as Draconian from the point of view of breathers.

Similarly, the U.S. Congress instituted a ban on smoking during airline flights of less than two hours' duration in 1984, and then in 1988 extended that ban to all domestic flights except those scheduled to require over six hours' flight time (which occurs only on flights to Hawaii and Alaska). Here, the logic for universal application of a regulatory regime seems somewhat thinner, because a market response to the question of whether smoking should occur on an airplane seems more likely to emerge. Indeed, before any regulations on the subject occurred at all, some airlines privately banned smoking on all flights, "specializing" in the market for those customers who preferred nonsmoking flights. In a city-pair market with many daily flights (such as between New York and Boston or Washington), one might well expect such specialization to emerge. It did not, however, take widespread hold, and the regulatory approach soon swept over the market.

The second and more serious externality arising from cigarette smoke has now become more carefully understood—even nonsmokers' health risks increase when they spend considerable time in close proximity with smokers. A series of epidemiology studies has emerged over the past several years showing substantially heightened risks of lung cancer, heart disease, and other lung diseases (such as emphysema) from nonsmokers who live in a house with at least one smoker. Indeed, one study demonstrated that the *dogs* of smokers had a 50 percent increased risk of dying from lung cancer, compared with dogs whose owners did not smoke.

The magnitude of secondhand smoke morbidity and mortality is yet to be fully determined, but evidence continues to accumulate that this is a more serious externality than had previously been recognized. A recent study compares the health outcomes of nonsmokers who live with smokers with that of nonsmokers who live with nonsmokers (Sandler et al., 1989). This 12-year study of nearly 28,000 individuals found that the age-adjusted risk of death was 15–17 percent higher for nonsmokers who lived with smokers than for those who lived with nonsmokers. Such a study will necessarily understate the actual health consequences of secondhand smoke, however, because it ignores the exposure out of the home, which will tend to increase the risks of those living with smokers and nonsmokers similarly.[17]

More recently, a case-control study investigated the risks of lung cancer for those exposed to secondhand smoke in the household (Janerich et al., 1990). Household exposure to 25 or more "smoker years" during childhood and adolescence doubled the risk of a person's later contracting lung cancer. (If two parents smoke for 15 years, that creates 30 "smoker years" of exposure.) The study estimated that one out of every six cases of lung cancer among nonsmokers was due to secondhand smoke exposure acquired during childhood from parents.

These and other studies will continue to clarify the role of this particular type of externality. Even with incomplete information, it seems safe to say that tobacco consumption does create an important health externality, certainly for those within the household, and possibly in other surroundings as well (e.g., the workplace). We can analyze the consequences of this behavior just as for any other externality.

"FINANCIAL" EXTERNALITIES—TRANSFERS BETWEEN PEOPLE

An important distinction exists between two types of economic events, known generally as "transfers" and "externalities." A transfer takes place when the government (or some other agent) takes money (or something of value) and gives it to another person.

---

[17] To see why this causes an understatement of the actual health effect, consider the following hypothetical case: Suppose that all nonsmokers had a 10 percent increase in risk due to second-hand smoke acquired at work, restaurants, etc., and that the true effect of living with a smoker was a 20 percent increase in risk, compared to no second-hand smoke exposure at all. Then the actual increase in risk due to smoke exposure would be 32 percent ($1.2 \times 1.1 = 1.32$), but the apparent effect would be only the 20 percent increase.

Using an income tax to raise government resources and then giving welfare payments to low-income persons represents a classic "transfer." In theory, but in theory only, one can accomplish a transfer without distorting anybody's behavior. In such cases, total social welfare rises or falls only to the extent that the utility produced by $1 differs from person to person. Normal practice of "welfare economics" generally presumes that $1 matters just as much, on the margin, to each member of society. Using per capita income or GNP as a measure of social welfare makes this assumption implicitly, and the usual methods of cost-benefit analysis make the same assumptions (Harberger, 1971).

In real-life events, however, one cannot either raise or distribute money without distorting people's behavior somehow. If the government raises income taxes, it alters people's incentive to work. If it distributes welfare payments to people on the basis of income or assets, it similarly distorts people's incentive to work, to accumulate assets, and to give away assets to family and friends.[18] In the health care system, if the government provides medical care for free (or at reduced rates) to repair the damage people inflict upon themselves through their own life-style choices, it increases the rates at which people will engage in those behaviors. Should we call these kinds of behavioral responses "externalities"?

These issues can matter considerably. For example, when calculating the apparent "externality" from consumption of tobacco and alcohol, considerable differences emerge in the estimate of how much of an externality these activities create, depending on whether one counts a large number of "transfer payments" made or received by alcohol and tobacco users. The issue, ironically, hinges on the mortality-raising effects of these substances. In our society, where Social Security, pensions, and other income transfer accrue automatically to individuals who survive, heavy consumers of tobacco and alcohol confer a "fiscal externality" (benefit) on others by dying sooner than they otherwise would (and hence not collecting on as much of their Social Security and pension payments). In overall calculations of whether tobacco and alcohol users "pay their own way," such transfer payments hold large sway in the arithmetic. By the usual definition, some people would not consider these as true externalities, but "merely" transfer payments. However, once the legal entitlement to a pension or Social

---

[18] The discussion in Chapter 13 about "spending down" to create Medicaid eligibility provides a good example.

Security payment is established, money flows from one person to another much in the same way as with "normal" externalities. These matters hinge on choices that lie outside of pure economic logic. We need to remain aware of the distinction, and to understand how and why transfers and pure externality costs arise, but it probably remains a good idea to keep them separate in analyses of activities like this.

INFORMATION AS AN EXTERNALITY

Topics such as alcohol and tobacco consumption raise an important issue about the economics of information. It is widely understood that the production of knowledge creates a beneficial externality, because the marginal costs to disseminate the knowledge are small compared with the marginal costs to produce it. Once produced, the logic goes, knowledge should receive widespread distribution, limited only when the marginal costs of dissemination finally match the incremental benefits. In addition, the production of knowledge itself is likely to be undertaken too little in a society with less than perfectly functioning property rights to knowledge, creating a government role for the subsidy and/or production of knowledge.

Another issue arises in the categorization of externalities from such things as alcohol and tobacco use: Should (say) the deaths of persons drinking alcohol "count" as an externality, or should they get counted as a purely private cost, presumably taken into account by people when they decide to engage in behavior that harms their health (such as smoking tobacco or drinking alcohol)?

The pure economic model views consumers as fully informed about the risks and benefits of consuming any commodity, and making their decisions accordingly. The person's demand curve for tobacco or alcohol automatically incorporates those risks. If this is true, the logic for government intervention to reduce smoking and drinking rests *solely* on the external damage produced by these activities.

Another way of thinking about the problem suggests that at least some of the so-called private costs of these activities represent an externality. Consider a hypothetical demand curve for (say) tobacco or alcohol by a fully informed consumer. Now contrast that demand curve with that of a consumer who is the same except (to pose an extreme case) for having *no* knowledge of the risks of the activity. Obviously, the "uninformed" demand curve will exhibit (at every quantity) a higher willingness to pay—that is, the demand curve will be shifted outward. At any price, the uninformed consumer will consume more than a fully informed consumer with otherwise identical tastes and circumstances. Figure 15.3 shows this problem.

The welfare loss arising from the lack of information appears on this figure as the triangular area A. The uninformed person consumes at a rate X_2, but would consume only X_1 if fully informed. The dotted area between the informed demand curve and the cost line C, bounded by X_1 and X_2, represents expenditures exceeding the value of incremental consumption. Adding up welfare triangles like these across the entire population represents the maximum possible value of the information necessary to move all persons from the uninformed to the informed demand curve. (Note that a similar logic arises when information causes the demand curve to shift outward, rather than inward.)

In the case of tobacco, the case can be made that information arising over the past quarter-century has produced just this sort of change in behavior. The precipitating event here was the publication of a major study in 1964 by the Surgeon General of the United States (Luther Terry, M.D.) on the health risks of smoking. Prior to that year, smoking rates had increased steadily from at least 1930 onward. Because of that initial study—"merely" the publication of new information—and the many subsequent changes in

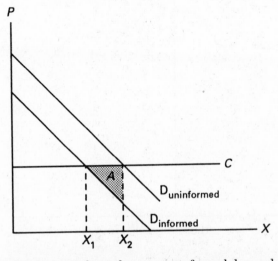

Figure 15.3 Informed versus uninformed demand.

public policy and personal attitudes toward the use of tobacco, smoking rates began a steady decline that continues apace today. Figure 15.4 shows the actual per capita cigarette consumption in the United States over time, and projections of what the smoking rates would have been in the absence of the information campaign.

To evaluate the economic gains from this information, one could readily apply the type of model portrayed in Figure 15.3, and estimate the annual gain from consumers' obtaining the information.

Note also that information, once produced, creates a long stream of benefits, the present value of which must be offset by the cost of producing the information. If the information has a very long life, then one can approximate the present value of the long stream of benefits by using a "perpetuity" calculation. To do this, one divides the annual benefit by the interest rate. Thus, for example, if the discount rate is 5 percent then the perpetuity value of information is 20 times the annual benefit.

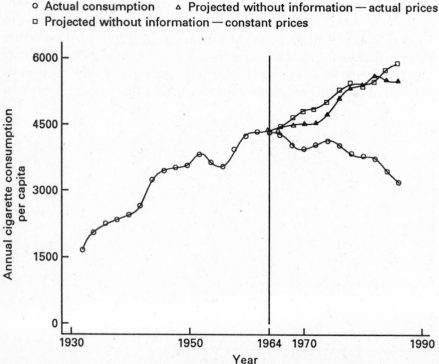

Figure 15.4 Smoking behavior with and (projected) without information on risks.

TRANSFUSION-INDUCED AIDS AND HEPATITIS

Another important externality in the health care system arises through the transmission of dangerous and/or fatal diseases through transfused blood. Blood transfusion policy is complicated both in the biology and the public policy implications, but the problem remains of paramount importance, even though (as has been true since April 1985), screening tests have been available to test donated blood for HIV antibodies, a sign that the person has acquired the viral infection that ultimately leads to AIDS. Problems of transfusion-induced disease persist because (1) the AIDS test is imperfect, with some persons donating blood while infected but before their antibody levels reach detectable levels, and (2) other diseases, notably hepatitis, continue to pass undetected into the blood supply system and create major health problems.

Recent analyses of the risks of AIDS infection have some intrinsic variability, but suggest that the risks of acquiring AIDS through a single unit of blood lie somewhere between 1/300,000 and 1/100,000 (Cumming et al., 1989). The average transfusion recipient receives 5.4 units of blood, so the risks of acquiring AIDS from transfusion are approximately 1/20,000 to 1/60,000. If recent estimates of the value of reducing probabilities of death (consistent with "a value of life" of about $5 million) are correct (Viscusi and Moore, 1989), then these risks from AIDS correspond to a cost of about $17–$50 per unit of blood transfused.[19]

Thinking about blood supply policy in the context of property rights and liability has a long history, and the AIDS question has only served to remind us again of issues previously raised. An early and important work on this topic came from Richard Titmuss (1972), a British scholar who argued that the nature of a country's blood supply was in many ways a key indicator of its moral character. He strongly opposed "commercial" blood supply, at that time the majority supplier of whole blood in the United States, and sought to promote a wholly voluntary (unpaid) donor supply. Many have argued for a voluntary supply on the grounds that it is allegedly safer; Titmuss made the argument as well on moral grounds.

Titmuss struck a raw nerve in the ranks of the economics profession in this country. A barrage of commentary on Titmuss's

---

[19] This number comes from multiplying the $5 million figure by the incremental risk of 1/300,000 to 1/100,000.

thesis emerged, including responses from Reuben Kessel (1974, of the "classic" Chicago school of thought) and two subsequent Nobel Laureates of substantially different political persuasion than Kessel, Kenneth J. Arrow (1972) and Robert Solow (1971), the latter even before the book was officially published.[20]

The essence of the argument appeared in Kessel's original blast at Titmuss: Kessel asserted that the *real* problem with blood supply in the United States was not that it was *over*commercialized, but rather that it was *under*commercialized. The discussion turns closely on the question of the "liability" side of property rights law. In donated blood, the donor never acquires any liability for the blood's damage (if any), and in fact, neither does the supplying agency (most predominantly, the American Red Cross and its affiliate agencies). Because of this failure to acquire liability, Kessel argued, agencies that collect and distribute blood have too small an incentive to worry about the safety of that blood, and hence we have too much infection through the blood supply.

This issue shows precisely the relationship between the system of property rights and the creation of externalities. If, as Kessel and many others have argued, property rights were complete, suppliers of blood (like manufacturers of stereos and cars) would have the strongest possible incentives to guarantee the quality of their product, to promote information about that quality, and (the other side of the coin) to take all necessary steps to prevent a legal liability suit if tainted blood were administered under their auspices.

The legal system has not conferred property rights (and liability) on the blood system, and the externality persists. Steps remain to be taken that would reduce the incidence of transfusion-related iatrogenic illness, yet the current legal and organizational incentives lead to their being ignored. Most prominently, tighter donor screening and exclusion provide potentially powerful tools to reduce transfusion-induced disease (Eckert, in Eckert and Wallace, 1985).

We should note here that the altruistic motives of blood-banking organizations are not under attack from this line of argument. Indeed, the complex issues associated with blood safety and

---

[20] An extended discussion of this topic appears in a written debate featuring Ross Eckert, an economist, on one side and management professor Edward Wallace on the other. See *Securing a Safer Blood Supply: Two Views* (Eckert and Wallace, 1985).

supply provide a rich portrait of organizational purposes, and show how such purposes must sometimes end in conflict. In the case of blood supply and safety, the conflict arises through the protection of both recipients and donors of blood—banking organizations wish to preserve the "right" of donors to donate, and hence are reluctant to exclude them from donating, even if their blood is "implicated" in a previous illness.[21] Because of these conflicting objectives, even an altruistic organization will end up placing patients at risk; the question is one of balance. Altering the legal structure would, of course, alter the choices made by participants in this market. With the current legal incentives and structures, however, some blood donors create serious and potentially fatal health hazards for others and do not account fully for these costs in their own actions, thus creating another externality.

SUMMARY

Externalities occur when Person A's behavior imposes costs (or benefits) on other people (Persons B, C, . . .) that Person A's private decision making does not account for. In almost all cases (perhaps all), these externalities arise because of the failure of our legal system to define fully property rights, and hence legal liability. Health-affecting externalities arise in numerous ways, including (trivially) sneezing when you have a cold, and (more importantly) vaccinations for infectious diseases, alcohol-involved driving, tobacco use, and the donation of blood, among others. These examples serve to illustrate, but do not provide a complete catalog of the types of health-affecting exernalities that we might consider.

PROBLEMS

1. What is the key economic characteristic of an externality? What legal feature is most likely to be present when externalities occur?

---

[21] Most transfusion recipients receive multiple units of blood. For example, if a patient becomes ill after receiving 10 units of blood (each from a different source), then each of those 10 donors has a 1/10 chance of being "contaminated." One blood-banking organization excludes donors only when the probability of implication reaches 0.3 to 0.4 (AABB Technical Manual, no date, chap. 18). Clearly, tighter restrictions are possible, and would increase blood pool safety at a higher cost of obtaining blood.

2. Many people have criticized the U.S. blood supply system as being "too commercial" and prefer a more altruistic method of collecting and distributing blood for transfusions. The issue is whether donated blood contains viruses that might cause disease to the recipient (such as hepatitis or AIDS). Reuben Kessel, an economist from Chicago, argues to the contrary that the problem with blood supply is that it was not sufficiently commercial. How might more commercialization improve the safety of blood supply in the United States? Are there risks to taking such an approach?

3. "Herd immunity" describes a beneficial externality arising from vaccination against contagious disease. Describe "herd immunity" precisely. What public policies might lead to the proper amount of vaccination? What might affect the ability of your proposed policies to achieve their goals? (Hint: Think about transactions costs.)

4. Regarding the problem of drinking and driving, two approaches have been widely discussed. The first would raise the price of alcohol to reduce drinking (and thus, by presumption, drinking and driving). The second approach would rely on enforcement of current drunk driving laws, punishing the dangerous event (drinking *and* driving) rather than just "drinking" alone. What do we know about current enforcement of drunk-driving laws to suggest which strategy might have a better chance of success?

5. Many "externalities" have some aspect of "privateness." On a scale of 0 = purely private to 1= purely external, rank the following consequences of cigarette smoking in terms of how "external" they are, and discuss the logic underlying your answers. In your answer, think broadly about mechanisms of insurance, and the possibility of "financial" externalities. (Note: There is no purely "correct" answer here.)
   (a) Lung disease to the smoker.
   (b) Heart disease to spouses and children of smokers.
   (c) Emitting smoke into the air in a restaurant or commercial airplane.
   (d) Emitting smoke into the air of a private airplane or automobile.

APPENDIX

Value of Life

The "value of life" turns out to affect importantly a number of questions in health economics and even, in some cases, health and safety regulation. The "value of life" is really a misnomer for two reasons: First, all of us eventually die, so we could use "value of preventing premature death" as a more proper phrase. Second, one seldom confronts the question of what any single individual's life is actually worth, or at least, one can almost never measure that value.

Rather than attempting to measure the value of "lives" on a one-by-one basis, economists have turned to a related question: How much are people willing to pay for a small reduction in their probability of dying? The idea then takes on a different slant. If 1000 people (say) each have something happen to them that decreases their chances of dying by 1/1000, then statistically we can say that 1 life has been saved (1000 people each have 1/1000 reduced chance of dying). Thus the *aggregate* willingness to pay by those 1000 people for this risk reduction reflects the value of 1 statistical life saved.

Economists have turned to several sources to estimate the value of life following this approach. First, one can think of settings in the labor market where some jobs are riskier than others, but otherwise the jobs have similar demands in terms of physical effort, mental skills, training, and so on. Economists use regression models to estimate wage rates in different occupations including explanatory variables such as age, education, and so forth, and also a measure of the *on-job* occupational risk. The relationship between on-job risk and wage rates provides a measure of how much extra compensation people require in order to accept the extra risk. For example, consider the recent work by Moore and Viscusi (1988). The average hourly wage in the sample they studied was about $7, equivalent to an annual wage (assuming 2000 hours of work per year) of $14,000. (The data came from a 1982 study of persons' income, occupation, education, and so forth.) To each person's occupation, they matched data from two sources, the U.S. Bureau of Labor Statistics (USBLS) and the National Institute of Occupational Health and Safety (NIOSH), on occupational mortality rates. The average fatality rate in the NIOSH data was just under 8 per 100,000 full-time equivalent years, and in the BLS data, about 5 per 100,000. For a variety of reasons, Moore and Viscusi prefer the NIOSH data. Their estimates show that each increase of 1 death per 100,000 workers (e.g., from 8 to 9 per 100,000) increases the hourly wage by about 2.7 cents. This corresponds to $54 for a year's exposure to the risk, or correspondingly (adding up for 100,000 workers) $5,400,000 per statistical life in 1982.

Other approaches to the same problem yield quite different results. Merely shifting to the BLS measures causes the estimated value of life to fall by half, as Viscusi and Moore show. Alternative approaches have looked at quite different types of behavior to try to learn the same thing (e.g., how much people will pay for added safety in automobiles), or in some cases have turned completely away from market-related measures. One study, for example, looked at court cases for wrongful death to determine the magnitude of awards in civil trials when one person's behavior caused another person's death (Perkins, Phelps, and Parente, 1990). Others have looked at the implied value of life represented in government decisions to embark on various safety and regulatory programs (Graham and Vopel, 1981).

Even within a single type of study (such as the labor-market studies by Moore and Viscusi), wide variability exists in the estimated results.

Moving across methods (for example, comparing labor market studies with court awards) gives an even wider dispersion of estimates. All that we can say at present is that the subject of placing a monetary value on "life" has some considerable importance, but that the estimates now available contain a wide degree of disagreement. One can easily find credible studies that disagree with one another by as much as a factor of 10 in their estimates of the value of life. As is true of many empirical problems in health economics, this one begs for further work.

Chapter
16

Regulation in the U.S. Health Care Sector

*T*he U.S. health care system, while among the most "market oriented" in the industrialized world, remains the most intensively regulated sector of the U.S. economy. The health care sector has been under continuous price controls imposed by both federal and state governments since at least 1971; entry of new products into the market (drugs and medical devices) has been under stringent regulation for much longer, with requirements of a proof of efficacy of drugs since 1962, and of drug safety for far longer. Licensure of providers—physicians, dentists, nurses, therapists, pharmacists, and others—has been a part of state laws for most of this century. More recently, restrictions on the entry and size of hospitals creates a new form of controls, called "certificate of need" (CON) rules, still in place in some states, but recently abandoned by others.

Figure 16.1 sets out the possible domains of regulation in the health care sector. It separates "input" from final product markets, and it isolates four separate areas in which regulation might apply: price, quantity, entry by new providers, and quality. While this figure is not exhaustive, it covers the major areas in which local, state, and federal regulation affect the U.S. health care system. As Figure 16.1 shows, some types of regulation actually have effects in more than one area. For example, licensure affects both entry into the market by various types of "labor" (each an input into the production of final products) and also their quality. Price control rules administered by the Nixon Administration's Economic Stabilization Program (ESP) during the 1970s were typical of the

	Price	Quantity	Entry	Quality
Input	Wage controls, Antitrust	CON	Licensure, CON	Licensure, Voluntary "boards"
Product	ESP State regulations Medicare Antitrust law	—	FDA	Health Department Tort law FDA Peer review

Figure 16.1 Domain of health care regulation.

genre. A later section on price controls provides details. CON rules affect entry into the market by new providers, and also limit the quantities of inputs used by existing providers. Thus, it makes more sense to talk about each type of regulation separately, rather than to talk about each of the "cells" in Figure 16.1 one at a time. The set of regulations appearing in Figure 16.1 and the discussion that follows are not intended to be comprehensive, but rather to review some of the main forms of regulation and the economic issues associated with them.

LICENSURE

Licensure of professionals has a long history, substantially preceding any other form of regulation. The idea of licensure of "professionals" probably arose originally from various guilds in Europe, as they attempted to use the power of the state to their own advantage. Debates about licensure continue to carry the flavor that arises from this historical background: proponents of licensure (usually, existing practitioners of a field) praise licensure as a protection of citizens against fraudulent or unsafe providers of care. Opponents of licensure decry the entire idea as a tool to restrict entry and to limit competition. Like the mythical "Force" in the Star Wars movies, licensure has a "good side" (quality enhancement) and a "dark side" (limitation of entry and competition).

Licensure as we commonly think about it applies to *labor inputs* in the provision of health care, but seldom to the organizations (firms) that produce the final products of health services.[1] In health care, virtually all licensure comes from the states, rather than the federal government.[2] Licensure spans professionals ranging from physicians, dentists, and psychologists to nurses (RNs and LPNs), pharmacists, physical therapists, social workers, and dental hygienists. Virtually all licensure of individuals is limited to areas of the economy where the final product is a "service" rather than a

---

[1] An obvious exception is the licensure of hospitals and nursing homes. This sort of licensure is usually limited to obvious fire and safety questions, and has little to do with the quality of care provided.

[2] The only important exception is the license to use narcotic drugs, which comes from the federal government, arising through federal control of those drugs.

good, suggesting that the consumer's inability to exchange the "service" is an important component in the logic of licensure.[3]

If one set out the logic of licensure as a safety- and quality-enhancing regulation, it would contain the following premises:

1. Important variation exists in the quality of individual inputs (e.g., doctors, nurses).
2. Low-quality inputs translate, at least partly, into poor outcomes of the final product.
3. Managers of the firms that produce the final product either cannot or do not wish to measure the quality of labor inputs they hire. (Tort liability makes firms liable for damage caused by their employees, so the idea that managers would not wish to measure quality of labor inputs could arise only because of a failure of the tort system to produce sufficient incentives for product quality and safety.)
4. Consumers lack full information about product quality.
5. The product cannot be readily exchanged (it is a "service" tailored to the individual) and/or use of a defective product/service would endanger the consumer.

The idea that consumers cannot perceive product quality accurately has a reasonable statistical basis, because of the variability in outcomes that occurs even among "competent" providers. Health (H) is produced by medical care (m) with a random outcome—that is, $H = h(m) + v$. In plain words, sometimes people get better even with minimal or poor medical care, and sometimes people get worse or die even with the best medical care. The random term v captures these "random" outcomes, and the production function $h(m)$ captures the systematic effects of medical care. Medical care (m) is in turn produced by combinations of inputs, say doctors (D) and capital (K), so $m = m(D, K)$. Finally, doctor quality can vary, so $D = D^* + u$, where D^* is average quality of doctors. Thus, the quality of m will vary with the random doctor quality u, and the final outcomes will vary with the random component v in the production of health.

The consumer seeks to infer whether the doctor is "good" or not, but the inference problem is compounded by the various sources of "noise" in the system. Does a "good" health outcome

---

[3] We do not license farmers, although the food they produce is crucial to our survival, nor do we license the people who make automobiles. We do license, however, airline pilots, taxi drivers, and barbers.

guarantee that the doctor is "good"? (No; the doctor may merely have been lucky.) Does a "bad" health outcome guarantee that the doctor is "bad"? (No; it may have been a freak of nature, despite the best care.) The consumer must try to draw information from a very limited number of events in some cases, and may have almost no information available before making a choice. (How many times can you sample a surgeon's ability to remove an appendix?) In statistical terms, the problem is that when one attempts to infer whether u is large or small (the doctor is "good" or "bad), the noise sent up by the random component v may overwhelm the signal about u. If the variance in v is large compared with the variance in u, then it may be almost impossible for a single patient to infer anything about the quality of a single doctor, even after a series of encounters.[4]

Licensure can provide two types of information about quality, at least in concept. First, it can certify that the licensed person has sufficient command of the body of relevant knowledge to pass an examination on that material. (This does not test, for example, the manual dexterity of a surgeon, but it can test for knowledge about the indications for surgery, or the proper procedures to follow if a surgical incision becomes infected.) Second, the licensure authorities are in a position to collect information about poor outcomes by an individual doctor (no matter what physician-firm the doctor works for or in what hospital that doctor practices). This gives the licensure authorities a much larger potential sample of events, which can in concept help identify an individual doctor's competence far faster than any single patient (or physician-firm) can.

Most state licensure authorities have relied almost exclusively on the first type of quality indicator; an examination passed when the doctor graduated from medical school may often suffice for an entire lifetime.[5] Licensure revocation turns out to be quite rare, and most commonly occurs for events that have little bearing on what we would normally consider "medical quality." Previous studies find three reasons that dominate the list of actual revocations: (1) drug use by the physician, (2) fraudulent billing prac-

---

[4] This explains why some doctors who lose their licenses can produce a list of patients who will testify about the marvelous quality of their doctors, even in the midst of the licensure-revocation proceedings.

[5] Some states now have periodic relicensure examinations, but the idea of periodic reexamination is much more common in voluntary certification of quality, as will be discussed below.

tices, and (3) repeated sexual encounters between the doctor and his patients, most commonly in this case among mental illness healers. Revocations for "bad quality" turn out to be exceedingly rare.

Voluntary Quality Certification

Most areas of medical care that have licensure requirements also have voluntary indicators of quality. For many providers, these are private quality-certifying groups, typically with the word *College* or *Board* in their titles. These organizations perform two functions: They designate certain training programs in their specialty as "approved," and they administer written and oral examinations to applicants for their quality certification. Typically, before becoming eligible to take the examination (to "sit for the board" in the usual jargon) the doctor must complete a certified training program, thus becoming "board eligible." These training programs are "residencies" in medicine, surgery, pediatrics, and so on that "teaching" hospitals provide. Some of these subspecialty programs require additional specialized training beyond a residency. Thus, for example, one can become board eligible in pediatrics after a three-year residency, but to subspecialize in neonatology (the care of sick newborn children) requires an additional three-year "fellowship" in the subspecialty area, also at an approved site. Fellowship training usually takes place only at hospitals closely allied with medical schools; residency training is more diffusely provided, often at hospitals quite remote from medical schools both geographically and organizationally.

Voluntary quality certification has some distinct advantages over mandatory licensure. First, it is more difficult for a voluntary "board" to inhibit competition, because it does not have the compulsion of the state at its disposal. Second, the boards have produced indicators of varying degrees of quality, providing more information to the market about quality than a single-quality license can provide. (Nothing, of course, prevents mandatory licensure from also providing more specific information about quality, but in practice, one only knows that a provider is licensed.) Third, again in contrast to the usual practice of licensure, most medical specialty boards and many other quality-certifying private bodies have adopted requirements for repeat examinations periodically, along with requirements for "continuing medical education" (CME) for the successful renewal of a board certification. This puts the specialty "boards" in one more line of business, namely the certification of the quality of producers of this CME activity.[6] Many states now have CME requirements for relicensure by physicians and other providers—for example, that the doctor acquire 100 hours of CME credit over a three-year period—but few have yet adopted the reexamination rules that some specialty boards have now adopted.

Once an organization gains a market reputation for high quality, it has strong incentives to promote itself and its members. In other areas, franchises of businesses (like McDonald's Hamburgers, Midas Mufflers, Holiday Inn, and Better Homes and Gardens Realtors) all provide quality certification and national advertising for their constituent firms. Specialty boards provide a comparable certification of quality that they promote widely, including enabling patients to call and determine whether any individual doctor is "board certified."[7]

Specialty boards and voluntary certification have their own problematic features, just as does mandatory licensure. First comes the question of the quality of the certifier: It may be quite difficult for individual patients to understand what a certification really means. For example, the American Board of Internal Medicine is a classic "specialty board," but the American College of Physicians offers an alternative certification of quality, and to many patients, membership in the American Medical Association has a similar connotation of quality. Which of these implies a higher standard of quality, if any?[8] This problem blurs into other areas as well; voluntary certification of scuba instructors exists from at least four separate national organizations (National Association of Scuba Diving Schools, National Association of Underwater Instructors, Professional Association of Diving Instructors, and the YMCA), each of which proclaims its own certification as the best.

---

[6] CME courses commonly take place at pleasant locations such as Aspen, Colorado; Kaanapali Beach, Hawaii; Carmel, California; and Naples, Florida. However, there are now strict requirements about attendance at these events that are closely monitored, probably in response to early versions of CME courses that amounted to little more than tax-deductible vacations.

[7] The American Board of Medical Specialties, a consortium of 23 specialty boards, has a toll-free telephone line (800-776-CERT) that it advertises widely in Yellow Pages sections of most telephone directories, allowing patients to determine whether a doctor is certified by any of those 23 boards or an additional 50 subspecialty boards operated by the 23 primary boards.

[8] The ACP has the most vigorous reexamination requirement of any of these bodies, requiring exams every five years. The AMA requires only payment of a membership fee.

The second question that arises is the ability of a certifying body to restrict entry into the specialty. If an organization achieves a very strong market position in terms of quality certification, it can acquire a *de facto* ability to limit entry. Some of the various medical specialty boards may have achieved this position by limiting the number of approved residency positions throughout the country. (Recall that a trainee must complete an approved residency— approved by the specialty board—before sitting for the board exam itself.) Thus, limiting the number of approved residencies provides a convenient vehicle to limit the number of practitioners of a given specialty.

Some evidence suggests that an effective restriction to entry has been achieved by a number of specialty boards, most notably those in surgical specialties and subspecialties, and more recently in radiology, pathology, and anesthesia. These specialties have two common markers: First, practitioners of those specialties have unusually large economic returns to receiving the specialty certification. Evidence presented in Chapter 7 about returns to specialization provide the basis for this observation. It is of particular interest that the economic returns to surgical specialization have persisted from the early 1960s (when Sloan, 1975, studied the market) through the 1980s (when Marder and Wilke, 1990, made similar calculations).

Second, specialties where *de facto* entry restriction exists should have extensive queues from doctors trying to enter the specialty. The difficulties in obtaining residency positions and fellowships in these specialties and subspecialties match these economic returns; competition for these positions is notoriously fierce, and training positions are often committed four to six years in advance.

Quality Certification of Organizations

The previous discussion has focused on the quality certification of individual providers of care—doctors, dentists, nurses, and the like. There are also important certifications of quality for hospitals, nursing homes, and even medical schools. Licensure plays a role here as well, since every state licenses hospitals, clinics, and nursing homes. This licensure commonly limits itself to fire safety, food preparation processes, and so on, seldom venturing into realms of "medical quality." As with individual providers, voluntary certification has proven more of a degree of quality assurance.

With hospitals, the voluntary certification comes through several groups. Any hospital may apply for certification from the Joint Commission on Accreditation of Health Care Facilities (commonly referred to as the Joint Commission), a cooperative body sponsored by the American Hospital Association and the American Medical Association (hence the term *joint* in its title). This certification focuses on a wide variety of hospital activities, with specific reports given to hospitals in their on-site inspections that note specific "deficits" in performance. This process focuses on organizational structure and the process of operating the hospital, with no attempt to measure outcomes (e.g., patient survival, infection rates, readmission rates, etc.). Accreditation from the Joint Commission is important to hospitals, because many insurance plans limit their payments to accredited hospitals. Medicare allows approval from the Joint Commission as an alternative to a specific Medicare approval, even though the Joint Commission approval is generally viewed as "easier" than Medicare approval.

Even medical schools have an accreditation program—the American Association of Medical Colleges (AAMC)—that serves the same role with medical schools that the boards do for residency training. As with other certification bodies, the importance of accreditation depends in part on the externally perceived quality of the schools, and partly on how other groups (including governments) adopt the private certification in their own behavior. For example, in order to receive a license from most states, the doctor must (as one precondition) graduate from an approved medical school, and the list of schools approved by various states almost universally reflects the approval list of the AAMC. Thus, just as Medicare delegates the certification process of hospitals to the Joint Commission, many state governments delegate the process of approving medical schools to the AAMC. This type of implicit or explicit delegation of regulatory authority obviously blurs the line between private voluntary certification and mandatory state licensure.

Quality Certification and Consumer Search

Another important issue arises when we consider quality certification in health care markets: How does certification affect consumers' incentives in searching for lower-priced providers? In many markets, low price has a connotation—often correct—of low quality. When quality is intrinsically difficult to measure, sellers may be able to send a false "high quality" signal by raising their prices, a strategy that will probably work only when the customer has relatively few encounters with the seller. (We would not expect this strategy to work as well for grocers and barbers as we would for vacation resorts and divorce lawyers.) Providing quality "assurance" about a provider of medical care may actually promote competition in the final product market (for medical services) even if it inhibits entry into the factor market (the market for physician labor). The models of market price and consumer search set forth in Chapter 7 show the importance of search in markets that are intrinsically monopolistically competitive, a model that seems to fit medical service markets (like physicians, dentists) quite well. However, consumers may be quite reluctant to search for a lower-priced provider if they cannot measure quality well. Thus, establishing a base-floor for quality, as licensure does, or gradations of quality, as voluntary certification does, may move final product markets to a more competitive level.

DRUG AND MEDICAL DEVICE REGULATION

We now turn to a completely different type of regulation, namely that of pharmaceutical drugs and medical "devices." Federal regulation of prescription drugs began in 1938, when Congress established the Food and Drug Administration (FDA). This regulatory agency originally had purview over both the food supply and the supply of drugs in this country, but the "food" part of their work has evolved more to an activity of the U.S. Department of Agriculture.[9] The FDA establishes which drugs can enter the market, the purposes for which the drug can be used (for which diseases doctors can prescribe the drug), and even the information contained in the drug package when it is sold. The FDA must (by law) restrict drugs from the market that have proven in tests to be unduly unsafe, and for those with potentially risky side effects and hazards must select a set of illnesses for which the risks of side effects are deemed worth the risk. Virtually NO drug is truly "safe" in the

---

[9] This accounts for the ubiquitous stamp appearing on inspected meat in the store: USDA, standing for U.S. Department of Agriculture. The USDA both inspects meat for safety and grades its quality.

sense that large overdoses would cause no harm, but of course the same is true of most food substances, including salt, sugar, protein, caffeine, many vitamins, and even oat bran, and the FDA rules recognize this by comparing risk with potential benefit.

In 1962, an amendment to the FDA law added a separate requirement: Drug manufacturers were required to establish not only the safety of a new drug before it could reach the market, but also its *efficacy*. That is to say, they had to show not only that the drug was safe, but that it worked as described.

The testing mechanisms required by the FDA include three phases of tests. Phase I tests take place in research animals, once the basic chemical properties of the drug have been established in the chemical laboratory. These tests often use massive doses of the drug, compared with the intended use in humans, on the belief that any bad effects will appear sooner and with a smaller sample of animals.[10] Researchers also strive to find animal "models" of comparable disease processes, and attempt to show that the drug "works" in those animal models. Once the efficacy has been established in animal models and the level of dose that is "safe" has been found, the company may move on to Phase II tests in controlled human trials. These trials use the paradigm of the randomized controlled trial (RCT), widely held as the best possible paradigm for the analysis of a drug's effect. Box 16.1 sets out the basic parameters of a classic RCT and discusses why it is so widely admired.

If the RCT demonstrates that the drug is both safe and effective in Phase II trials, it continues on to Phase III studies, in which certain clinical investigators use the drug, but continue to report on effects on each patient.[11] (This phase greatly expands the sample available to detect side effects.) Finally, once past a prescribed amount of this "postmarketing" testing, the drug may be released for routine use by all physicians.

The rigorous tests of drugs on the U.S. market have produced a drug supply with safety unsurpassed throughout the world. On very few occasions, drugs even past the postmarketing phase are discovered to have serious side effects, and are subsequently with-

---

[10] Among other things, these tests establish the dose known as LD50, the lethal dose (per unit of body weight) for 50 percent of the animals receiving the drug.

[11] The physician must obtain an Investigative Clinical Drug number that is used in all subsequent reports on the drug's use by that doctor.

Box 16.1 **RANDOMIZED CONTROLLED TRIALS**

In a pure randomized controlled trial (RCT), patients agree to sign up for the "experiment" without knowing whether they will get the "new treatment" (here, dubbed T) or the customary therapy for the disease, known as the "control" (here, dubbed C). Patients are randomly assigned to receive either T or C, commonly (if possible) without either the patient or the treating doctors knowing which patient receives which. When both the patient and the doctor cannot "see" the treatment, this is known as a "double blind" study. (Research colleagues provide the drug or the "control" in identical form. If the drug has no obvious manifestations such as creating a rash, then nobody except the statistical analyst on the project has any idea which patients received T and which received C.) Sometimes it's impossible to keep the "secret" from the doctor, in which case one has a "single-blind" study, and sometimes even the patients know. (It's awfully difficult to run a randomized trial on whether back surgery "works," for example, without having both the patient and the doctor know whether the patient had received an operation. On rare occasions in the past, patients have been deluded with a "sham operation" when they are put to sleep, an incision is made in the appropriate place in their bodies, and they are sewn up. These types of experiments are no longer conducted in this country.)

In a pure randomized trial, it's quite easy to measure how well the drug, device, or procedure "works": One looks at the outcomes for the treated group and compares them with those in the control group. The difference is the "treatment effect" and is the standard measure of treatment efficacy. For example, if 70 percent of the T group recover from their disease, but only 50 percent of the control group (i.e., those receiving existing therapy), we would commonly accept the idea that the treatment "works" if the sample size were sufficiently large. Studies like these, incidentally, form the basis for some medical decision analysis "decision trees" as described in the Appendix to Chapter 4, since they provide one type of estimate of the outcomes for people who do and don't receive treatment for a disease.

The beauty of the RCT is its simplicity: Since patients are randomly assigned to T or C, we have a strong presumption that "other factors" possibly affecting their recovery are also randomly distributed across the T and C groups. Thus, there is no need to measure other "covariates" or to worry about oddities of individual patients. If the sample is sufficiently large, the test is robust and we can assert with strong confidence that "T is better than C" (or vice versa).

We should recognize that an RCT never "proves" that a drug works, in the sense that one can *never* "prove" anything with a statistical estimate. Take the previous example of 70 percent recovery versus 50 percent recovery. If the sample in the RCT included only 10 patients each in the T and C groups, then one could observe 7 "healed" patients in the T group by some chance, even if it worked no better than C. To see this for yourself, flip a coin 1000 times and divide it into strings of 10 "outcomes." On average, about 1 out of every 6 of such groups will have 7 or more Heads. This means that an RCT with sample sizes of 10 patients each in T and C stands 1 chance in 3 of "finding" that T is better than C, even though they really have the same effect. (To see this, string out another set of 1000 coins, divided into groups of 10, and call the first group T and the second group C. Now count the number of groups of 10 coins each in which T has more heads than C. This will happen about a third of the time.) Elaborate statistical designs help to decide how large the samples should be in RCTs, but the problem persists with 100, 1,000, and even 1,000,000 patients in each "arm" of the trial. There's always a chance, albeit very small once the sample size grows to several hundred per arm, that the true effect is zero when the sample shows a benefit. This is why RCTs can never "prove" that a treatment works. More precise language would say that the RCT increases our confidence that it works, or we can even say that it works "with 99 percent confidence" (or whatever the appropriate statistical confidence turns out to be).

drawn.[12] If this were the end of the story, it would be a happy ending, and indeed would probably not appear in this textbook. However, the FDA rules have come under vigorous attack from various quarters as widely dispersed as the drug companies themselves, the economics profession, and various AIDS patient activist groups. The sources of concern all focus on the effect of having the FDA rules keep drugs off the market when various groups of patients (and their doctors) would prefer that they were allowed into the market. These patients fall into three categories:

1. Some patients cannot obtain the drug in the United States, even though it has sometimes been in common use for years in other countries, including Canada, European nations, and Japan.

2. Some patients have such rare diseases that the overall prospective sales for the drug are not sufficient to pay for the costs of the required FDA tests. These are commonly called "orphan drugs," and the FDA rules have been modified considerably to ease the testing and marketing of such drugs.

3. Finally, some patients have presumably fatal diseases with no known cure, such as cancers and (more recently) AIDS. For these patients, the question of "safety" of the drug has only limited meaning—they will die anyway if something isn't done—and the issue of "efficacy" is also something they may have only limited interest in learning about. De-

---

[12] The most dramatic example was the tranquilizer Thalidomide, which when prescribed to pregnant women created a high incidence of badly deformed babies. Another prominent example was the drug diethylstilbestrol, known more commonly as DES. This drug, widely used by women, appeared in later epidemiologic studies to have heightened the risk of their *daughters'* having cervical cancer.

lays until FDA-required standards of proof appear are necessarily delays in arrival of the drug on the market. For those with rapidly progressing and eventually fatal diseases, it often seems worth the risk of paying for a drug that has no proven efficacy, since the consequences of waiting are so permanent.

Foreign-Marketed Drugs First consider the question of foreign availability of some drugs. U.S. citizens are occasionally confronted with stories about a "wonder drug" that has been available in Europe for years, and is used widely there. They wonder why they cannot receive the benefits from the drug. To analyze this question, we need to ask what welfare loss occurs from restricting the drug. This, in turn, depends on how well potential "substitutes" work. Figure 16.2 shows the question in standard economic terms, where we now interpret "demand curves" in the same way that we did in our discussions of variations in medical care use—namely, as measures of incremental value to consumers as the rate of use increases, or "willingness to pay" curves (WTP for short).

The panel on the left shows the WTP curve for European consumers for the drug they have readily available at a price P_E. Their consumer surplus appears as triangle A. In the United States, with only a less efficacious drug available at price $P_{U.S.}$, U.S. consumers' WTP is not as high, and they get consumer surplus shown in triangle B from using this drug. Prohibition of the European drug from U.S. markets reduces U.S. consumer surplus by an area equivalent to A minus B.

Of course, the "risk" the FDA worries about is that European testing of drugs may not be as complete as that in the United States, so that some damage of value D might occur to patients with

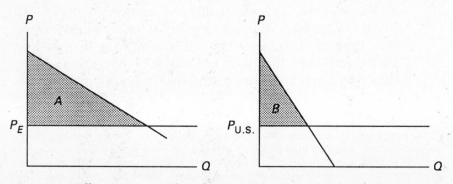

Figure 16.2 Willingness to pay for a drug based on price and efficacy.

probability π that has gone undetected in the European markets. Continuing to exclude the drug makes sense if $A - \pi D < B$, which we can easily solve to say that we should exclude the drug if $\pi > (A - B)/D$. The point of using the European (and other) experience, of course, is that when a drug has been in common use for several or many years, π presumably becomes very small, so continued exclusion makes no sense. We must remember, however, that π is wholly "subjective" in the sense that no laboratory experiment can help us measure the risk. We can only attempt to learn from others' experiences. Nevertheless, the U.S. FDA continues to ignore the experience of such drugs in foreign markets, and insists on complete testing before allowing U.S. marketing.

The extra costs that this approach creates for U.S. patients can sometimes be quite large. Many patients hear about such drugs, of course, and often establish foreign contacts to purchase them.[13] These added transactions costs represent an additional cost of the FDA rules as now implemented. The more important problem, however, is the forgone consumer surplus in the markets for these drugs. Consumer surplus—that ephemeral economists' concept of "value received above and beyond the price paid for a commodity"—measures the net social value of having the drug on the market. (Review the discussion of consumer surplus in Chapter 4 if the terms seem unfamiliar to you now.) Several attempts have been made to measure the lost consumer surplus from keeping drugs off the U.S. market "too long." Peltzman (1973) estimated that the lost consumer surplus represented $0.42 billion per year in 1970 ($1.4 billion in $1990), equivalent to 8 percent of annual drug sales.[14]

Orphan Drugs   So-called "orphan" drugs are those for whom the market is so small that it can never be profitable for the drug firm to conduct the required FDA testing. In the extreme case, suppose that the drug were produced by only one seller, who establishes a (perfectly legal) monopoly and prices accordingly at the profit-maximizing price p_M. (Review Box 7.2 on monopoly pricing if you need to.) Figure 16.3 shows this case.

---

[13] Some people even fly to foreign countries to purchase their own supplies of such drugs.

[14] The methodology for this study came under attack by three economists (McGuire, Nelson, and Spavins, 1975) with a rebuttal by Peltzman (1975). This debate makes interesting reading for people who want to learn more about the economic issues associated with FDA regulation.

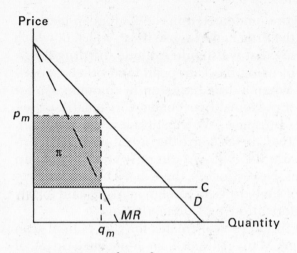

Figure 16.3 Monopoly profits.

The monopoly profits *per period* (Π) are shown as the box of height $P_M - C$ (the cost of producing each dose of the drug) times the quantity sold, Q_M. The company gets this profit each year, once the drug begins to be sold. It must compare the *discounted present value* of all such future profits against the cost of the FDA-required testing procedures. For many "rare" diseases whose market is intrinsically small, it would never be profitable for the drug company to engage in testing, and hence the drug would never reach the market.

This problem has several possible solutions. One would be to have the FDA agree to subsidize the testing costs for drugs like this. Another (which the FDA chose to do several years ago) allows drug companies to file for "orphan drug" status for such drugs. Once the company establishes to the FDA that the market is too small to support full FDA testing costs, the FDA allows substantially revised testing procedures that create notably less cost. This solves some but not all of the problems created by the FDA rules, for no matter what set of rules is devised, some groups of patients will find their diseases "too rare" to merit the attention of the drug companies. One irony in this process, of course, is that when the disease is quite rare, the number of patients who might be harmed by the side effects of the drug is also quite small, hence presumably reducing society's concerns about harmful side effects.

"Fatal Illnesses"   A final set of diseases of immediate concern are those that have increasing severity and eventual death as the only known outcomes for patients. For patients with such diseases, delay in introducing an efficacious drug is fatal. Many persons argue—with considerable success in the case of AIDS drugs—that the usual rules about safety and efficacy should be set aside in such cases, since the alternative of death cannot be worse than taking the drug. Again, we can place this argument in straightforward economic terms: Suppose the drug has a chance ρ of curing any patient's disease, and costs C_D per unit to buy. Even if *nothing* is known scientifically about the efficacy of the drug, it is rational to use it immediately in some settings. Suppose that we think "saving a life" (or more properly, preventing a premature death) is worth some specific amount, which we can call V_L. (Numerous economic studies put this value somewhere in the range of $1 to $3 million or more. A discussion about economic valuation of life and of extending life by a year appears in the appendix to Chapter 15.) Suppose that it takes N doses of a drug to determine whether it works (say, a year's supply, when the person would normally be expected to die within a year under normal treatment). Then it makes sense for the patient to use the drug if $NP_D < \rho V_L$, or equivalently, if $\rho > NP_D / V_L$.[15] Of course, if some researcher monitors the drug use and records what happens, society can often learn a considerable amount about the efficacy of the drug, even though it would not be administered through a classic RCT. This creates an additional social benefit beyond the private benefit (to the patient) already discussed.

The logic of using such information arises from alternative designs to the classic randomized controlled trial. In this case, "historic" controls provide the basis of comparison. If doctors have high confidence about the "natural history" of a disease like this, then those historic patients provide a meaningful "control group" for comparison with a group receiving an experimental drug to try to help them. A risk to such a study, of course, is that the new "experimental" patients really don't have the same fatal disease that had previously been observed and measured, but if their disease is well documented, then this type of model can importantly add to the stock of knowledge.

---

[15] The cancer drug Laetrile represents one famous example like this. This drug, widely acclaimed as a miracle cancer cure in the 1960s and 1970s, was widely debunked by most cancer scientists. However, a few scientists continued to assert their belief in the drug's efficacy, and it became widely available in other countries. Many U.S. citizens traveled to Mexico to receive Laetrile treatment there, including the famous movie actor Steve Mc-Queen, who eventually died from his cancer.

Under normal FDA rules, even if this situation occurred, the drug company could not legally market the drug in question. However, under extreme political pressure from AIDS activists, the normal rules have been widely modified in ways that come much closer to following the economic logic set out above. Whether this political activism spills over into groups of patients with other diseases remains to be seen, but it seems plausible that the changes in drug-testing strategy in AIDS drugs could alter strategies for many other patient groups as well, particularly those confronting fatal illnesses with no known effective treatment.

FDA Rules and Market Competition

FDA rules about testing for safety and efficacy have the potential for affecting incentives for R&D and for entry of new producers in pharmaceutical markets. As with any invention, U.S. patent law provides 17 years of the exclusive right to produce the patented product for its inventor. (Of course, the inventor can license the patent to other producers.) Under the 1962 law, however, a drug company could not begin testing the drug until after the patent was filed, so that the "effective" patent life was shortened by the time necessary to obtain FDA marketing approval. This approval took so long that the effective patent life was shortened from 17 years (the statutory term) to about 7–9 years, according to studies of drug development. Taken alone, this aspect of the FDA laws would surely reduce the incentives for invention of new drugs.[16]

The same FDA laws had another important effect. Even after a patent expired, they required other drug companies that wanted to manufacture the same drug to go through the same process of testing before they entered the market, even if the new competitor-drug was chemically identical to the original product. This had the strong effect, of course, of reducing willingness of competitors to produce "generic" drugs that might compete with the brand-name

---

[16] The inventor gains monopoly rights for 17 years, so the economic profit from invention is the discounted present value of an annuity for 17 years of the annual monopoly profit Π. Shortening the effective patent life to 9 years (by delaying marketing of the drug 8 years for testing) allows the inventor only an annuity of 9 years' duration, and that annuity only begins 8 years from the initial date of investment. This makes a considerable difference in the economic gains to invention. For a discount rate of 10 percent, for example, the present value of a 17-year annuity is 8.02 times the annual profit Π. Eliminating the first 8 years of the patent for testing makes the patent worth only 2.69 times Π. Thus, the rule eliminates about two-thirds of the economic profits associated with a patent.

drug previously marketed. It also, of course, reversed some of the original erosion of the 17-year patent previously discussed. If the requirement for new testing of a competitive "generic" drug effectively eliminated future competition, the original inventor would gain an "effective" perpetual monopoly, but (because of the delay for testing) would not be able to exercise it until testing was completed. A problem at the end of this chapter shows that at discount rates of 10 percent, the incentives to invent are still blunted by these rules, compared with a straight 17-year patent protection, followed by open competition.[17]

The old FDA rules had the unhappy consequences both of reducing the incentives for investment and also of reducing the incentives to enter competitively after expiration of a patent for those drugs that actually reached the market. In 1984, a new revision of the FDA rules attacked both of these problems.[18] Under this revised law, if the "generic" drug is actually equivalent both in chemical content and route of administration, it can proceed with minimal delay. In general, the approval process is highly accelerated even for drugs that are not perfectly equivalent. This should greatly boost entry of generic competition for those drugs with patents expired or soon to expire. (Such drugs account for about half of the dollar volume of drug sales in the United States.)

The 1984 act also allows for extension of patents to replace time lost in the FDA approval process, up to five years of extension for all drugs patented after 1984. This will increase the incentives for invention of new drugs, possibly considerably. Finally, the 1984 law allows up to five years of "market exclusivity" for drugs even if they do not have patent protection. This will likely increase the number of European and other drugs brought into the United States, even if they are not eligible for U.S. patent protection.

A "Non-Event" in Regulation

It seems worth noting at this point the considerably different treatment of drugs (under the FDA law) and "medical interventions" in

---

[17] However, at a 5 percent discount rate, this combination of delay plus "permanent" monopoly increases the incentives above a straight 17-year patent protection. Most corporations use a discount rate of at least 10 percent, often 20 percent or more, in evaluating investment decisions.

[18] The act was called the Drug Price Competition and Patent Term Restoration Act of 1984. For a useful summary of the provisions of this act, see Mattison (1986).

general. In contrast to the considerable regulation of drug safety and efficacy, no comparable regulation exists for the promulgation of new "treatments" or the assessment of previously used treatments. Here, "treatments" means both surgical procedures and also "strategies" for treating patients that do not involve surgery. This creates an odd schizophrenia in the production of knowledge and also in the likelihood that a new treatment innovation will reach the market. Drug regulation in general will inhibit market entry, particularly the pre-1984 rules. Nothing inhibits entry of a new surgical technique, however. This will distort the economic incentives regarding these alternative forms of therapy, probably tipping us toward "too much surgery" and not enough "pharmaceutical treatment." This is a highly relevant comparison for many diseases, ranging across a wide spectrum of illnesses for which both surgical and nonsurgical approaches are feasible.

"CERTIFICATE OF NEED" (CON) LAWS

A completely different type of regulation has been present in U.S. health care markets for decades—namely, limitation on construction of new "capital" facilities in order to try to limit total hospital use. These regulations prohibit the building of new hospital bed capacity (and often the addition of any "expensive" equipment in the hospital) without having government approval in advance. The architects of such laws attempt to determine how many hospital beds (and possibly how many MRI units, etc.) a given geographic area "needs," and then allow new construction only if available supply does not meet that "need." Because of this underlying logic, these rules are called "certificate of need" rules.[19]

CON-like programs have existed for many years, but until 1974, they had (with only a few local exceptions) advisory capability. In 1974, the National Health Planning and Resources Development Act institutionalized the idea of health planning, provided federal funds to support planning agencies, and (to put some teeth into the law) limited Medicare capital payments to those facilities that had been approved by the relevant planners.

---

[19] A 1981 book by economist Paul Joskow examines CON and other hospital regulations in detail. He offers a relatively complete history of CON laws, and reviews broadly the literature on their effectiveness. His conceptual approach to CON laws differs somewhat from that discussed below, but much of the empirical evidence he summarizes can be interpreted equally well in light of the models used in this book.

This type of regulation raises several conceptual issues, which we can first consider, and then we can turn to a number of empirical studies to learn whether CON laws have had any measurable effect on actual hospital costs and prices. These questions include (1) How might we interpret "need" economically, and how is the concept interpreted in the regulations? (2) Is there any logical basis for believing the CON laws might increase or decrease costs systematically, or is the issue completely an empirical one? (3) If such laws are put in place, what types of economic behavior should we expect in response by hospitals?

"Need"

Consider first the idea of "need." One approach, of course, asks what level of a commodity is necessary to sustain life itself—for instance, how much water or how many calories of food, and so on a person must consume daily in order to stay alive. Even this question has no clear answer, however; the minimum quantity of food and water depends in part on the activities a person undertakes. In normal sedentary life, for example, an average-sized person burns about 1800 calories of food each day. However, the same person riding a bicycle 100 miles per day will burn an added 3000 calories or so. How much does the person "need"? It depends on other activities.

The same holds true for hospital care and for any other "necessary" commodity. Certain hospital procedures are "necessary" to save lives when otherwise-fatal medical conditions occur—for example, operable cancers, profuse bleeding following a vehicle crash, and so forth. Others will save lives at least with some probability, if not with certainty—for example, coronary artery bypass surgery for some patients. One approach to a community's "need" for medical resources is to add up all of the uses for such procedures that the area's population might have. But this also ignores other behavioral questions, just as the example about "needed" calories to stay alive. Lung cancer will occur more frequently in communities that smoke more cigarettes, making more lung surgery "necessary"; more liver transplants will seem "necessary" in communities with more heavy drinkers; vehicle crashes will injure more people in communities that allow drunk driving more, that have poorer roads, or that invest less in enforcement of speeding laws. "Need" cannot have a predetermined, biologically defined meaning.

In practice, CON laws typically center the state's definition of need on something like the average beds per capita for the state at the time the law was passed, commonly adjusting the averages at least for differences in age and gender mix of various regions. Once the "standard" is chosen, then regions with "excessive" beds will find CON certificates for new construction difficult to obtain, while those areas with relatively low bed supply (often dubbed "underserved areas") will be allowed to grow to the regulatory standard, if any hospital wishes to enter the market or expand. Box 16.2 summarizes the CON rules for one particular region, to provide a flavor for the types of rules that CON regulators devise.

Entry Restriction

The fundamental logic of CON controls rests on the premise that natural market forces will lead to "too much" hospital capacity, with undesirable economic consequences. Does any economic rationale exist for this belief? Normally, of course, in a for-profit industry, we would hold the belief that entry restrictions were a bad idea, helping to sustain monopoly pricing. Numerous studies of government rules restricting entry in industries ranging from airlines to taxicabs have reached the conclusion that entry restrictions harm consumers, and indeed the fundamental premise of the substantial decontrol of such industries as airlines, trucking, and others rests on the idea that free entry creates the best possible market conditions for consumers.

Tools developed in Chapters 8 and 9 provide a basis for understanding the potential effects of CON rules. Return to the discussion surrounding Figure 9.2 and we can see how CON laws might work. Recall that the EE curve in Figure 9.2 represents the set of possible quality-quantity combinations that are both feasible for the hospital because of market conditions and "legal" because of the zero profit constraint. Recall also that when one hospital enters or expands its capacity in the market, the demand curves (and hence the EE curves) of all other hospitals in the same market shift leftward. In general, however, we would not expect the *cost* curves of the hospital to shift with such entry.[20]

---

[20] Cost curves might shift upward if the new entry created higher costs for inputs (such as nurses) for all hospitals in the market. This would only emphasize further the types of changes that we can discuss while presuming that cost curves do not change any.

Box 16.2 **CERTIFICATE OF NEED (CON) REGULATIONS**

Many states have CON laws that specify the number of hospital beds, nursing home beds, or large hospital-based "devices" (like MRI, cobalt therapy units, burn units, etc.) allowed in any region. In order to expand a region's capacity, an organization must receive permission from the state's regulatory authorities. These vary from state to state—sometimes a special commission, sometimes the State Health Department—and commonly delegate much of the overall process to a regional or local organization. Previous federal law supported these local organizations—called "Health System Agencies" (HSAs) in their most recent manifestation—until the mid-1980s, when political support for regulation in the health care field diminished. During this period, many states completely abandoned the CON process, essentially allowing free entry into the areas previously regulated. Some states continue to fund these HSAs, and use them in ongoing regulatory processes.

Where they persist, these regional organizations receive applications from hospitals (or nursing homes) that wish to expand their bed capacity or add equipment or special "units" covered by the state law. They then determine if the capacity is "needed" according to predetermined formulas, and then make recommendations to the state authority, which makes the final decisions. In reaching decisions about "need," regulators often rely on specific formulas, commonly devised at the state level, but with some local "flavor." As an example, consider the following calculations, which come from the regional HSA for a county in upstate New York. These calculations show how the bed "need" for long-term care beds in the county was determined for 1993.

The analysis begins with the population of the county of age 65+ and under-65. In this county, the over-65 population was 4,926, and the under-65 population was 133,833. (This county, at 3.55 percent, has a much lower proportion of over-65 persons than the United States as a whole. The corresponding U.S. population has about 12.5 percent of the population over age 65—see Table 2.2.) The rules apply statewide "use rates" for various services for the under-65 and over-65 populations. For nursing home beds, the "rule" (i.e., the statewide average) allows 173 beds per 1,000 population over age 65, and 0.453 beds per 1,000 persons under age 65. Thus, for this county, the "allowed" number of nursing home beds is $(173 \times 4.926) + (133.833 \times 0.453) = 852 + 61 = 931$. A similar calculation for "supportive housing" allowed a total of 192 units, and finally, for community-based long-term care services, the rules allowed a total of 686 units.

The next step (you didn't think that a bureaucratic rule could be *that* simple did you?) adds these allocations up: $(913 + 192 + 686) = 1,791$ persons in "need" of some type of long-term care. Next, it calculates the *local* patterns of provision of long-term care to those actually receiving services in the county in some previous period (in this case, 1986). In this county, there were 967 persons receiving nursing home care, 205 receiving supportive housing, and 426 receiving community-based long-term care, for a total of 1,598 persons. These lead to a local "mix" of services of 60.5 percent nursing home, 12.8 percent supportive housing, and 26.7 percent in community-based long-term care.

The final step applies the local patterns (60.5 percent, 12.8 percent, and 26.7 percent) to the total "need" predicted for 1993—1,791 persons for this county. This total gets apportioned as 1,084 nursing home beds (60.5 percent of 1,791), and similarly, 229 supportive housing units and 478 community-based long-term care units.

The local providers in this county have 980 nursing home beds actually constructed and operating. The forecast "need" for the community in 1993 is 1,084 nursing home beds. Thus, the community will have the legal right to build an additional 104 (1,084 − 980 = 104) nursing home beds. Applicants for the right to build those beds may file a CON request, and the authorities will likely grant permission to build those 104 beds on the basis of the previous calculations.

Capacity expansion decisions, as one might expect, have the possibility of provoking considerable dispute, particularly if a regulatory group has determined that (say) 100 new nursing home beds may be constructed in a region, and several nursing homes each seek permission to build those 100 beds. In these situations, the disputes take on very formal legal structures, often including an administrative law judge and sometimes an actual legal trial. Each applicant will attempt to show why it is best qualified to build the new bed capacity instead of other applicants. Issues of competition and quality of care often appear, and the disputes commonly take on much of the same tone as an antitrust case involving mergers. (In such cases, definitions of the relevant market and market concentration measures such as the Herfindahl index become the weapons of battle.) A substantial source of income for some health economists and health planning experts—testifying in such CON proceedings—largely disappeared with the widespread decontrol of the hospital and nursing home industries in many states. Some of these people have turned to other activites, including writing textbooks in health economics, to replace the income no longer available from these legal/expert-witness activities.

Consider now a not-for-profit hospital that (initially) holds a monopoly position in a market, so that market demand curves and the hospital's demand curves (for differing qualities) are the same thing. Figure 16.4 shows such a setting for *one* quality level, shown as D_M. (The complete story about the hospital's decision will actually reflect demand curves at numerous possible qualities, but that would clutter up the figure too much.) The $E_M E_M$ curve traces out all of the intersections of each quality's demand and cost curves as if the figure showed them. (A review of Chapter 9 material on how EE curves get constructed may prove beneficial.) Now consider the consequences of entry; the hospital's demand curves will shift leftward, possibly as far as D_C, as will its EE curve. Unconstrained entry will continue to the point where the hospital's EE curves consist of points of single tangency between AC curves and de-

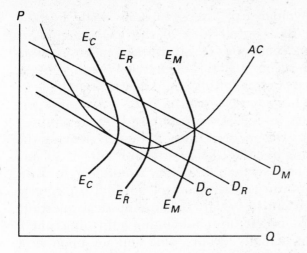

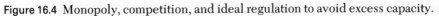

Figure 16.4 Monopoly, competition, and ideal regulation to avoid excess capacity.

mand curves. Figure 16.4 shows this for the one level of quality appearing in that figure, and the "competitive" opportunity set $E_C E_C$ shows the way the EE curve shifts inward in general. At that point, neither entry nor capacity expansion is feasible for the hospital.

It is obviously the case that the hospital in Figure 16.4 would choose an operating point above minimum AC for all quality levels if it had a monopoly, and also that the hospitals would operate above minimum average cost *on the other side of the AC curve* once unconstrained entry had shifted the EE curve to $E_C E_C$. No matter what quality selected, either the monopolist or the "competitive hospital" would operate above minimum average cost.

It is easy to see that some combination of entry restrictions *must* exist that will create a set of demand curves, including D_R for the quality we are looking at here, and a "regulated" $E_R E_R$ curve somewhere between the $E_M E_M$ curve of the monopolist and the fully competitive $E_C E_C$ curve also shown in Figure 16.4. This would happen if the regulator allowed "some" but not "complete" entry to erode the pure monopoly shown by $E_M E_M$. An exquisitely careful regulator might even be able to construct a set of hospital capacity rules such that each hospital had an $E_R E_R$ curve that went *exactly* through the minimum average cost curve for each possible level of quality. An all-knowing and beneficent regulator could actually achieve this goal, thereby reducing social cost of providing hospital care.

This basic idea—avoidance of "excess capacity" in the hospital industry—represents the central idea of the CON regulatory model. The underlying economics of the situation makes it *possible* that regulation can accomplish this objective. We can also see, however, that regulators in this setting could do considerable economic harm. Nothing in logic says that the $E_C E_C$ curve must produce average costs that exceed those produced by an arbitrary regulator. A regulator could restrict entry too much (leaving the industry near something like the $E_M E_M$ curve in Figure 16.4), and actually produce higher costs than free entry would lead to. How successfully regulators can actually control hospital capacity to reduce costs will turn out to be strictly an empirical question, because the theory tells us that they could either help things out or goof things up. What actually occurs will depend upon how adroitly regulators know the cost curves and demand conditions for the hospitals they regulate. As the next section shows, this discussion also ignores an important question about substitution between "beds" and other inputs, a phenomenon that will actually cause cost curves of hospitals to rise. This will have the effect of reducing hospital output, but in a socially inefficient way. We turn next to this question.

Input Substitution with a Capacity Constraint

The question of substitution in production emerges when one considers the actual structure of these rules: They do not really control "output" of a hospital (as Figure 16.4 implies) but rather place limits on a specific input (beds). If hospital production functions have no ability to substitute other inputs for beds, then the previous discussion stands complete. However, if hospitals have the ability to substitute one type of input (say, nursing intensity) for another (beds), then they can in part "defeat" the intent of the regulation by expanding output further than the regulators intended, even if the capacity constraint on beds is effective. The logic for this rests on the simple rules of cost minimization, as Figure 16.5 shows.

Consider a hospital producing at output level Q_1 and confronting costs of two inputs ("beds" and "nurses") of amounts w_B and w_N respectively, reflected in the slope of the iso-cost line $I_1 I_1$. If the hospital faces a constraint on the number of beds it has available, say to B_R, then that constraint will "bind" the hospital only at relatively high levels of desired output—for example, Q_3 and Q_4 in

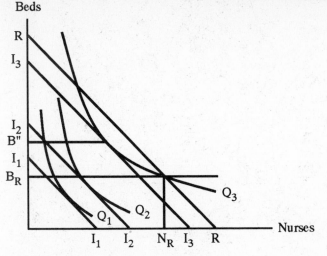

Figure 16.5 Cost minimization.

Figure 16.5. Note that at output levels Q_1 and Q_2, the cost-minimizing solution, shown at tangencies of I_1I_1 to Q_1 and I_2I_2 to Q_2, would use fewer than B_R beds, so we would say that the constraint would not be binding at those levels of production.

Now look carefully at output level Q_3. Without the constraint B_R, the hospital could produce at minimum cost with B^* beds and N^* nurses, but this solution is not allowed with the CON rules; the isoquant's tangency with the constant-cost line I_3I_3 shows this solution. With the constraint binding, in order to produce Q_3, the hospital must slide down the isoquant Q_3 until the bed constraint B_R is just met, using B_R beds (and therefore abiding by the law) and N_R nurses. The constant-cost line going through *that* point is the line I_RI_R, rather than the more efficient line I_3I_3 that the hospital would freely choose if it could. A similar effect occurs, but more so, for output level Q_4, and all higher levels of output. As the hospital reduces the number of "beds" to conform to the regulatory constraint, other inputs (here, "nurses") are increased. The net consequence of this constraint is to alter the cost curve of the hospital for each level of quality it might produce, at least for outputs like Q_3 and Q_4 in Figure 16.5

Since it costs more to produce Q_3 (and other quantities similarly) with the bed capacity constraint B_R than without it, we need to redraw the cost function of the hospital. Figure 16.6 shows the original and modified cost curves for several levels of quality that the hospital might select. The original EE curve shows the set of possible quality-quantity points that the hospital might select

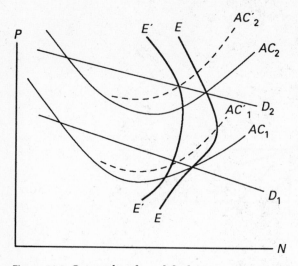

Figure 16.6 Original and modified cost curves for several levels of quality.

without regulation. The higher cost curves (shown as AC'_1 and AC'_2) induced by the regulatory constraint create a modified opportunity set, shown as $E'E'$ in Figure 16.6. The higher cost curves cause the EE curve to shift to the left to $E'E'$, so the optimal quantity to produce will fall, holding constant the location of the demand curves confronting the hospital.

Summarizing the Effects of CON Rules

We can now combine these two phenomena to understand fully the potential consequences of CON laws. First, we can say without ambiguity that the hospital will want to substitute toward "nurses" and other inputs when the bed constraint becomes binding. The cost curves for each level of quality shift upward because of this behavior. We had also seen previously that the regulatory constraint alters the family of demand curves facing each hospital, shifting them inward as overall market capacity expands. CON laws inhibit that inward shift of demand curves for each hospital before they reach the monopolistically competitive curve E_cE_c shown in Figure 16.4, by reducing the number of competitors in the market. On balance, we can see that CON laws have numerous economic effects, some of which *must* raise costs, and some of which *may* reduce and *may* raise average costs, depending upon the adeptness of the regulators.[21]

Empirical Evidence

This conceptual ambiguity forces us to turn to real-world behavior to understand the consequences of CON laws. Several studies of the effect of CON laws on hospital costs have appeared in the literature. The first, and perhaps most widely cited, is the study by Salkever and Bice (1976), using data from the early 1970s. Their study took advantage of the fact that various states across the nation adopted CON laws at different times, so they could compare CON and non-CON states' hospitals directly. They found that CON laws had succeeded in reducing the rate of hospital bed growth, but that hospitals had increased "other assets" per bed in response, a form of substitution comparable to that discussed above for "nurses." They also found that, if anything, costs had increased slightly *faster* in CON states than in other states.

Sloan and Steinwald (1980) conducted a similar study for data from 1970–1975, and, not surprisingly, reached similar results. They found insignificant effects in general for CON laws, but if anything, concluded that the laws had increased (rather than decreased) average costs of hospital care. They also found substantial and significant increases in nurses per bed, the type of substitution discussed in the previous sections.

Several observers of CON laws have hinted that these laws became more effective through time, either because added time gave the laws more time to "bite" or because the regulators became more savvy about the behavior of the hospital industry. Sloan (1983) studied hospital costs, length of stay, and "profitability" for the time frame 1963–1980, including measures of both mature and new CON programs. If anything, he found that the more mature programs had *less* effect on hospital costs than did younger programs, and his broad results indicate that CON laws were ineffec-

---

[21] A final issue remains: If hospitals are profit-maximizing firms, rather than not-for-profit firms, their behavior will differ. CON constraints that block entry of competitors will merely enhance the profit-making opportunities of the for-profit hospitals, in the standard industrial-organization view of how entry restriction enhances profitability of monopolists. Of course, the effects from substitution on the hospital's cost function would persist no matter what the organizational goals of the hospital, since those results come from cost-minimizing input selection, which we can presume for both for-profit and not-for-profit hospitals.

tual in reducing per-day or per-admission hospital costs. They did have more significant effects on hospital length of stay (modest reductions in LOS), suggesting that substitution for other inputs was working through the process of speeding up each patient's stay in order to "open up" the beds for new patients. Hospital "profitability" also fell in CON states, perfectly consistent with the idea that the EE curve for the hospital would shift inward on net, reducing its opportunities to accomplish its many goals.

On net, it would appear from the empirical studies of CON laws that they were generally ineffective in controlling hospital costs, and possibly had perverse effects on overall costs, primarily by leading to cost-increasing substitution of other assets and personnel for the beds that regulation restricted.

CON laws began to fall out of favor politically in the 1980s during the Reagan administration, probably in part because the accumulated evidence showed that they had no beneficial effect on hospital costs, despite the beliefs of their (sometimes) fervent admirers. Most state CON processes were supported by local planning agencies that make studies of hospital use and "need" and then make recommendations to the state regulatory authorities.[22] As enthusiasm for the CON laws diminished, so did the federal and state funding to support these planning agencies. Most states do not now have any formal CON process, although some states have persisted with these rules, both for hospitals and for nursing homes.

PRICE CONTROLS

The prices charged by health care providers constitute another realm of common governmental regulation. Setting aside general economywide price controls during World War II and the Korean War, significant regulation of medical costs began in August 1971,

---

[22] This arrangement was actually required by law. The 1966 Comprehensive Health Planning Act created both state and local planning organizations, and required that the local planning organizations get half of their funding locally, with heavy dependence on the hospitals they regulated. In 1974, the CHP Act was replaced by the National Health Planning and Resource Development Act, and states had to put such planning into place or sacrifice important federal funds for health care. (This act precluded local planning agencies from taking funds from local hospitals, reversing the earlier CHP requirement for such funds.) This requirement for planning was eliminated during the Reagan administration in 1986.

with the overall wage-price freeze installed by the Nixon administration, using powers bestowed upon the president by the Congress; this program was called the Economic Stabilization Program (ESP).[23] The initial 90-day freeze was lifted in favor of "Phase II" controls on most of the economy that quickly went away, except in two sectors of the economy—petroleum products and the health care sector. President Reagan decontrolled petroleum prices in early 1981, leaving only medical prices subject to serious federal government controls, which persist to this day via Medicare's control of hospital and physician prices. In addition, numerous states control the prices of hospitals very strictly, adding further price regulation to that imposed by the federal government.

The ESP used standard "price control" methods from 1972 (postfreeze) until the actual decontrol of the hospital sector in 1974. However, by then, Medicare pricing rules had adopted the general flavor of the ESP program, and continued controlled prices for hospitals and doctors continually since then. (Recall that hospitals typically get about half of their revenue from Medicare and Medicaid, for whom these rules apply, and doctors typically get about a quarter of their revenue from these "controlled" sources.)

Typically, cost controls such as these have a "cost-passthrough" price feature, wherein the price in the base year becomes the basis for all future allowed prices. In the typical Nixon-era control, the price in time t was established as the base period price P^0 plus increases in "allowed costs." Generically, the allowed price $P^t = P^0 + \Delta P_A^t (A_0/Q_0)$, where input A represents the set of all inputs for which cost increases are allowed to be passed through, A_0/Q_0 represents the input-output ratio in the base period, and ΔP_A^t represents the change in price of input A from the base period to period t. Using the base-period price plus changes in input costs allows the price r‿gulators to distinguish (say) between hospitals in different markets (with different cost structures), for hospitals of differing qualities, and so on. (Review Chapters 8 and 9 regarding hospital pricing and quality decisions.)

The Medicare pricing rules (at least those before Prospective Payment, introduced in 1983) followed the same general approach.

---

[23] At the time, annual inflation rates were about 7 percent. By more modern standards, this inflation rate would seem modest at best, and would generate little if any demand for price controls. Of course, the interim years have seen annual inflation rates in excess of 12 percent.

The 1972 amendments to the Social Security Act authorized limits for routine operating costs, putting a direct cap on prices separately from the type of rule described above. In addition, the scope of "allowed" inputs was decreased, specifically by withholding payment on capital expansion that had not been approved by state CON regulators. The amendments also established rules for physician pricing that very closely corresponded to the rule described above—physician prices could be increased only according to changes in an input cost index defined by Medicare.

State regulations of hospital prices have followed a similar approach in general. Nine states have at one time had statewide hospital rate regulation, and eight persist with such regulation now—Connecticut (1974), Maryland (1973), Massachusetts (1971), New Jersey (1971), New York (1969), Rhode Island (1971), Washington (1973), and Wisconsin (1975).[24] The first, and widely regarded as the most stringent of these rules, is the New York system, with a very specific cost-passthrough type of formula based on an input-cost index determined each year by a panel of economists. At the other extreme, the state of Washington's regulation is better described as a budget review system with "negotiations" between the hospitals and the state. Other state programs fall between these extremes, but most involve a formula of the New York genre (Joskow, 1981).

We can use the analysis of not-for-profit hospital behavior previously developed in Chapter 9 to help understand the likely consequences of a price control imposed on hospitals. (For-profit hospitals would be expected to respond somewhat differently, of course, because of their emphasis on profit motives, but the general ideas remain the same.) Figure 16.7a and b re-creates comparable figures from Chapter 9, showing in Figure 16.7a the set of demand curves and cost curves for different qualities of output, and in Figure 16.7b the "decision maker's" preference function for various combinations of service intensity (S) and quantity of output (N). The unregulated hospital will pick a quality-quantity mix that maximizes the utility of the (phantom) decision maker. This choice appears as the point (N^*, S^*) in Figure 16.7b and has a corresponding demand curve, cost curve, and price associated with that choice in Figure 16.7a, labeled P^*.

Consider now what happens when regulators impose a price control lower than P^*. (A price control higher than P^* is possible,

---

[24] Colorado had a mandatory rate regulation program from 1977 until 1980.

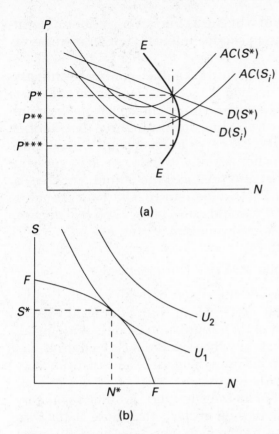

Figure 16.7 (a) Demand and cost curves for different qualities of output, and (b) "decision maker's" preference function for various combinations of service intensity (S) and quantity of output (N).

but uninteresting in general. A discussion below about New York State regulations shows what happens when some hospitals face a "binding" price control while others are unconstrained.) Three things should happen: First, the listed price will decline; the law requires this to happen. Second, service intensity (S), perhaps the same thing as quality of care, should decline. This may mean a decline in "amenities," or perhaps forestalling or elimination of new equipment that would otherwise increase the "medical quality" of care, or it might mean substituting lower-skilled workers (e.g., LPNs) for higher-skilled workers (e.g., RNs). Finally—and distinctly different from the classic analysis of price controls in a for-profit environment—quantities of output can actually increase. Indeed, if the control is only "mildly" binding, we expect this to happen, since the hospital will always operate in the preregulatory environment in the downward sloping portion of its EE curve. (See

Chapter 9 for a discussion of why this is true.) A price control set at P^{**} will produce a decline in service intensity but an increase in quantity of output.

Of course, it is possible to construct a price control so strongly binding that both S and N decline. To see this, look again at Figure 16.7a. The unconstrained quality-quantity choice has the hospital operating at (P^*, N^*). If we drop a line downward from that point, it will eventually intersect the EE curve in its upward-sloping portion. From there, go across horizontally to the price axis, and you will find the price P^{***} that also produces the output N^*. Thus, a regulation setting price at P^{***} would reduce S without changing N. Any price lower than P^{***} would cause both N and S to decline from the original (freely chosen) values of N^* and P^*.

Empirical Studies of Hospital Price Controls

Several studies have estimated the effects of these various regulatory programs on hospital costs. Sloan and Steinwald (1981), within the broader context of their regression models studying all types of controls, found that only the Nixon-era ESP controls had significantly affected hospital price increases, and that the state programs were essentially ineffective.

Dranove and Cone (1985) investigated state hospital regulatory authority while presuming in their analysis that those states with the "worst" price increase problems might have been the most eager to adopt regulations. They found that this was indeed true and that when one accounted for this issue statistically, there was a modest but nevertheless measurable effect of state regulations on the rate of hospital price increases.

Some of the more complex state regulations actually affect various hospitals differently, depending on how the rules get written, and how each hospital "comes out" can vary as state regulators change the "rules of the game." A recent study of the New York State pricing rules showed this in detail (Thorpe and Phelps, 1990). The rules—called the New York Prospective Hospital Reimbursement Methodology[25]—had complicated structures intended both to promote cost savings and to create a sense of "fairness" among the hospitals. As a result, they ended up imposing controls on some hospitals that were quite stringent—like P^{**}

---

[25] The acronym for this set of rules is NYPHRM, pronounced "nye-frum" when properly spoken, sounding approximately like the clearing of a throat.

or P*** in Figure 16.7a—and for other hospitals, imposing a control that was either quite close to or actually above the original choice of the hospital (like P*). Thus, some of the hospitals faced a binding constraint, and others did not.

These rules, like those described above, "allowed" some price increases that were "due to" increases in input costs. The real test of the rules' effectiveness therefore turns on rates of price growth, rather than on actual decreases in hospital prices. Thorpe and Phelps found that the differential impact of the rules on "constrained" and "unconstrained" hospitals was dramatically different. Unconstrained hospitals had an average annual price growth of 5.5 percent over a four-year period (1982–1985). Tightly constrained hospitals had a comparable annual growth of 1.9 percent. The differences in cost growth apparently occurred because of actual decreases in employment of "nonnursing" staff, such as laboratory technicians, phlebotomists (those who draw blood samples), and so on.

Medicare Prospective Payment as a Price Control

One can also interpret the Medicare Prospective Payment System (PPS) as a price control, although it applies legally only to patients admitted under the Medicare program. (Some private insurance plans have adopted similar pricing mechanisms in their contracts with hospitals and patients.) Two shifts in thinking are necessary to understand fully the consequences of the Medicare system. First, the price rules apply to specific categories of admissions, called "Diagnostically Related Groups" (DRGs). The Medicare PPS sets a different price for each admission, depending on which DRG category it falls into. (The Medicare rules list about 470 DRGs at present; each has a different price the hospital may charge.) The second important distinction is that the DRG system makes no allowance for each hospital's own cost conditions (at least within a given region). It is the *same* price for all hospitals. Note that this differs remarkably from the "cost-passthrough" types of rules that most price controls use. Those types of rules begin with each hospital's "base period" price and add "allowable cost increases." The DRG system disregards all of that and simply announces "a price" for each DRG. Figure 16.8 shows the effects for two prototype hospitals, one a relatively low-cost hospital, and the other a relatively high-cost hospital. (Note that these cost differences could arise from differences in input costs, or from differ-

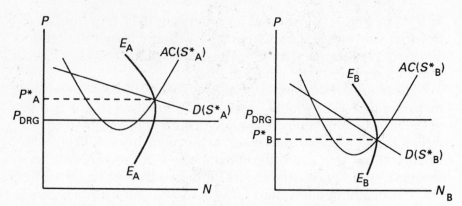

Figure 16.8 Effects for two prototype hospitals, high- and low-priced.

ences in the quality or intensity of care, labeled S in previous discussions.)

Hospital A, the relatively high-priced hospital, will necessarily reduce its quality, and could either contract or expand its output (N), depending on how much P_{DRG} was below its previously chosen P_A^*. Hospital B, by contrast, finds its optimal price P_B^* below P_{DRG}, and hence is "unbound" by the rules. It has no incentive to do anything differently, at least in this simple analysis.

A more esoteric analysis would include the following idea: As hospitals like Hospital A reduce their quality, this will shift the demand curves confronting Hospital B in predictable ways. Demand for lower-quality care will fall, because Hospital A will treat some of those patients previously treated at Hospital B. Demand for higher-quality care will rise, because Hospital A no longer produces at its previously chosen higher quality. This will cause the EE curve facing hospitals like Hospital B to shift, making higher quality relatively more attractive, and lower quality relatively less attractive. After final adjustments, Hospital B should end up producing at a higher quality than before.

Most analysts of hospital behavior expected that the DRG system would create strong incentives to reduce average length of stay in the hospital, since the hospital got paid the same amount whether the patient stayed 5, 6, or 10 days for a common surgical procedure.[26] This in turn created concerns about the health and

---

[26] A set of special rules actually allowed the hospital to receive more money once a patient had been in the hospital a very long time, a so-called "outlier" in the DRG system. For example, if the DRG allowed 10 days, once the patient had been in the hospital (say) 25 days, then the hospital's actual incurred costs would get paid.

safety of patients; the catch phrase became that patients would get discharged from the hospital "quicker and sicker." We can readily think of extra time in the hospital as an equivalent to "service intensity" (S) in the formal models of hospital behavior. The incentives to discharge patients "sicker and quicker" would thus seem to concentrate in those hospitals actually squeezed by the DRG pricing rules; those unconstrained by the rules (like Hospital B in Figure 16.8) would seem to prefer to maintain their previous level of quality, as discussed before, except for the pressures to shift quality that changes in other hospitals' quality would cause, as noted.

QUALITY CONTROL

Previous discussions have shown how the medical malpractice law (Chapter 14), licensure, and voluntary certification (earlier in this chapter) have potential effects on the quality of medical care services. A completely different type of regulation further attempts to control the quality of care rendered by providers of medical care, both "personal" care (like doctors, dentists, etc.) and "institutional" care (like hospitals, nursing homes), described here as "direct quality regulation." These activities differ from licensure and comparable quality-enhancing efforts by focusing directly on individual patients' treatment. Most of these interventions have come through the rules of Medicare and Medicaid, the two major federal governmental insurance plans discussed in Chapters 12 and 13.

The first of these federally mandated quality-control organizations were designated as Professional Standard Review Organizations (PSROs). These organizations reviewed individual treatments for Medicare and Medicaid patients to determine both the adequacy and appropriateness of care. Despite the overtone of quality control, many observers of PSROs consider these programs primarily as cost-containment efforts, attempting to deter "unnecessary surgery" and overly long hospital stays, and in this role, it seems clear that the PSRO system failed. One report, for example, concluded that the PSRO system had saved about $50 million in hospital costs in 1978, but the program itself cost $46 million, and a separate report was eventually abandoned by the federal government.

As a quality-controlling organization, the PSRO system was never very effective. The customary approach in the PSROs was to sample hospital records and compare the treatment (as recorded in the record) with a physician-determined standard. This leaves the obvious question of the scientific basis for the physician standards. The degree of variability across regions in medical practices (see Chapter 3) creates the strong suspicion that locally determined standards have no scientific basis; they cannot, in fact, and still have such widespread disagreement. Since these programs rely on retrospective record review, the concern also arises that the programs measure record-keeping patterns more than actual patterns of care.

When the Congress approved the Prospective Payment System (PPS) in Medicare, it became concerned with the obvious incentive to shorten hospital length of stay, and that quality of care would fall. Thus, despite the earlier dissatisfaction with the PSRO system, Congress mandated a similar quality-control system called the Peer Review Organizations (PROs). As with the old PSROs, the government contracted out with private organizations to conduct the peer review, and again, the system became wholly dependent on the medical record of the hospital. (The PROs abstract a large number of Medicare records from hospitals, and use that information as a basis for quality studies.) The PROs focus on premature discharge from the hospital (reflecting concern about the "sicker and quicker" problem), but also look for unnecessary hospitalizations, again creating the dual role for the PROs of quality control and cost reductions.

SUMMARY

Regulation pervades the health care sector, much of it aimed at reducing costs in the industry. These regulations have attempted to limit capital formation in the industry (CON laws), directly control prices (the federal ESP rules and Medicare pricing rules), and even individual screening of patient records for inappropriate use (PSROs and PROs). Empirical studies of these rules have found only the direct price controls "work" well, and then not universally.

A number of regulations also directly attempt to improve quality of treatment. FDA rules to control safety and efficacy of drugs and medical devices provide an obvious example, as does licensure of various medical personnel (doctors, dentists, nurses, etc.). In both cases, private provision of information (advertising of drug quality, certification of provider quality by "boards") provides a nonmandatory alternative that coexists with mandatory licensure. In most (if not all) cases, private certification provides evidence of a higher standard of quality than the mandatory licensure requires.[27]

Many regulatory activities, whether aimed at controlling cost or enhancing quality, have the obvious side effect of changing the nature of competition, primarily because of the implicit or explicit effects on entry by competitors. CON laws directly attempt to control hospital entry. FDA rules (until the 1984 revisions) considerably restricted competition, especially in the entry of "generic" products.

Other regulations that attempt to control costs also have consequences for quality of care. The effect of price controls on quality of care seems obvious, since not-for-profit hospitals can only meaningfully respond to a binding price control by reducing quality of care. (A for-profit hospital could reduce price and maintain quality simultaneously, although it would probably not choose to do so if confronted with a binding price control.)

Careful economic analysis of regulation in the health care sector remains an important realm of study; every regulation considered in this chapter has the *potential* for providing some social benefit, and none is unambiguously without merit. Each, in turn, has the *potential* for creating economic mischief, either through effects on competition, quality, or combinations thereof. Thus, we leave the discussion of regulation in the health care sector with the usual economist's two-handed evaluation: "On the one hand, they might be good, and on the other hand. . . ."

PROBLEMS

1. Calculate the present value of an annuity of 17 years' duration for an annual profit of $1 million from a new drug at discount rates of 5 percent and 10 percent. You can use "annuity" tables found in most business-math texts, or you can generate them in a spreadsheet or statistical program.

    Now calculate the present value of the last 9 years of the annuity—that is, the present value of the 17-year annuity minus the present value of an 8-year

---

[27] Indeed, it would seem strange if this were not true, because it would imply that the provider would be spending money and time in order to provide certification of quality that had no meaning to the consumer.

annuity. The difference exemplifies the loss in profits due to the (average) 8-year testing imposed by FDA rules.

2. Calculate the present value of a "perpetuity" of monopoly profits at 5 percent and 10 percent. (Hint: The present value of a perpetuity of $1 is $1/r$, where r is the discount rate.) Now calculate the present value of a perpetuity after "discarding" the first 8 years of profits. Compare this with the present value of a 17-year annuity. Which provides greater incentive for invention—a perpetuity delayed by 8 years, or a 17-year annuity? How does the answer change as you vary the discount rate from 5 percent to 10 percent?

3. The FDA regulates the sale of drugs in the United States by requiring evidence of safety and efficacy. What effects does this have that might improve the health of the U.S. population? What effects that might harm the health of the U.S. population?

4. For which disease is a drug company more likely to invest resources to develop a cure: high blood pressure (affecting millions of Americans) or cystic fibrosis (which affects about 1500 children born each year)? How, if at all, might their incentives differ if the FDA did not exist?

5. Regarding hospital regulation, the models of hospital behavior developed in Chapter 9 predict that the effects of DRG payments on hospital behavior will differ across hospitals, since that system offers the same price to all hospitals. What types of hospitals will likely have to change their behavior when confronting the DRG system, and what will their changes do to hospital output and quality, if anything? What types will likely do nothing?

6. "The free market has too much entry for hospitals, and regulators can potentially improve social well-being by restricting entry." Discuss.

7. "Licensure restricts entry into medical professions for such health care providers as doctors, pharmacists, and dentists. This can only benefit these professions, and will always harm patients by making doctors, dentists, and so on more expensive." Discuss.

8. What evidence do you know about that shows whether price controls and CON laws really do affect hospital costs?

Chapter
17

International Comparisons of Health Care Systems

*T*he preceding chapters have analyzed how the U.S. health care system works, emphasizing the interlocking roles of consumer demand, insurance coverage, tax policy, doctors, hospitals, and government regulation. It would probably be fair to state that, at least over the post–World War II era, U.S. health policy has emphasized a balancing of two concerns: access to care and control of costs. Massive growth in insurance coverage through employer-related groups left two areas of concern: retired persons and low-income persons. In the 1950s and 1960s, access issues dominated, culminating in the passage of laws creating Medicare and Medicaid. In the 1970s, issues of cost control began to emerge, with the growth of CON laws, state hospital rate regulations, and increasingly strict payment mechanisms for both doctors and hospitals under Medicare and Medicaid. By the early 1980s, cost control had become *the* dominant issue in health policy. Shifts of the entire Medicare hospital payment system (Part A) from a fee-for-activity basis to the diagnosis-related Prospective Payment System in 1983 and to an alternative "Resource-Based Relative Value System" for physicians, scheduled to begin in late 1992, provide good examples of these governmental concerns over unit costs and total spending within these programs.

As detailed in Chapter 2, growth in spending in the United States has proceeded almost without interruption, increasing the share of GNP devoted to health care from about 5 percent in 1950 to about 12 percent in 1990. Per capita costs, even after adjusting for large changes in the overall consumer price index, have increased steadily during this period. In constant dollars per capita, total spending has increased nearly sevenfold between 1950 and 1990, and hospital spending has increased twelvefold in that period. (Review the data in and discussion surrounding Table 2.5 for details.)

At the same time, access issues continued to create concern. Despite the widespread growth of employer-related plans, the universal coverage of over-65 persons under Medicare, the addition of over 1 million disabled under Medicare, and widespread coverage for the poor under Medicaid, large pockets of persons persist who have no health insurance coverage. Current estimates of the number of persons who have no health insurance coverage at all center now at about 30 to 35 million Americans, or about one out of every seven persons.

The dual concern about access to care and total spending growth has led increasingly to health policy analysts looking to other nations for alternative models of organizing the delivery of health care. As is commonly described in news and commentary on the U.S. health care system, we simultaneously have among the greatest spending per capita and the least comprehensive coverage of any major industrialized nation. We also have very high life expectancy, and by any objective standard, a broader and deeper set of medical technologies and interventions available than does any other nation. Nevertheless, many people continue to call for a rethinking of how we organize our health care system, commonly emphasizing the desirability of universal coverage for all persons and some sort of centralized cost control. The Canadian health care system is the current "favorite" for comparison, but in previous years, the British National Health Service, the German system, the Swedish system, and others have been held up as desirable standards for the United States to emulate. Surprisingly, when one looks carefully at these alternative models, one finds that they encompass a wide array of differences in cost growth, the way "coverage" is created, and the ways providers are paid. The next portion of this chapter decribes a (partly representative) handful of the health care systems in various countries, and seeks to draw from them any available lessons about the effects of "the system" on access, costs, and quality of care.

This review will necessarily omit many nations whose systems could doubtless offer useful insights for the United States, and the description of those discussed will necessarily remain both brief and (to knowledgeable persons) simplistic. For persons interested in more detail on these types of comparison, a number of recent publications (and citations therein) will provide an expanded picture of other nations' systems.[1] Here, discussion will follow on four countries of interest—Canada, Germany, Japan, and the United Kingdom. Each country has substantial similarity to the United States in some important ways, but important differences in its health care system. All are major industrialized nations with large populations; two (United Kingdom and Canada) have common legal heritage with the United States; two (Japan and Germany) rely heavily on employer-related insurance; three (all but the United Kingdom) primarily have private ownership of health care production. The primary major difference is that all of these countries (and, indeed, almost all other industrialized countries) have universal insurance of one form or another. These discussions explore briefly the differences between these countries and the United States.

SNAPSHOTS OF FOUR COUNTRIES

Canada

Many analysts use Canada's health care system as an important point of comparison to the U.S. health care system, in part because of Canada's obvious similarity to the United States in economic and political structure, and in part because the health care cost behavior in the two countries, while once quite similar, has now diverged considerably; Canada "costs less" than the United States, and cost growth is persistently lower there than in the United States. Before 1971, Canada spent 7.4 percent of its GNP on health care, and the United States spent 7.6 percent. Then the Canadians instituted the universal Medicare system, and its spending share

---

[1] Several sources seem particularly useful. First, John Igelhart (who edits *Health Affairs*) has published a sporadic but extremely useful series of articles describing the health care systems of various countries in the *New England Journal of Medicine*. The countries he has discussed include Canada (1986, and an update about "problems" in 1990), Japan (1988a, 1988b), and Germany (1991a, 1991b). A recent issue of *Health Care Financing Review* (1989 Supplement), published by the Health Care Financing Administration (the people who run Medicare), was devoted entirely to international comparisons, with extensive data and commentary; I recommend this volume highly. Finally, a 1990 supplement to *Advances in Health Economics and Health Services Research* (Rosa, Scheffler, and Rossiter, eds.) focused on international comparisons, with studies of 10 nations included. Full citations appear in the Bibliography.

has increased only to about 8.5 to 9 percent, while the U.S. system now consumes 12 percent of the U.S. GNP.

Canada's universal insurance plan is organized around its provinces (most analogous to states in the United States). Each province has its own Medicare system, with commonalities imposed by the federal government, but with some regional differences. The Canadian system retains some similarities to the United States, with important differences as well. Before the Canadian Medicare system was enacted, Canadian and U.S. systems were similarly organized, with employer-provided health insurance the most common vehicle for private insurance coverage, with some governmental insurance programs, and with both hospitals and doctors operating as private entities. Doctors' fees were determined quite similarly to those of U.S. doctors (i.e., market determination); hospital charges had evolved into a mixture of controls and bargaining, but hospitals were paid on a per-service basis.

The introduction of Medicare changed three important features of the Canadian system. First, insurance coverage became universal, provided by the provincial governments to all citizens of the province. Private insurance essentially vanished. Second, hospitals came under a central regulatory authority in each province, with a total budget cap for all provincial hospitals established by governmental authority. Within that overall cap, each hospital in the province receives a direct budget. Physicians now receive fees according to a negotiated schedule within the province, but generally continue to function as independent firms. Thus, Canada now has universal governmental insurance, mostly private production of medical care, and a strong governmental regulatory control on prices paid to both hospitals and doctors. Health care cost growth in Canada is thus almost entirely a political decision, since governments decide the rates of increase in spending in each province, with virtually no role for markets to set prices.

The consequences of this difference appear readily in Canadian cost data; since 1971, the proportion of GNP going to physicians in the United States has increased by 40 percent, where in Canada, by only 10 percent. Hospital spending has also increased at a lower rate than in the United States, although not as dramatically. These differences depend primarily on differences in payments for providers (doctors, hospitals) rather than on changes in utilization rates. The United States and Canada have maintained similar hospital admission rates (per capita) across the years. U.S.

average lengths of stay have declined steadily over the past decade, while Canadian ALOS has remained steady.[2] Doctor office visits per capita have remained similar in both countries across the intervening years since 1971, when Canada adopted Medicare.

Payments to doctors have constituted one important difference, as noted before. Another important source of difference is the rate at which the health care system has adopted new technologies. Canadian hospitals cannot (by law) use private capital markets, and must generally turn to provincial authorities for funds to support capital acquisitions such as more beds, MRI units, lithotripsy, and so on. The Canadian system has adopted new technologies at a notably slower pace than the U.S. system, accounting for an important part of the cost growth differential.

Germany

Germany is commonly touted as having the first "universal" health insurance plan, dating back to the social insurance plans of German Chancellor Otto von Bismarck in the 1880s. Bismarck's government established a spectrum of social insurance for workers and (partly) their dependents apparently in response to labor unrest and to quell growing labor union influence. Initially, the sickness funds hired doctors directly, but in the pre–World War II era, the physicians gradually separated to "panels" that negotiated with the sickness funds to provide care for patients.

After an interim when Hitler's National Socialist (Nazi) party turned the health care system into a true socialist (government owned and operated) plan, the postwar German system returned to a sickness-fund-based system, but with a prominent role for independent office-based (private practice) doctors. The sickness funds are now the prominent model, each operating as a not-for-profit entity (like the Blue Cross and Blue Shield plans in the United States), but with compulsory membership for workers and their families, with free choice of membership but choices commonly based upon occupation or geographic region (there are also numerous "local" funds independent of the regional or national funds

---

[2] The Canadian system has seen a considerable increase in the number of elderly persons using the hospital in lieu of long-term care facilities, accounting for the failure of ALOS to decline in Canada, as it has in the United States. These patients "clog up" the hospital system, creating a de facto constraint that prevents acute care hospitalizations from occurring.

with specific occupational connections). These plans account for seven out of eight German citizens. Workers and firms both contribute to these funds, with total contributions ranging from 8 to 16 percent of a worker's salary, averaging almost 13 percent. Unemployed persons have their premiums paid by the federal unemployment insurance fund. When a person retires, his or her pension plan pays mandatory insurance premiums, equalling (by law) a percent of their pension payment equalling the national average payroll contribution (now about 13 percent).

The link between the sickness funds and providers is much more formalized than in the United States. To be eligible for payment from a sickness fund, a doctor must belong to a regional association of physicians. The sickness funds pay lump sums to the physicians' association, which in turn pays doctors for services provided. These regional associations function similarly to the "physician groups" that IPAs and HMOs in the United States contract with, wherein the doctor belongs to the group practice, the insurance plan (HMO or IPA) pays the group itself, and the group pays doctors on a fee-for-service or salary basis, depending on the plan. The rates are negotiated between the sickness funds and the doctor associations.

Doctors in Germany almost always specialize either in hospital treatment or in ambulatory treatment; very few of the ambulatory-care doctors can admit patients to hospitals. In turn, hospital-based doctors are barred from practicing ambulatory medicine, and are paid on an annual salary by the hospital, the amount commonly depending on specialty and seniority. Hospital-based physician salaries come from the overall operating costs of the hospital, negotiated between the hospitals and the sickness funds. Thus the financial incentives confronting doctors are quite different than those in the United States, with a strong fee-for-service incentive in ambulatory care and an HMO-like incentive (straight salary) for hospital care.

Differences between the United States and Germany in medical care use correspond to these financial differences; German citizens, with universal and virtually full-coverage insurance, and with an active fee-for-service system in ambulatory care, consume about 11 patient visits per year, compared with 5.4 in the United States and 7.1 in Canada. Doctor visits are also much shorter in Germany, the average patient visit being about 9 minutes in Germany for a G.P. versus 15 minutes in the United States. Bed-days used by German citizens are quite high by international comparison, mostly due to the very long ALOS (18 days currently) in Germany, almost triple the U.S. ALOS.

One way the Germans achieve the high rates of service use while apparently not spending enormous sums of money on hospital care is through much lower staffing intensity in their hospitals than is typical in other countries. German hospitals have about 1.25 employees per occupied bed currently, versus about 2.8 per occupied bed in the United States, about 2.0 in Canada, and 2.5 in the United Kingdom.[3]

Japan

Japan's health care system has both elements familiar to the United States and some that differ remarkably. As in the United States and Germany, the large bulk of persons in Japan obtain their insurance via employer-related groups. In industries where "insurance societies" or "mutual aid societies" are established, these provide the insurance; in other industries, the national government sells the insurance. These societies were formed as early as 1922, when a law enabled such insurance plans to organize. Both the structure of Japanese medicine and the reliance on industry-related insurance plans were specific choices to emulate the German system at the time, widely regarded as the world's best system during the early 1900s. A separate national health insurance plan is available for persons not eligible for some employer-related group. Participation in some kind of insurance group is mandatory following legislation that passed the Japanese Parliament in 1961. Currently, the employment group plans insure about 75 million people, and the national health insurance plan insures the remaining 45 million. The employment groups contribute an average of 8.1 percent of an employee's pay (4.6 percent from the employer, 3.5 percent from the employee), although premium rates range from 6 to almost 10 percent of an employee's income. The governmental plans receive 8.9 percent of a person's monthly income, evenly split between employer and employee.

Most of the employer-group plans require copayments (20 percent for inpatient care, 30 percent for ambulatory care) for dependents, with 10 percent copayments for workers. These plans also have a "catastrophic cap" feature that limits monthly out-of-pocket

---

[3] But see discussion regarding Japan!

expenses to ¥54,000 (now about $400) per month. The community plans ("national health insurance") have premiums paid by the covered person, with a *per household* cap of ¥370,000 per year (about $2800 now), and a 30 percent copayment, both of which are related to family income. On net, premiums pay for about half of the cost of these plans, with the remainder from national and local government (tax) support. Retired persons are covered in the same plans with contributions from employment and community plans, plus funds from both national and local governments, with very small copayments for patients at the time of service (¥400 per hospital day, about $3) and ¥800 *per month* for ambulatory care.

Most hospitals in Japan are private, many organized as not for profit, but with a substantial number owned by physicians, but the government operates some hospitals also; the mix between private and public hospitals is similar to that in the United States. Most of the large, complex hospitals in Japan are public, most with affiliation to one of Japan's 80 medical schools.

Physicians primarily operate out of "clinics" rather than solo or small-group practice. Many physicians own their own clinics (and hire other doctors to work in them), but particularly among younger doctors, the preference has shifted to salary arrangements with hospital clinics.

As in Germany, "clinic" doctors cannot follow their patients into the hospital, but (in contrast to the German setting), the presence of large ambulatory-care clinics affiliated with the hospitals makes the free-standing clinic doctors very reluctant to refer their patients for hospitalization, since they could well lose the patient to the hospital clinic. Japan's hospital admission rates are by far the lowest of any major country in the world (7.5 admissions per 100 persons), half of the U.S., Canadian, and U.K. rates and a third of that found in Germany, Denmark, and Sweden (the countries with the highest hospital admission rates).[4]

The extremely low propensity to hospitalize patients in Japan is offset partly by extremely high rates of use of physician visits and prescription drugs, the two often going hand in hand. Japanese citizens see doctors an average of 15 times per year, triple the U.S.

---

[4] The Japanese rate, low as it is, represents a major increase over the past several decades. In 1960, the hospital admission rate in Japan was only 3.7 per 100 persons, half of the current rate. The increase is (as in other countries) due in part to the aging of the population, but it is also partly explicable with the shift to hospital-based clinics, where hospital referrals are not as financially dangerous to the referring doctor.

rate, which in turn stands as one of the highest per capita physician visit rates of any nation (5 per year) except Japan. These visits are apparently extremely short in duration, and almost always occur without an appointment, rationed by waiting time (about an hour's wait on average), since the universal insurance provides virtually full coverage for all physician visits.

Ohnuki-Tierney (1984), an anthropologist studying the Japanese medical care system, reported typical workloads for physicians. Across several specialties, those physicians she studied commonly treated 40–50 patients in one three-hour period, making the average visit length about 3.5–4.5 minutes, in contrast to the average visit for G.P.s in the United States of 15 minutes, with longer visits for specialists. Thus, while the number of visits in Japan is hugely different than the U.S. rate, the number of minutes a typical patient spends annually with a physician is similar to if not less than in the United States.

Japanese citizens are also among the most heavily medicated in the world, a practice that also corresponds closely to the economic incentives in Japanese society. In a recent study, Japanese citizens apparently were dispensed over 20 drugs per person per year, compared (say) with the U.S. average of about 6 prescriptions. This issue is of intrinsic economic interest, because it reflects both the role of insurance on drug use and also sheds some light on the question of "supplier-induced demand." U.S. visitors to Japan after World War II found it very surprising that Japanese doctors both prescribed and sold drugs, quite in contrast to the U.S. norm of separating prescribing and dispensing.[5] In 1955, the Japanese Parliament passed a law requiring separation of prescribing and dispensing, but the law was apparently so riddled with exceptions that it had little effect. The law was supported by the health insurance plans (who paid for the drugs) and the pharmacists' association, whose members would stand to benefit hugely from a change in the market, and strongly opposed by the Japanese Medical Association, whose members would of course lose by the change. Virtually all of the 27,000 free-standing "clinics" continue to dispense drugs, as of course do the hospital-clinics.

Insurance plan reimbursement for prescription drugs in 1980 constituted nearly 40 percent of all insurance benefits, although

---

[5] In earlier times, U.S. doctors functioned similarly; the "traveling" doctor in the 1800s both prescribed drugs and then sold them, often out of the wagon in which the doctor traveled.

this had dropped to 28 percent by 1987. (By contrast, pharmaceutical expenses paid by all sources constitute 8 percent of the U.S. health care bill.) Most of the reduction in spending has been due to reductions in the governmentally decreed price list for drugs; during the 1980s, these regulated retail prices for drugs fell by over 60 percent.[6] The rates of prescribing do not seem to have changed over this period, and the Japanese continue to consume prescription medications at a rate hugely above that in most modern societies.

The overall balance of these events has allowed Japan to operate its health care system with one of the lowest spending rates in the industrialized world. Per capita spending is about half of that in the United States[7] and the Japanese spend only about 6.5 percent of their GNP on medical care. One obvious source of this difference is hospital spending, with very much lower hospitalization rates in Japan than in the United States. When patients are hospitalized there, they remain for nearly 40 days, on average, but much of that care is more like hotel care than the intensive hospital care in the United States, so that overall spending for hospital services remains low. The combination of infrequent admissions and very long stays leads to a days/year use by the Japanese of twice the U.S. rate. However, the intensity of hospital care is so low that overall costs remain much lower than U.S. costs. Hospital staffing ratios reflect this low intensity; in Japan, hospitals hire about 0.75–0.8 employees for each occupied bed, while in the United States, staffing ratios are 3.5 times higher.

Physician costs are much more similar between the United States and Japan than one might expect, given the 15-visit per person rate in Japan versus the 5-visit U.S. rate. As noted, the much shorter time spent with a physician on each visit makes the total physician-contact time per year similar for patients in the two countries. Japanese doctors have gross revenues of about $350,000 per year—roughly twice the rates of doctors in Germany and Canada, and 1.5 times those of U.S. doctors—but net incomes appear similar in Japan and the U.S. (Sandier, 1989, Figure 6). Finally, the

---

[6] This creates an interesting issue of incidence; it appears in some discussions that the incidence of these reduced payments fell primarily on drug manufacturers and wholesalers, rather than on the retailers (clinics and doctors).

[7] Remember that such calculations depend heavily on the exchange rate between currencies that is used to make such calculation. As the U.S. dollar declines against other nations' currencies, calculations of their spending will decline when converted to U.S. dollars.

relative supply of active physicians in Japan is lower than in the United States—one doctor for each 640 people in Japan, and one doctor for each 425 people in the United States—so that total spending on doctors' services is quite similar between the two countries, despite the higher per-doctor gross income in Japan.[8] ("Gross incomes" reflect all payments to "physician-firms," and represent best the whole scope of services associated with "physician visits." Using net physician income would be analogous to trying to understand spending in hospitals by measuring nurses' incomes.)

The other remarkable feature of the Japanese system is that, despite the substantially lower spending on medical services than in the United States, Germany, or even Canada, the Japanese people appear to have among the best health outcomes of any major industrialized nation. Life expectancies at birth for males (75.6 years) and females (81.4 years) are the highest of any industrialized country. The United States, by contrast, has life expectancies of 71.5 years for males and 78.3 years for females. The contrast with earlier Japanese outcomes makes these results all the more remarkable: At the end of World War II, life expectancy in Japan was only 50 years for males and 54 years for females.

Japanese perinatal mortality (infant deaths and stillborn children) is the lowest in the world—0.66 per 100 births, compared with about 0.85 in Canada, 0.75 in Germany, 0.9 in the United Kingdom, and 1 in the United States. Two factors account for these excellent results, in addition to the obvious effect of universal access to prenatal care: The Japanese have an active maternal and child health program that includes issuing a handbook of information to every pregnant woman in Japan, and the highest abortion rate in the world (24 per 1000 women of childbearing age). As has been shown in the United States (Joyce, Corman, and Grossman, 1988), increased accessibility of abortion reduces infant mortality in the United States, and we could expect a comparable result in Japan.

---

[8] These aggregate data on physician supply and visit rates provide an independent confirmation of the very short time spent on each visit in Japan. On a per capita basis, the Japanese have about 80 percent of the physician supply that the United States does, but use three times the number of physician visits. If doctors in Japan and the United States work about the same number of hours each year, then the typical Japanese doctor must see 3.75 times as many patients per hour as does the typical U.S. doctor (3/0.8 = 3.75). If the typical U.S. visit lasts 15 minutes, then the typical Japanese visit lasts exactly 4 minutes (15/3.75 = 4). Ohnuki-Tierney's estimate was 3.5–4.5 minutes.

Nonmedical care factors probably play a very important role in these Japanese health outcomes. The traditional Japanese diet is almost devoid of fat, concentrating heavily on vegetables and fish. The role of diet on heart disease appears most prominently in the rates of heart surgery in Japan—1 bypass surgery per 1000 people per year, compared with 26 per 1000 in Canada and 61 per 1000 in the U.S. Rates of gallbladder removal (cholecystectomy) are even more remarkable, however—2 per 1000 in Japan, versus 219 per 1000 in Canada, 203 per 1000 in the United States—and the biomedical link between fat intake and gallbladder disease is well established.[9] The Japanese avoid surgery generally compared with the United States, but the coronary bypass surgery and cholecystectomy rates stand uniquely in their differences from other countries' rates. The difference seems due to Japanese persons' preferences to eschew the fat.

Great Britain

The British health care system is the best known to U.S. citizens of any socialized system, although the health care systems of many other countries are also owned and operated by the state. The British National Health Service (BNHS hereafter) was formed in 1948, following a long evolution from voluntary and then mandatory social insurance programs that the British had developed during much of the twentieth century. The BNHS provides a pure model of a socialized health care system: The vast bulk of hospitals are owned by the state; with few exceptions, all health care workers are employees of the state. A very small private insurance market exists in Great Britain, and a handful of hospitals provides service to persons with such insurance. The primary distinction between the Canadian, German, and Japanese systems (on the one hand) and the U.K. system (on the other) is the ownership of the resources used to produce health services. In the United Kingdom, the state owns the resources of production, while in the other nations, most of the resources are owned by private organizations, with some public ownership of hospitals, and broad controls over prices and investment common in the hospital sector.

The direct control over resource investment provided by the BNHS has led to considerable differences in their choices about

---

[9] Gallstones, which clog up the gall bladder duct and lead to its surgical removal, are formed as a combination of bile salts (created by the gallbladder) and cholesterol.

the organization of health care, compared with the United States or Canada, two societies with similar cultural heritage. Part of the differences arise from income differences: Using Organization for Economic Cooperation and Development (OECD) purchasing power parity measures, per capita income in Britain is only 70 percent of that in the United States, and systematic evidence across nations shows that aggregate spending closely relates to per capita income. As the next section shows, this income difference alone accounts for much (but not all) of the low per capita spending in the U.K. The mix of services chosen by the BNHS is much more hospital- and nurse-intensive, and relatively low in doctors.

The other feature—relatively low density of doctors—reflects a market force that not even a socialized health care system can control: British doctors emigrated to the United States and Canada in large numbers during the 1960s and 1970s, particularly after the economic returns to a medical license rose dramatically in the United States after Medicare was introduced in 1965. In 1960, the United States had 1.44 doctors per 1000 population, and the United Kingdom had about 1 doctor per 1000 population. Over the decades, physician supply rose greatly throughout the world, but the United States absorbed a disproportionate share of this increase; by 1987, the U.K. physician supply had increased to 1.36 doctors per 1000, but the U.S. supply had increased to 2.33 per 1000. Put differently, in 1960, the United Kingdom had a physician: population ratio about 70 percent of the U.S. figure, and by 1987, that ratio had slipped to 58 percent. (Review also Chapter 7 about immigration of physicians to the United States in the post-Medicare era; many of those physicians came from Britain.)

AGGREGATE INTERNATIONAL COMPARISONS

Health Care Spending

This brief excursion through four nations' health care systems masks many important differences among the systems. Yet surprisingly, stepping back to an even higher level of aggregation suggests common behavioral patterns in health care spending among virtually every industrialized nation of the Western world: Aggregate per capita spending on health care turns out to have a strong relationship with per capita income—so strong, in fact, that some analysts have suggested that little else remains to be explained by other factors such as the ownership of resources, cost controls, or the nature of the health insurance system in the various countries.

Kleiman (1974) and Newhouse (1977) first showed this strong relationship, using data from industrialized countries in the early 1960s (Kleiman) and the early 1970s (Newhouse). They converted other countries' expenditures to U.S. dollars using currency exchange rates from the same period, and estimated a simple linear model of the form Expenditures = Constant + β (per capita income). Several subsequent studies have expanded the data set to include more than 20 countries (Culyer, 1980; Parkin, McGuire, and Yule, 1987; Gerdtham et al., 1987). Some important differences emerge when one uses currency exchange rates versus an alternative measure called "purchasing power parity" developed recently by the OECD. Box 17.1 discusses these issues in more detail.

Box 17.1 INTERNATIONAL FINANCIAL COMPARISONS

A difficulty when making international spending comparisons arises when selecting the exchange rate with which to convert other currencies to dollars (or some other common unit). For example, since 1960, the exchange rate with the German mark has varied from 4.2 deutsche marks per dollar (in 1960) to 1.8 deutsche marks in 1987, with several fluctuations around systematic trend. Within a single country, the exchange rate doesn't matter, but when we seek to learn whether the Germans spend "more" than the United States on medical care, we must convert deutsche marks to dollars, and changes in the exchange rate alter our perception of how much Germans spend on health care. For example, suppose the average German spent 5000 deutsche marks on medical care in a year. At an exchange rate of 4.2 deutsche marks to the dollar, we would describe that as $1190. If the exchange rate is 1.8 deutsche marks per dollar, we would describe that as $2778. The currency exchange rate is highly relevant if you intend to travel to Germany, but not very relevant in comparing how much the Germans spend on medical care compared with how much is spent in the United States.

An alternative system uses "purchasing power parity" rates, a system designed to standardize each country's currency by comparing its prices to the OECD average for a fixed basket of goods. This system provides a much more stable picture of international spending. Between 1960 and 1987, the "purchasing power parity" conversion rate between the United States and Germany has varied only between 3.37 and 2.47 deutsche marks per dollar, much less than the variation in the currency exchange rate.

The exchange rate problem accounts for differences in various writers' perceptions about where the United States "stands" in per capita spending. Maxwell (1981) reported that the United States was third in 1977 per capita medical care spending in the world ($769), behind Germany ($774) and Sweden ($928), based on average exchange rates over calendar year 1977. Comparable calculations using the OECD purchasing power parity index would place the

1977 German spending at just under $600, and the Swedish spending at $625 per person. By these calculations, the United States is an easy first-place winner in the per capita medical spending race.

Most analysts now prefer to use the purchasing power parity index when assessing international spending differences, because it more closely resembles the kind of price index that we commonly prefer to use for adjusting spending data across time and space. All of the data shown in this text comparing various countries use the OECD purchasing power parity index.

A recent publication of a rich data set on 24 OECD countries (Poullier, 1989) gives us the opportunity to explore some of these issues directly. Figure 17.1 shows the relationship between per capita income (measured as gross domestic product per person) and per capita health care spending. The size of the bubbles in Figure 17.1 shows the relative sizes of the countries, with the United States the largest (at the upper right), and tiny Luxembourg and Iceland showing as small dots.

The figure shows an obvious nearly linear relationship between income and health care spending, across a wide array of

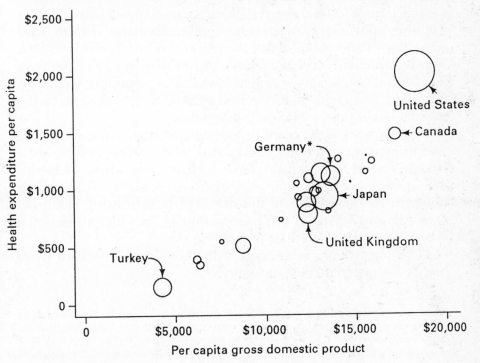

*Formerly West Germany.

Figure 17.1 Per capita income and per capita health care spending.

nations of different sizes, health care delivery systems, forms of government, geographic and climatic features, and ethnic basis of the population. A regression through these points shows a very tight fit indeed. (Review Box 5.2 on regression analysis.) The exact model depends in part on whether one weights the data by each country's population (the preferred approach) and whether one ignores or includes "outliers,"[10] but the main idea persists broadly: Per capita income explains most of the international differences between spending across countries, with the following as a typical equation estimated:

health care expenditure
per capita = $\qquad -290 \quad + 0.1$ (per capita income)
$\qquad\qquad\qquad (t = 3.5) \quad (t = 14.3)$
$\qquad\qquad\qquad N = 24; \quad R^2 = 0.91$

If one calculates the income elasticity of demand from this regression (review Box 4.1 on elasticities if you need to), the estimated income elasticity is 1.35, a result very similar to those found by the other studies cited above.

In Figure 17.1 (and statistically in the model), the United States has the appearance of having abnormally large health care spending—that is, lying above the regression line fitted to the data. This is particularly true if one fits a line to all of the data other than the United States and "predicts" what U.S. spending would be from that line; with that approach, the U.S. spending is three-eighths larger than would be predicted.

Alas, in seeking to learn from these international income and spending patterns, nothing tells us that we should necessarily see a straight-line relationship to show how income relates to health care spending. For example, the relationship might just as well be linear in the logarithms of the data, rather than using "natural" data. Figure 17.2 shows this relationship, again with the countries shown proportional to their population size.

The logarithmic regression with these data shows an even higher estimated income elasticity:[11]

---

[10] Outliers are identified as those single observations that create significant differences in the estimated regression parameters. In the regression of health care spending versus per capita income, the United States is an outlier, pulling the regression line up to itself. In a regression using logarithmic forms (see below), Turkey is an outlier that pulls the regression line down to itself. In both cases, eliminating the outliers reduces slightly the apparent relationship between income and health care spending.

[11] In logarithmic data, the estimated coefficient on income is the income elasticity itself.

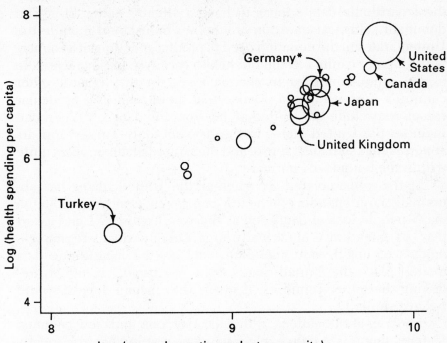

*Formerly West Germany.

Figure 17.2 Per capita income and per capita health care spending (logarithmic relationship).

$$\begin{aligned}
\log \text{ (per capita} \\
\text{medical spending)} = -10.2 & \quad + 1.8 \text{ (log (per} \\
& \quad \text{capita income))} \\
(t = 10.0) & \quad (t = 17.0) \\
N = 24 & \quad R^2 = 0.93
\end{aligned}$$

Here, the income elasticity is estimated at 1.8, and the United States is not nearly as meaningful an "outlier" in this model. If we fit the same model to all other countries except the United States, however, the United States continues to appear unusually large in health care spending.

To know which model is correct, we need to conduct a very expensive experiment: We need to increase other countries' incomes so that some of them reach the per capita income of the United States, and then see how they behave. In these times of restricted research budgets, this experiment is not likely to become funded. An alternative approach seeks to use the relationships in the data to tell which is the "best" model; in this approach, the logarithmic data appear to have a slightly better fit.[12] The dominant statistical question is whether one should include the United States in the equation used to predict spending of a country with U.S. per capita income but health care systems characteristic of the other countries in the data set. In that regard, other countries' systems are so diverse that there seems no particular reason to exclude the United States from the data set. When one includes the United States in the data set, only Turkey appears abnormal in a statistical sense, and the United States appears quite within the bounds of "normality."

Statistics cannot tell us whether the United States has abnormally large spending for health care, given its income, although many people look at data such as those in Figure 17.1 and assert that U.S. spending is abnormally large. Other perfectly reasonable models of health care spending (such as the logarithmic data model) place the United States right "on trend," in effect predicting that other countries, if given U.S. income levels, would behave similarly.

Looking at the nations in this data set, one finds few common features that might help explain why one country has abnormally large or small spending on medical care. Socialized medical care systems seem to have little effect: Great Britain has unusually low spending (below the regression line), but Sweden has abnormally large spending. Japan is abnormally low, but not by much, despite the apparently large effects of their diet on the use of at least some surgical procedures. As Newhouse (1977) pointed out, the information on per capita income explains so much of the data that little else remains for other variables to affect systematically. Despite the obviously major differences in health care systems, nothing emerges at least from this type of comparison to suggest how one approach to organizing a health care system might dominate another.

Health Outcomes

The other major question confronting health analysts is how these differences in health care spending affect health outcomes. This turns out to be a question that we cannot answer cleanly with

---

[12] One cannot just compare the R^2 in these models, since the dependent variables differ. The proper approach uses a Box-Cox transform to test whether the log or linear model is best. In these data, the logarithmic data have a somewhat better fit.

available data, but we can see some suggestive patterns in the data. The question is of obvious interest because if we could answer it, we would know more than we do now about the marginal productivity of medical care. The difficulty that analysts confront when attempting to judge such matters is the high correlation between income and medical care spending that the last section displayed. We suspect that medical care affects health outcomes (favorably) but we also suspect that income by itself affects health outcomes— for example, by affecting the amount of good (and bad) consumption items that people consume. A third variable, education, probably affects both income and health outcomes favorably, lending further confusion to things.

Finally, if we compare current medical care use (or income) in various countries with the life expectancy of their adult populations, we might find very little relationship, because life expectancy will be affected both by current and past income, consumption patterns, and medical care use. A more sensitive indicator—because it happens "now"—is the infant mortality rate or (correspondingly) the perinatal mortality rate.[13]

Because of the high correlation between per capita income and per capita medical spending, one can look at the effect of either variable on life expectancy or infant mortality and see essentially the same thing. Figure 17.3a and b shows the relationship (respectively) between per capita income and male life expectancy (a) and per capita medical spending and male life expectancy (b). (Data for female life expectancy show a very similar pattern, but females generally live longer.) The statistical relationships are very weak in both cases, whether or not one includes the U.S. data point in the estimates.

The relationships between income (and medical spending) and perinatal mortality, as expected, are stronger. Figure 17.4a and b shows the relationships for perinatal mortality rate and per capita income (a) and medical spending per capita (b). The statistical relationship is modestly tight, if one excludes the U.S. data point

---

[13] Infant mortality measures the proportion of all live-born children who die early in their lives. Perinatal mortality measures all stillborns plus infant mortality, relative to all live births plus stillborns. Nations differ in their reporting practices regarding the definition of a live birth; some count only infants weighing more than (say) 0.5 or 1 kilogram, while other countries count births of all weights. The perinatal mortality rate provides a more uniform measure of outcomes.

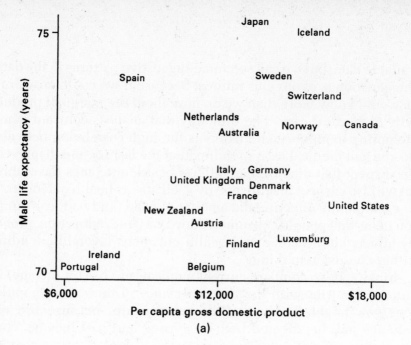

(a)

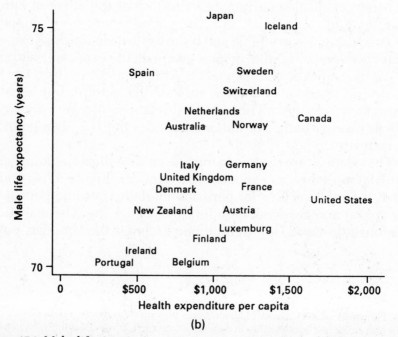

(b)

Figure 17.3 Male life expectancy and (a) per capita income and (b) per capita
medical spending, by country. (*Note:* Germany refers to the region that was
formerly known as *West Germany.*)

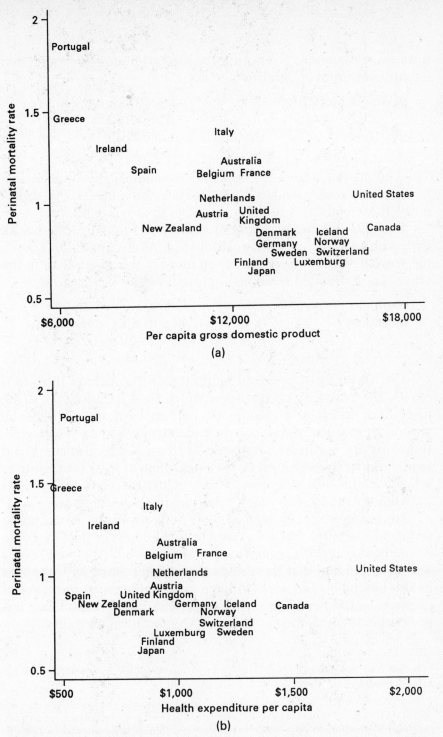

Figure 17.4 Perinatal mortality rate and (a) per capita income and (b) per capita medical spending, by country. (*Note: Germany* refers to the region that was formerly known as *West Germany*.)

from the estimation. Excluding the United States, the estimated equation for infant mortality is[14]

perinatal mortality = 1.90   − 0.0000755 (per capita income)

$$(t = 6.53) \quad (t = 3.28)$$
$$N = 21 \quad R^2 = 0.35$$

The United States and Italy appear to have unusually large perinatal mortality rates compared with their income levels. Put differently, these countries seem to lie above the regression line.

The data relating perinatal mortality rates to per capita health expenditure show an even relationship. As Figure 17.4b shows, the United States "sticks out" even more in this graph than in the other. The estimated relationship (omitting the U.S. data) is

perinatal mortality =   1.43   − 0.000517 (per capita health care spending)

$$(t = 6.74) \quad (t = 2.30)$$
$$N = 21 \quad R^2 = 0.21$$

The elasticity of perinatal mortality with respect to income is −0.98; the elasticity with respect to medical spending is −0.49, just half of the effect of income. (In both cases, if one includes the United States, the statistical relationship vanishes.) While relationships are not nearly as strong as those between income and medical care spending itself, it is clear that, at least for countries other than the United States, a strong statistical relationship exists between infant mortality and either medical spending or income.

Both of these relationships show "effectiveness" in reducing perinatal mortality, but we cannot tell whether it is "just" income or medical care, or a combination of the two.[15] We have every reason to believe that the things that income buys will improve health outcomes, but we also have good evidence at the level of individuals that, even holding income constant, prenatal care improves health outcomes as measured by perinatal mortality.

---

[14] In this case, the statistical models say that the "natural" data are a better fit than the logarithmic data.

[15] In econometric terms, since we believe income affects both the health outcome and medical care use, we cannot identify the separate relationship of medical care use on infant mortality. If we attempted to estimate an equation including both income and medical care at the same time, it would be underidentified. Indeed, in pure econometric terms, the equation estimating health outcomes as a function of medical care is not "legitimate" because medical care is jointly determined with illness levels, which we cannot observe.

In any case, these data clearly show the United States as a statistical outlier, with substantially higher perinatal mortality than one would expect, given the per capita income and the relationship between the two that exists in other countries. The causes of this discrepancy in outcome remain unresolved at this point. Critics of the health care system point to such data as evidence of the inferiority of the health care financing arrangements in the United States, noting that almost all other industrialized countries have universal health insurance, and apparently achieve better outcomes than does the United States while spending less of their income on medical care. Others respond that the U.S. health care system must deal with a wide heterogeneity of persons that other systems do not confront, with persons of different ethnic backgrounds who speak diverse languages, and in many cases of infant deaths, with immigrants (both legal and illegal), and so on. As one writer notes, "The U.S. population, often referred to as a 'melting pot,' cannot be readily compared with those of Iceland or Japan, with their much more homogeneous populations. . . . The degree to which poor health outcomes reflect social causes vs. the inadequacy of the health system . . . is hard to quantify" (Davis, 1989).

GROWTH IN COSTS AND HEALTH OUTCOMES

The final question we can address with these international comparison data is whether or not the U.S. health care system has experienced abnormally large cost growth and/or different changes in health outcomes through time. This is a different question than the ones posed in the previous section: It asks whether the *change* in costs and the *change* in health outcomes over time seem unusual in an international perspective.

A provocative article by Evans et al. (1989) lays down the gauntlet: They argue that the U.S. and Canadian health care systems were very similar in 1971, when Canada adopted universal health insurance and budget controls through their Medicare system, and that in the interim, U.S. health care spending has increased at a much faster pace than has Canadian spending. The spending comparisons were discussed above in the description of the Canadian system. Evans et al. did not discuss health outcomes. This section, briefly (and certainly not definitively!) addresses both the question of cost growth and health outcomes in various nations.

Table 17.1 describes the annual per capita spending on health care in each of the five countries previously discussed, and Table 17.2 (top) shows the annual rates of growth in spending in these countries over the three available time periods. Table 17.2 (bottom) converts these data to annual growth rates in spending after correcting for the internal inflation rate in each country, so that each of its components show *real* rates of spending increase. Note that these increases could reflect changes in productivity in the medical care sector, changes in payment associated with changes in market power of factors (doctors, nurses, etc.), or both.

The data in the second part of Table 17.2 seem to refute the general thrust of Evans et al. After adjusting for within-country inflation rates, the patterns in the United States and Canada seem quite similar, except for the 1960–1970 decade, easily explicable because of the introduction of the U.S. Medicare and Medicaid programs in that decade. German growth rates are much lower than U.S. rates in the 1980s, but comparable in previous decades. U.K. growth is commonly lower than that of other countries, and Japan's growth is commonly higher, although declining rapidly through time. The unusually low growth in medical spending in the 1970s is in no small part due to the U.S. price controls instituted in 1971–1974, and subsequent fee controls in Medicare. The overall real growth in the United States and Canada are extremely similar since 1970.

Data in Table 17.3 show perinatal mortality rates in the corresponding periods, and Table 17.4 shows the annual decline in perinatal death rates in each decade of these data. The United States started out at the top of the pack in 1960 among these countries, and ended up in 1987 at the bottom of the pack.

The relationships between changes in spending and changes in perinatal mortality rate within countries is very weak, if any

Table 17.1 PER CAPITA SPENDING ON MEDICAL CARE[a]

	1960	1970	1980	1987
Canada	$117	$282	$819	$1515
Germany	$84	$189	$704	$1073
Japan	$26	$128	$521	$1282
United Kingdom	$77	$149	$455	$751
United States	$148	$365	$1089	$2050

[a] Converted to U.S. dollars using OECD Purchasing Power Parity index; see Poullier, 1989.

Table 17.2   ANNUAL GROWTH RATES IN NOMINAL PER CAPITA SPENDING

	1960–1970	1970–1980	1980–1987	Overall
Canada	9.2%	11.2%	9.2%	9.95%
Germany	8.4%	14.1%	6.2%	9.89%
Japan	17.3%	15.1%	13.7%	15.55%
United Kingdom	6.8%	11.8%	7.4%	8.80%
United States	9.4%	11.6%	9.4%	10.22
Corrected for internal inflation rate in each country				
Canada	5.73%	4.06%	5.06%	5.12%
Germany	5.48%	6.66%	1.86%	4.94%
Japan	15.34%	7.14%	3.96%	9.25%
United Kingdom	3.81%	5.02%	3.02%	4.05%
United States	6.44%	4.20%	5.08%	5.25%

exists. Over the entire period shown in these tables, a modestly weak association exists between annual changes in spending and perinatal mortality rates, but the result is driven entirely by the Japanese experience, where there are both high growth rates in spending and large reductions in perinatal care. Given the large changes in per capita income in postwar Japan, it is difficult to know what to make of such data.

These data suggest the following hypothesis: The United States with higher medical spending than any other country, is closer to the "flat of the curve" than other countries in the sense that additional spending on medical care is less likely to produce increases in health outcomes. Ignoring the Japanese data, the United States experienced modestly higher growth in health care spending in the postwar era, but less decreases in perinatal mortality. Japan, by contrast, started with the lowest investment in health

Table 17.3   PERINATAL MORTALITY RATES

	1960	1970	1980	1987
Canada	2.84%	2.18%	1.09%	0.82%
Germany	3.58%	2.64%	1.16%	0.73%
Japan	3.73%	2.03%	1.11%	0.66%
United Kingdom	3.36%	2.38%	1.34%	0.88%
United States	2.86%	2.30%	1.32%	1.00%

Table 17.4 ANNUAL PERCENT CHANGES IN PERINATAL DEATH RATES

	1960–1970	1970–1980	1980–1987	Overall
Canada	−2.6%	−6.7%	−4.0%	−4.5%
Germany	−3.0%	−7.9%	−6.4%	−5.7%
Japan	−6.0%	−7.3%	−5.8%	−6.2%
United Kingdom	−3.4%	−5.8%	−5.8%	−4.8%
United States	−2.2%	−5.5%	−3.8%	−3.8%

care per capita and had the largest increases in spending and the largest gains in perinatal mortality. We might well characterize Japan as having been on the "steep of the curve," if that forms a legitimate contrast to the "flat of the curve."

We still confront a puzzle in analyzing health care spending and outcomes: Does more spending increase health outcomes? Contrasting the United States and Japanese experiences suggests that the answer is "yes" if you begin with little health care, and "not so much" if you begin with lots of health care. In all countries, spending increased and health improved, using both the perinatal death rate and overall life expectancy as measures of improved health. Alas, we still confront as well the added puzzle: Does the added health gain come from the increased medical spending, from increased income in general (which leads to added health spending), or to increases in education, which fueled both the increased income and the improvements in health? We cannot answer such questions with the aggregate data provided by the OECD.

SUMMARY

International comparisons of health care systems can provide several types of useful information for those studying the U.S. health care system (and probably conversely as well). As comparisons of aggregate spending on medical care and its relationships to health outcomes show, many regularities exist across countries despite the very large differences in organization of health care systems in those countries. The relationship between per capita income and spending almost certainly shows that, at the level of societies, medical care is a luxury good. (This stands in large contradistinction to cross-sectional results for individuals within the United

States, where income elasticities are very small. Health insurance should produce that kind of income-related equality within a country, but it will have nothing to do with cross-national comparisons.) While the United States has much larger health care spending per capita than any other nation, much of it (nearly all, if one looks at the logarithmic relationships between health care spending and income) is quite "in line" with what one would expect when looking at choices made by other nations. The very strong relationship between per capita medical care spending and per capita income is all the more remarkable, given the wide diversity of health care systems that various countries have chosen.

Either higher income or additional medical care spending (or both) seem to improve health outcomes, both for infant mortality and for adult life expectancy. On these comparisons, however, outcomes in the United States look particularly unfavorable, since U.S. mortality and life expectancy results are notably worse than one would expect for a country with U.S. income and medical care spending.

Careful study of the health care delivery systems also reveals many other ways to achieve the same goals that U.S. institutions have sought to achieve—for example, broad access to care, cost control, and so forth. The roles of incentives in the use and provision of care are especially open to study across nations, since the diversity of choice makes a far richer field of study than one could find within any single country. Insights about drug-prescribing habits in Japan offer but a single example of this type of study.

PROBLEMS

1. The Japanese health care system has two unusual features compared with many other nations' systems: (a) hospitalization rates are very low, but once a patient is admitted to the hospital, lengths of stay are very high; and (b) the population seems to receive prescription drugs at a very high rate. What organizational features of the Japanese system are distinct to it, and which also could plausibly lead to the aspects of behavior described above?

2. "The United States has higher medical care expenses per person than any other country, and this proves that the U.S. system is wasteful." Discuss.

3. What evidence do you know about from international comparisons of hospital use that points dramatically to the role of life-style (e.g., dietary composition) on illness rates and hospital use? (Hint: Think about demand for specific hospitalizations where diet might have an effect, e.g., heart disease.)

4. The U.S. health care system costs more per person than those in other countries. How much of the difference is explained by higher incomes in the United States than in other countries?

5. Cross-national comparisons show a quite large income elasticity of demand for medical care, perhaps exceeding 1.0. (a) What evidence supports this assertion? (b) How can you "square" this information with the very low income elasticity found in many cross-sectional (individual data) studies of demand for medical care, e.g., the RAND Health Insurance Experiment? (Note: You should presume in your answer that both the cross-national and the individual cross-sectional estimates are correct.)

Author's Postscript

*F*or many readers, finishing this book completes their formal study of health economics. For a few, it may represent only the beginning of a long excursion into the subject. Economists everywhere are now finding that the study of subjects in the domain of "health economics" offers a fascinating opportunity to pursue and expand other specialized economic tools and interests. Persons with separate interests in the economics of uncertainty, game theory, cost function estimation, demand theory and modeling, labor economics, industrial organization and regulation, econometrics, public finance and taxation policy, and other "traditional" fields in economics are now turning to the vast array of unsolved problems in health economics for new and interesting applications in their own special fields. Indeed, no "health economist" can hope to master all of the areas of expertise that one might bring to bear on the subject of health economics. As knowledge expands, specialization invariably increases. (Remember Adam Smith's dictum: "Division of labor is limited by the extent of the market." The market is growing larger!) This trend is quite natural and desirable, for there is very little in the study of health economics that is so "special" that others cannot enter the field and make important contributions. As will be painfully obvious to persons who have completed this textbook, a vast array of unsolved problems remains as fertile ground for future study. Yet, I hope that this textbook also points out the importance of scrutinizing carefully the special circumstances that often surround the study of health economics. Careful modeling of the novel institutional arrangements in health care will pay considerable dividend, as will studies of why particular institutions emerge in some settings but not others. (International comparisons of institutions may provide very fruitful paths for some studies.) As I noted in the preface, I believe that careful attention to the role of uncertainty will often, if not always, yield considerable payoff in studying behavior and institutions in the area of health care.

I also believe that we will find increasingly that the fruitful study of health economics reaches further into the medical detail of problems than has been true in the past. Biology has moved from the broad classification of plants and animals to the study of organ systems, and then to the study of cells, and now of molecules. Physics has moved from the study of objects to the study of atoms and now subatomic particles. So also has the study of economics reached for increasing detail. Early studies in health economics used highly aggregated data—for instance, average spending by persons in the 48 (or 50) states per year. This work charted the way for studies with much finer data, gathered at the level of individual annual spending on medical care, for example. Studies using household surveys and the RAND Health Insurance Study represent this class of work at its best. Studies of providers similarly shifted from aggregate data to data on hospitals, departments within hospitals, physician-firms, individual physicians, nurses, and so forth.

We can and should now turn to even more disaggregated data— for instance, patients' behavior within single clinical episodes and the corresponding behavior of providers. Work on medical practice variations, for example, has highlighted the importance of turning to specific clinical conditions to understand behavior more broadly. One reason I included the appendix on decision analysis in Chapter 5 is because that technique focuses attention on single clinical episodes. I believe that when we make this step in health economics, it will allow us to improve our understanding of much behavior that now remains puzzling and unresolved. As we successfully complete such studies, we will be able to build back up to a much better understanding of the macro-behavior in health care systems that we strive to understand. Both the data sets and the computational power now exist to make such studies feasible, but gathering new and highly detailed data on apparently "small" problems seems prerequisite to making many important advances. It would seem that any new researcher in health economics should become highly adept in the methods of statistics and econometrics, including those skills necessary to study discrete choices. I also believe that collaboration with clinician-researchers will have a growing payoff for at least some economists, because of their detailed knowledge of both clinical events and institutions involved in the production of health care.

For those readers who end their study of health economics at this point, I hope that what you have learned will assist you in your chosen work or study. For some, I also hope that this book will be only the beginning of an extended excursion into this fascinating realm.

Answers to Selected Problems

1. a. Uncertainty about illness creates financial uncertainty that in turn creates demand for insurance.
   b. Uncertainty in the production of medical care leads to large differences in the way doctors recommend therapy for patients.
   c. Uncertainty about illness creates an opportunity for doctors to deceive patients and hence to overtreat them.
   d. Uncertainty about patients' proclivities for illness creates risks to insurers that the patients may spend more than average.
   e. Uncertainty about provider skills leads to licensure and regulation.

5. The model of production of health (transforming medical care and life-style choices into an "asset" called "good health") allows us to analyze the demand for medical care using standard approaches. "Health" becomes the "good" in the utility function, and the demand for medical care is then treated as a *derived demand*. This is similar to the idea that "tires" and "gasoline" themselves don't create utility, but rather that "transportation" does, and tires and gasoline are inputs into the production of transportation.

6. The stock of health can't be observed very well, and it certainly can't be marketed in the same way that other durable goods can be marketed. Thus, we have to turn to "household production" models to understand how health is produced. In addition, "health" is highly random. Finally, at least for some illnesses, the body can self-repair without medical intervention. Few houses or automobiles have this characteristic.

CHAPTER 2

1. Income, age, insurance coverage, medical care prices, beliefs about the efficacy of medical care. Age is a good predictor of demand, but the random variations around age are also very important. Individual illnesses, if we could

measure them well, would probably be the strongest predictor of individual medical spending. When aggregated into groups, however, the individual variations in illness would wash out, and the systematic trends of age would likely be the strongest predictor of medical care use.

3. Government and private research; regulations regarding who may produce medical care and under what circumstances; tax laws that affect the organization of medical care firms, especially hospitals; government activities in the training of health care workers, including nurses, dentists, doctors and so forth.

8. The person's age. Medical spending increases systematically with age.

CHAPTER 3

1. This doesn't seem likely. First, some studies have held the insurance coverage constant across regions, and large variations still appear. More directly, studies of medical practice variations within countries like Great Britain (with a nationalized health care system) and Canada (with a nationalized health insurance system) still exhibit large variations in medical care use across regions. These types of studies virtually eliminate insurance coverage as an important determinant of the large variations that we observe (say) in the rate of use of tonsillectomy or surgery for low back injuries. Medical disagreement remains a more potent explanation.

4. Degree of training doesn't appear to have a lot to do with observed regional variations. First, most regions have similar patterns of generalists and specialists per 1000 population (except for rural areas), yet they have important differences in use of services. Second, the Boston versus New Haven comparisons show that even in cities where medical care is produced predominantly through university-affiliated hospitals by medical-school-affiliated doctors, large differences in behavior can emerge. Third, one can see considerable variation even within procedures that are carried out *only* by highly trained specialists, for example, coronary artery bypass grafts.

CHAPTER 4

2. It will create a direct upward and parallel shift in the demand curve by an amount (measured on the price axis) equal to the payment per visit (e.g., $25 per visit).

3. Yes, they could cross, if the price sensitivity differed considerably for illnesses of similar severity. This *could* happen in theory, but it seems implausible intuitively.

5. We have to think about demand curves in the same terms as markets supply medical services. Thus, we need to think about the demand for things like office visits, penicillin, and so on. The demand for penicillin represents the

aggregated demand for all people (see Problem 4) across all illnesses where the doctor might recommend penicillin (e.g., sore throats, sexually transmitted diseases, and wound infections).

7. Consumer surplus is the extra amount you would be willing to pay to consume something (e.g., a hamburger) above and beyond the amount that it costs you to acquire it (e.g., the price charged by McDonald's, Wendy's, or Burger King). If you would willingly pay $5 for a hamburger and McDonald's will sell you one for $1, then you have $4 consumer surplus from consuming that one hamburger.

8. Five visits is totally irrational. In order to get to five visits, the consumer must forgo the consumer surplus triangle traced out by the upper price line ($20), the demand curve, and the vertical portion of the price line. (Call this triangle A.) It would be better to stop at two visits and not lose this consumer surplus. Once having gotten to five visits, the consumer could increase the overall consumer surplus by moving to seven visits. The amount of extra consumer surplus is the triangle traced out by the lower price line ($20 × 0.2 = $4), the demand curve, and the vertical portion of the price line. (Call this triangle B.) Thus, either two or seven visits must be better than five visits. In order to know whether two or seven visits is better, one must compare the loss (A) with the gain (B). If B > A in area, then seven visits is best, and if B < A in area, then two visits is best. To consume exactly five visits would be stupid and irrational.

CHAPTER 5

1. The study did give everybody a fixed sum of money at the beginning of the year covering maximum possible expenditure, but that amount came independently of any medical care use. Thus, this "participation incentive" adds income, and we can expect that it would be used just as any other income is. The experimental plans varied the price of care, and should provide a pure price effect. The distinction between income effects and price effects makes the experimental evidence valid. Also, if the assertion in the second sentence was correct, then there would be no differences in spending between the plans, and there was.

3. The arc-elasticity of demand provides the basis for one answer (see Table 5.3). The definition of arc-elasticity is $E = [(q_2 - q_1)/(q_2 + q_1)] / [(p_2 - p_1)/(p_2 + p_1)]$. Here we can set $p_1 = 1$ and $p_2 = 0.2$. Table 5.3 shows (using dental care as an example) that $E = -0.39$ in the price range that's relevant here. Thus $[q_2 - q_1)/(q_2 + q_1)] = E \times [(p_2 - p_1)/(p_2 + p_1)]$. We can calculate $(p_2 - p_1) = -0.8$ and $(p_2 + p_1) = 1.2$. Thus $[(q_2 - q_1)/(q_2 + q_1)] = -0.39 \times (-2/3) = 0.26$. Then we can solve for $q_2 = 1.7q_1$. This means that dental care use would increase by 70 percent with that increase in insurance coverage. Similar calculations show (for example) that the increase in acute outpatient care use would be 54 percent, and for hospital use would be 21 percent. Comparison of the raw inpatient costs (Table 5.4) across plans gives a very similar answer, showing the comparison between a 95 percent coinsurance plan ($315 per year) and a 25 percent coinsurance plan ($373 per year), or an 18 percent difference.

The change in coinsurance between these two plans is slightly smaller than the ones used in this problem, so the change in spending should also be a little smaller.

CHAPTER 6

3. The firm competes in the final product market with other firms that may have different numbers of doctors, staff, and so on. The firm assembles a variety of inputs (doctors, nurses, office space, etc.) to produce the product (physician visits). The physician laborer may individually have a backward-bending supply curve, but in part because of the opportunities for substitution, there is no particular reason to believe that the supply curve of the physician-firm is backward bending. In sum, physicians are a factor input, combined with other factors, to create the final product of the physician-firm.

4. Going to medical school and specializing in surgery *is* profitable, but the high annual incomes are not appropriate evidence on this point. Rather, one must look at the internal rate of return for the income flow from specializing in surgery versus that from other alternatives. Surgical training requires many years (four to eight, depending on specialty) after medical school, in which earnings are lower than the market alternative, not to mention four years of medical school, entailing outlays of up to $100,000 total, *and* the forgone earnings of the alternative occupation. Any free market will show that those who do enter surgery will have higher earnings than the average person *during the years that they practice,* in order to make the investment of time worthwhile. Thus, the high annual earnings of doctors in practice tells us *nothing* in particular about the profitability of the decision.

5. The cost share of medical malpractice insurance is far too small to allow it to explain the large increases in physician prices over recent decades. (See Box 6.2 on physician cost structures.) Overall, malpractice insurance costs about 10 percent of all nonphysician costs in a typical physician-firm, and are well under 5 percent of total costs including physician-time payments. Thus, one could (for example) quadruple the malpractice insurance costs and not much would happen to the total costs of a physician-firm. For example, suppose that malpractice insurance costs $5,000; total non-M.D. costs were $50,000; and the physician payments were $80,000, making total costs of the firm $130,000 and insurance costs 3.8 percent of total costs. Now let malpractice insurance costs go up by a factor of 4 to $20,000 and all other costs remain the same. Then total costs rise to $145,000, or an 11.5 percent increase in total costs, and in most market settings, a smaller proportional increase in final product price.

7. At least four sources might arise, three in the short run and one in a longer run. First, doctors can always work longer hours than they did before. We have good evidence from HMO versus fee-for-service performance that hours of labor are sensitive to marginal rewards to physician time. Second, some doctors might decide not to retire when they might otherwise have chosen retirement. Third, we can import doctors from foreign countries, as happened dramatically in the 1960s and 1970s. Finally, medical school and residency training positions can increase, creating a long-run increase in the stock of physicians.

CHAPTER 7

4. With 15 doctors, the average is about 10,000 patients per doctor, and City C just can't compete for a doctor, with only 6,000 people. The 15 doctors will divide themselves with 10 in City A and 5 in City B, and every doctor will have nearly the same proportional demand. Nobody has any incentives to move at this point, so this allocation is stable.

    With 30 doctors, the average is about 5,000 patients per doctor, so City C is now an attractive location with 6,000 patients, and 1 doctor will locate there. The tricky puzzle is then whether City A or City B will come up a little short in order to supply the doctor to City C. Consider two alternative patterns of location:

    Pattern I: City A = 19 doctors, City B = 10 doctors, City C = 1 doctor.
    Pattern II: City A = 20 doctors, City B = 9 doctors, City C = 1 doctor.

    Take Pattern I first; A doctor in City B might think that City A looked appealing, because there are 5,210 patients per doctor (99,000/19) compared with 5,100 (51,000/10) in City B. However, if a doctor moves from City B to City A, the number of patients per doctor in City A will fall to 4,950, less than the number available now in City B. Thus Pattern I is stable. By the same logic, Pattern II is not stable, since a doctor in City A would find it attractive to move to City B. Thus Pattern I will emerge. The doctor in City C will earn unusually large returns, but nobody else can afford to move there, because it would split a small market into two parts that would both be smaller than the market remaining in either City A or City B in Pattern I.

6. Doctors are central to discussions of demand inducement because of the large informational asymmetry between doctors and patients, and also because the law has put doctors in the special position of being the only ones who can do certain things (like prescribe drugs or perform surgical operations).

CHAPTER 8

3. Nothing, to be honest. The larger hospital will almost certainly have more complex patients with more comorbidities, and so on, and will undertake more complicated procedures. Both of these phenomena will raise the cost per patient in a way that has nothing to do with "economies of scale" in the usual sense. In order to understand hospital efficiency, one must carefully control for the case mix of the patients in the hospital, and even after doing that, one must be certain that other characteristics of the patients don't affect hospital costs. Only then would an "efficiency" interpretation have any validity, and this almost never happens in real hospital data.

CHAPTER 9

4. If the hospital has any monopsony power, it will more likely emerge in the more specialized fields. Thus, intensive care nurses are more specialized than "general" nurses, who in turn are much more specialized than janitors. Even a

single-hospital town probably has no meaningful monopsony for janitors, since such general-skilled labor can easily ebb and flow from and to other industries as hospital demand changes. General nurses are more specialized, but many of them work in other settings than the hospital, and are probably more mobile across regions as well. Specialized nurses will exist only in the hospital setting, however, increasing the chance of monopsony power. In the standard monopsony setting, there will always exist "unfilled positions," and thus, we would more expect to see ads by the hospital for those professions.

CHAPTER 10

2. The group (any group, in fact) provides great economies of scale in sales costs. The work group in particular also protects the insurance company against adverse selection, since people presumably chose where they work mostly on the basis of other job characteristics, not the generosity or specific nature of the health insurance. Thus, the work group should mostly represent a random selection of people from the labor force, rather than (say) people who have high cancer risk or who are planning to have some elective surgery. (By contrast, people who seek individual insurance are more likely to be high cancer risks or planning elective surgery.) This reduces or eliminates the necessity of trying to learn a lot about individual health risks for the insurer, providing another efficiency of the employer-related group.

3. Heterogeneity of the work force probably is the major reason. Consider a fictitious group with two young people and one older worker in the work group. The two young people may prefer very comprehensive insurance, but the premium they pay in the work group—the same premium for every worker— will include the costs of the older person's insurance. Since older people spend much more on medical care than younger people (see Chapter 2, for example), adding additional coverage can become quite expensive for the younger workers. They may well prefer a less complete package, just because they have to carry the added costs of the older worker's premium as well as their own.

4. The tax subsidy to work-group insurance has probably increased the scope and generosity of private insurance considerably. This in turn has increased the demand for medical care through the usual mechanisms (see Chapters 4 and 5), which may well in turn drive up medical costs if the supply of medical care is upward sloping in the long run. All of these factors cause the size of the health care system to expand considerably as the tax subsidy increases. Thus, the tax-code provision making health insurance premium payments (by employers) exempt from all forms of income taxation has probably affected the health care system more than most other policy choices made by the government.

CHAPTER 11

2. Both contain an element of asymmetric knowledge. Adverse selection occurs when the patient knows more than the insurer about potential health risks of the patient (or spending plans, e.g., with pending pregnancy). Demand induce-

ment occurs when the doctor convinces the patient that some treatment is needed when it really isn't. The doctor's superior knowledge and the patient's trust in the doctor allow the demand inducement.

5. The most common uninsured person is young, and (to many people's surprise) employed. Typically, the worker has low education and hence low market wages, and commonly works for a small firm (e.g., a dishwasher in a restaurant). Such small firms face much higher loading fees for health insurance, and hence are less likely to offer insurance as a fringe benefit than are larger firms. The low earnings of these individuals also reduce their demand for insurance, since health insurance is a normal good (i.e., the income elasticity exceeds zero), and in fact is moderately large. Finally, their earnings are large enough to make them ineligible for public insurance plans such as Medicaid. Thus, they fall in between the two major sources of insurance in the United States—public insurance for low-income people and work-group insurance for higher-income people. Some people call this group " 'tweeners" because of their in-between position.

CHAPTER 12

4. It accords poorly. The model predicts that people would prefer to cover rare but very expensive events, and leave "trivial" events uncovered. The insurance demanded as supplemental to Medicare does just the opposite. Probably the best thing to conclude from this is that the current model of demand for insurance is incomplete. (One might also conclude that people are irrational or stupid, but economists don't usually find that a very fruitful approach to understanding human behavior.)

6. Payment to procedure-intensive specialties will fall, and hence the economic returns to specializing in those areas of medicine will fall. Conversely, the economic returns to geriatric medicine should rise, because it mostly involves diagnosis and counseling with patients. Everything we know about physician specialty choice says that physicians-in-training respond to economic incentives. Thus, demand for residency training in orthopedics should fall, and it should rise for geriatric medicine. Demand for pediatric residencies should be unaffected, since Medicare doesn't cover children. If a great number of private insurers adopted the same payment system, then pediatric training might become more desirable, since pediatrics, in general, involves "thinking," not doing procedures. Hospital use should decline, if anything. When doctors' incentives to recommend invasive procedures (e.g., surgery, some diagnostic interventions) declines, and they turn more to "watchful waiting" options, then hospital use should fall in parallel.

CHAPTER 13

4. The programmatic feature is the general requirement that medical care in the "Aid to Families with Dependent Children (AFDC)" category is allowed

only for families with no male head of household present. (This apparently comes from the historical welfare-program concept that families with a male head of household present should be able to "fend for themselves" with that person working.) This means automatically that the program will pay for the medical care of more females than males. (The children will presumably reflect a nearly 50/50 balance.)

The biological phenomenon is the longer survival of women compared with men, interacting with the programmatic feature that Medicaid pays for long term nursing home care for the elderly *after* they have depleted their assets. This means that most women outlive their male spouse, hence increasing the likelihood of asset depletion and eventual coverage by Medicaid for long term care.

6. Doctors in the military health care system may find it advantageous to induce their patients to return often for trivial matters, which the doctor can then treat readily and use the visit to count against the day's quota of required visits. Scheduling patients for many follow up visits for routine care offers one such opportunity. Another observed in some military health clinics: Prescribe medications for persons with chronic diseases (such as diabetes) with a relatively few number of days of medication (e.g., one month instead of every three or four months), so that the patient must return for a prescription refill more often.

CHAPTER 14

4. It has no major role. Malpractice law is wholly the territory of state governments. This is advantageous for economists wishing to understand how different malpractice legal arrangements affect doctor and patient behavior, since the various states have different laws, and change them over time, thus creating natural experiments for the economist to study.
6. A doctor's history of malpractice strongly predicts the chances that a future malpractice event will occur. In one study of doctors in Los Angeles, 46 doctors (out of 8000) accounted for 10 percent of all malpractice verdicts, and 30 percent of the damages awarded. Each of these doctors had lost four or more lawsuits in the period of study. Another study of insurance pricing showed that doctors' history of malpractice could predict malpractice awards in the future just as well as a doctor's specialty. Most insurance companies use doctors' specialty to set premiums, but few use past malpractice claims history.

CHAPTER 15

1. Externalities occur when Person A's action imposes costs on (or creates benefits for) other persons, but in such a way that Person A does not take account of these when making economic choices. A common feature of such problems is that property rights (and hence liability) are poorly or incompletely defined. Another common feature is that, even when property rights are well defined, transactions costs are large relative to the amount of damage. Thus, sneezing in a crowded room creates external costs (spreading of colds).

CHAPTER 16

4. The bigger market is always more attractive financially, since one has the prospect of more sales to recoup development costs. The burden of FDA testing, however, adds a "lump sum" cost to the development and testing of drugs, making it even more likely that the companies will ignore the small markets. This is the problem of "orphan drugs."

8. Since regulations vary by state, one can look at the rate of hospital cost increases in states with and without various forms of regulation, and infer the effects of those regulations. Studies by Sloan and Steinwald, Salkever and Bice, and others, for example, show that CON laws have little effect on hospital costs, despite the potential for "saving" unnecessary bed construction costs. Several studies, including Sloan and Steinwald and also Dranove and Cone, show that price controls do have some effect on cost growth. The most effective price control was the "freeze" imposed by the Nixon administration during the mid-1970s.

CHAPTER 17

1. a. Doctors in free-standing clinics have no rights to admit patients to the hospital (as in Germany and elsewhere), but doctors in hospital-affiliated clinics do have the right to carry out an ambulatory care practice (unlike Germany, where the two functions are completely separated). Thus, when a doctor in a free-standing clinic sends a patient to the hospital, the patient might get "trapped" by the doctors there. This could generate a reluctance to hospitalize patients. The long lengths of stay would also help the hospital-affiliated doctor to "take over" the referred patient by extending considerably the length of contact with the new patient, rather than sending the patient back to the referring doctor.

   b. The prescription drug rate might be high because, despite numerous attempts at reform, most doctors' offices not only prescribe but dispense pharmaceutical drugs.

4. Almost all of it. If one plots spending per person versus average per capita income, United States lies almost exactly on the line, perhaps slightly above it, depending in part on whether one fits a straight line or a logarithmic curve (see Figures 15.1 and 15.2).

Bibliography

Aaron, H. J., and Schwartz, W. B., *The Painful Prescription: Rationing Hospital Care,* Washington, DC: The Brookings Institution, 1984.

Acton, J. P., "Nonmonetary Factors in the Demand for Medical Services: Some Empirical Evidence," *Journal of Political Economy* 1975; 83:595–614.

Adamanche, K. W., and Sloan, F. A., "Competition between Non-Profit and For-Profit Insurers," *Journal of Health Economics* December 1983; 2(3):225–243.

Alchien, A. A., *The Economics of Charity: Essays on the Comparative Economics of Giving and Selling, with Applications to Blood,* London: Institute of Economic Affairs, 1973.

American Association of Blood Banks, *AABB Technical Manual,* Arlington, VA. No date.

American Hospital Association, *Hospital Statistics,* Chicago: The American Hospital Association, annual.

American Medical Association, *Profile of Medical Practice, 1981,* Chicago: American Medical Assn., 1981.

American Medical Association, *Current Procedural Terminology (CPT),* 4th ed., Chicago: The American Medical Association, 1990.

Anderson, O. W., *Blue Cross Since 1929: Accountability and the Public Trust,* Cambridge, MA: Ballinger Publishing Company, 1975.

Arnould, R., and Eisenstadt, D., "The Effects of Provider Control of Blue Shield Plans: Regulatory Options," in M. Olsen, ed., *A New Approach to Economics of Health Care,* Washington, DC: American Enterprise Institute, 1981, pp. 337–358.

Arrow, K. J., "Uncertainty and the Welfare Economics of Medical Care," *American Economic Review* 1963; 53(5):941–973.

Arrow, K. J., "Gifts and Exchanges," *Philosophy and Public Affairs* 1972; 1:343–362.

Auster, R., and Oaxaca, R., "Identification of Supplier-Induced Demand in the Health Care Sector," *Journal of Human Resources* 1981; 16:124–133.

Axelrod, R. C., *The Evolution of Cooperation,* New York: Basic Books, 1984.

Becker, G. S., "Theory of the Allocation of Time," *Economic Journal* 1965; 75:493–517.

Becker, E. C., Dunn, D., and Hsiao, W. C., "Relative Cost Differences Among Physicians' Specialty Practices," *Journal of the American Medical Association* 1988; 260(16):2397–2402.

Becker, E. R., and Sloan, F. A., "Hospital Ownership and Performance," *Economic Inquiry* 1985; 23(1):21–36.

Beitel, G. A., Sharp, M. C., and Blauz, W. D., "Probability of Arrest While Driving under the Influence of Alcohol," *Journal of Studies on Alcohol* 1975; 36(1):109–116.

Benham, L., "The Effect of Advertising on the Price of Eyeglasses," *Journal of Law and Economics* 1972; 15(2):337–352.

Benham, L., and Benham, A., "Regulating Through the Professions: A Perspective on Information Control," *Journal of Law and Economics* 1975; 18:421–447.

Benham, L., Maurizi, A., and Reder, M., "Migration, Location and Remuneration of Medical Personnel: Physicians and Dentists," *Review of Economics and Statistics* 1968, 50(3):332–347.

Berman, H., "Comment," in W. Greenberg, ed., *Competition in the Health Care Sector,* Germantown, MD: Aspen Systems Corp, 1978.

Blair, R. D., Ginsberg, P. B., and Vogel, R. J., "Blue Cross-Blue Shield Administrative Costs: A Study of Non-Profit Health Insurers," *Economic Inquiry* 1975; 13(2):55–70.

Boardman, A. E., Dowd, B., Eisenberg, J. M., and Williams, S., "A Model of Physicians' Practice Attributes Determination," *Journal of Health Economics* 1983; 2(3):259–268.

Booten, L. A., and Lane, J. I., "Hospital Market Structure and the Return to Nursing Education," *Journal of Human Resources* 1985; 20(2):184–196.

Brook, R. H., Ware, J. E., Rogers, W. H., et al., "Does Free Care Improve Adults' Health? Results from a Randomized Controlled Trial," *New England Journal of Medicine* 1983; 309(24):1426–1434.

Buchanan, J., and Cretin, S., "Fee-for-Service Health Care

Expenditures: Evidence of Selection Effects Among Subscribers Who Choose HMOs," *Medical Care* 1986, 24(1):39–51.

Buchanan, J., and Hosek, S., "Costs, Productivity, and the Utilization of Physician Extenders in Air Force Primary Medicine Clinics," Santa Monica, CA: The RAND Corporation Report R-2896-AF, June 1983.

Bunker, J. P., and Brown, B. W., "The Physician-Patient as Informed Consumer of Surgical Services," *New England Journal of Medicine* 1974; 290(19):1051–1055.

Burstein, P. L., and Cromwell, J., "Relative Incomes and Rates of Return for U.S. Physicians," *Journal of Health Economics* 1985; 4:63–78.

Cady, J. F., "An Estimate of the Price Effects of Restrictions on Drug Price Advertising," *Economic Inquiry* 1976; 14:493–510.

Cafferata, G. L., "Private Health Insurance Coverage of the Medicare Population," Data Preview 18, NCHSR National Health Care Expenditures Study, September 1984, Washington, DC: U.S. Dept. of Health and Human Services Pub. No. (PHS) 84-3362.

Campbell, D. P., and Stanley, J. C., *Experimental and Quasi-Experimental Designs for Research*, Chicago: Rand McNally, 1963.

Chamberlin, E. H., *The Theory of Monopolistic Competition*, 8th ed., Cambridge: Harvard University Press, 1962.

Chassin, M. R., Brook, R. H., Park, R. E., et al., "Variations in Use of Medical and Surgical Services by the Medicare Population," *New England Journal of Medicine* 1986; 314(5):285–290.

Coase, R., "The Nature of the Firm," *Economica* 1937, New Series, 4:386–405.

Coase, R., "The Problem of Social Cost," *Journal of Law and Economics* 1960; 3:1–45.

Colquitt, M., Fielding, P., and Cronan, J. F., "Drunk Drivers and Medical and Social Injury," *New England Journal of Medicine* 1987; 317(20):1262–1266.

Cook, P. J., and Graham, D. A., "The Demand for Insurance and Protection: The Case of Irreplaceable Commodities," *Quarterly Journal of Economics* 1977; 91(1):143–156.

Cowing, T. G., and Holtman, A. G., "Multiproduct Short-Run Hospital Cost Functions: Empirical Evidence and Policy Implications from Cross-Section Data," *Southern Economic Journal* 1983; 49(3): 637–653.

Cowing, T. G., Holtman, A. G., and Powers, S., "Hospital Cost Analysis: A Survey and Evaluation of Recent Studies," *Advances in Health Economics and Health Services Research* 1983; 4.

Cretin, S., Duan, N., Williams, A. P., Gu, X., and Shi, Y., "Modeling the Effect of Insurance on Health Expenditures in the People's Republic of China," RAND Corporation manuscript, 1988.

Cromwell, J., and Mitchell, J. B., "Physician-Induced Demand for Surgery," *Journal of Health Economics* 1986; 5:293–313.

Culler, S. D., and Bazzoli, G. I., "Moonlighting Behavior Among Young Professionals," *Journal of Health Economics* 1985; 4(3):283–292.

Culyer, A. J., "Cost Containment in Europe," *Health Care Financing Review* December 1989; Annual Supplement, 21–32.

Cumming, P. D., Wallace, E. L., Schorr, J. B., and Dodd, R. Y., "Exposure of Patients to Human Immunodeficiency Virus Through the Transfusion of Blood Components That Test Antibody Negative," *New England Journal of Medicine* 1989; 321(14): 941–946.

Danzon, P. M., "An Economic Analysis of the Medical Malpractice System," *Behavioral Sciences and the Law* 1983; 1(1):39–54.

Danzon, P. M., "Liability and Liability Insurance for Medical Malpractice," *Journal of Health Economics* 1985a; 4:309–331.

Danzon, P. M., *Medical Malpractice: Theory, Evidence and Public Policy*, Cambridge, MA: Harvard University Press, 1985b.

Danzon, P. M., "Liability for Medical Malpractice: Incidence and Incentive Effects," University of Pennsylvania, working paper, 1990.

Danzon, P. M., and Lillard, L. A., "Settlement Out of Court: The Disposition of Medical Malpractice Claims," *Journal of Legal Studies* 1982; 12(2):345–377.

Darby, M. R., and Karni, E., "Free Competition and the Optimal Amount of Fraud," *Journal of Law and Economics* 1973; 16(1): 67–88.

Davis, K., "Comment on 'What Can Americans Learn from Europeans?' " *Health Care Financing Review* December 1989; Annual Supplement, 104–107.

Davis, K., and Russell, L. B., "The Substitution of Hospital Outpatient Care for Inpatient Care," *Review of Economics and Statistics* 1972; 54(1):109–120.

Department of Health and Human Services, *International Classification of Diseases (ICD-9-CM)*, 2nd ed., Washington, DC: U.S. Government Printing Office (PHS)-80-1260, September, 1980.

DeVany, A. S., House, D. R., and Saving, T. R., "The Role of Patient Time in the Pricing of Dental Services: The Fee-Provider Density Relation Explained," *Southern Economic Journal* 1983; 49(3): 669–680.

Dionne, G., "Search and Insurance," *International Economic Review* 1984; 25(2):357–367.

Dranove, D., "Demand Inducement and the Physician/Patient Relationship," *Economic Inquiry* 1988a; 26(2):281–298.

Dranove, D., "Pricing by Non-Profit Institutions," *Journal of Health Economics* 1988b; 7(1):47–57.

Dranove, D., and Cone, K., "Do State Rate-Setting Regulations Really Lower Hospital Expense?" *Journal of Health Economics* 1985; 4(2):159–165.

Dranove, D., and White, W. D., "Agency and the Organization of Health Care Delivery," *Inquiry* 1987; 24:405–415.

Eckert, R. D., and Wallace, E. L., *Securing a Safer Blood Supply*, Washington, DC: American Enterprise Institute, 1985.

Eddy, D. M., *Screening for Cancer: Theory, Analysis and Design*, Englewood Cliffs, NJ: Prentice-Hall, 1980.

Eddy, D. M., "The Value of Mammography Screening in Women Under Age 50 Years," *Journal of the American Medical Association* 1988; 259:1512–1519.

Eggers, P., and Prihoda, R., "Pre-Enrollment Reimbursement Patterns of Medicare Beneficiaries Enrolled in 'At Risk' HMOs," *Health Care Financing Review* 1982; 4:155–173.

Eisenstadt, D., and Kennedy, T. E., "Control and Behavior of Non-Profit Firms: The Case of Blue Shield," *Southern Economic Journal* July 1981; 26–36.

Epstein, A. A., Stern, R. S., and Weissman, J. S., "Do the Poor Cost More? A Multihospital Study of Patients' Socioeconomic Status and Use of Hospital Resources," *New England Journal of Medicine* 1990; 322(16):1122–1128.

Evans, R. G., "Supplier-Induced Demand," in M. Perlman, ed., *The Economics of Health and Medical Care* London: MacMillan, 1974, pp. 162–173.

Evans, R. G., Lomas, J., Barer, M. L., et al., "Controlling Health Expenditures—The Canadian Reality," *New England Journal of Medicine* 1989; 320(9):571–577.

Evans, R. G., Parish, E. M. A., and Scully, F., "Medical Productivity Scale Effects, and Demand Generation," *Canadian Journal of Economics* 1973; 6:376–393.

Farley, P. J., "Theories of the Price and Quantity of Physician Services: A Synthesis and Critique," *Journal of Health Economics* 1986; 5:315–333.

Feldman, R., "Price and Quality Differences in the Physicians' Services Market," *Southern Economic Journal* 1979; 45:885–891.

Feldman, R., and Begun, J. W., "The Effect of Advertising: Lessons from Optometry," *Journal of Human Resources* 1978; 13(Supp.):253–262.

Feldman, R., and Dowd, B., "Is There a Competitive Market for Hospital Services?" *Journal of Health Economics* 1986; 5:277–292.

Feldman, R., and Greenberg, W., "The Relation Between the Blue Cross Market Share and the Blue Cross 'Discount' on Hospital Charges," *Journal of Risk and Insurance* 1981; 48:235–245.

Feldstein, M. S., "Hospital Cost Inflation: A Study of Nonprofit Price Dynamics," *American Economic Review* 1971; 61:853–872.

Feldstein, M., "Quality Change and the Demand for Hospital Care," *Econometrica* 1977; 45(7):1681–1702.

Filkins, L. D., Clark, C. D., Rosenblatt, C. A., et al., "Alcohol Abuse and Traffic Safety: A Study of Fatalities, DWI Offenders and Alcoholics and Court-Related Treatment Approaches," Ann Arbor, MI: Highway Safety Research Institute, University of Michigan, 1970.

Frech, H. E., and Ginsberg, P. B., "Competition Among Health Insurers," in W. Greenberg, ed., *Competition in the Health Care Sector*, Germantown, MD: Aspen Systems Corp, 1978.

Frech, H. E., "Market Power in Health Insurance, Effect on Insurance and Medical Markets," *Journal of Industrial Economics* 1979; 28:55–72.

Friedman, M., *Essays in Positive Economics*, Chicago: University of Chicago Press, 1953.

Friedman, M., *Capitalism and Freedom*, Chicago: University of Chicago Press, 1962.

Friedman, M., and Kuznets, S., "Income from Independent Professional Practice," New York: National Bureau of Economic Research General Series No. 45, 1945.

Friedman, B., and Pauly, M. V., "Cost Functions for a Service Firm with Variable Quality and Stochastic Demand," *Review of Economics and Statistics* 1981; 63(4):620–624.

Frymoyer, J. W., "Back Pain and Sciatica," *New England Journal of Medicine* 1988; 318(5):291–300.

Fuchs, V. R., *Who Shall Live? Health, Economics, and Social Choice*, New York: Basic Books, 1974.

Fuchs, V. R., "The Supply of Surgeons and the Demand for Operations," *Journal of Human Resources* 1978; 13(Supp.):35–56.

Fuchs, V. R., "Comment," *Journal of Health Economics* 1986; 5(3):367.

Fuchs, V. R., and Hahn, J. S., "How Does Canada Do It? A Comparison of Expenditure for Physicians' Services in the United States and Canada," *New England Journal of Medicine* 1990; 323(13):884–890.

Fuchs, V. R., and Kramer, M., "Determinants of Expenditures for Physicians' Services," Washington, DC: U.S. Dept. of Health, Education and Welfare, 1972.

Gerdtham, U., Anderson, F., Sogaard, J., and Jonsson, B.,"Economic Analysis of Health Care Expenditures: A Cross-sectional Study of the OECD Countries," CMT Rapport 1988: Linkoping, Sweden: Centre for Medical Technology Assessment, 1988.

Gertman, P. M., Stackpole, D. A., Levenson, D. K., et al., "Second Opinions for Elective Surgery: The Mandatory Medicaid Program in Massachusetts," *New England Journal of Medicine* 1980; 302:21:1169–1174.

Ginsberg, P., "Altering the Tax Treatment of Employment-Based Health Plans," *Milbank Memorial Fund Quarterly* 1981; 59(2): 224–255.

Glover, J. A., "The Incidence of Tonsillectomy in School Children," *Proceedings of the Royal Society of Medicine* 1938; 31:1219–1236.

Goldberg, G. A., Maxwell-Jolly, D., Hosek, S., and Chu, D. S. C., "Physician's Extenders' Performance in Air Force Clinics," *Medical Care* 1981; 19:951–965.

Goldstein, G. S., and Pauly, M. V., "Group Health Insurance as a Local Public Good," in R. N. Rosett, ed., *The Role of Health Insurance in the Health Services Sector*, New York: National Bureau for Economic Research, 1976, pp. 73–110.

Gould, J., "The Economics of Legal Conflicts," *Journal of Legal Studies* 1973; 2(2):279–300.

Graham, J. D., and Vopel, J. W., "Value of a Life: What Difference Does It Make?" *Risk Analysis* 1981; 1(1):89–95.

Grannemann, T. W., "Reforming National Health Insurance for the Poor," in M. V. Pauly, ed., *National Health Insurance: What Now, What Later, What Never?* Washington, DC: American Enterprise Institute, 1980.

Grannemann, T. W., Brown, R. S., and Pauly, M. V., "Estimating Hospital Costs—A Multiple-Output Analysis," *Journal of Health Economics* 1986; 5(2):107–127.

Greaney, T. L., and Sindelar, J. L., "Physician-Sponsored Joint Ventures: An Antitrust Analysis of Preferred Provider Organizations," *Rutgers Law Journal* 1987; 18(3):513–589.

Green, J., "Physician-Induced Demand for Medical Care," *Journal of Human Resources* 1978; 13(Supp.):21–33.

Greenberg, W., "Provider-Influenced Insurance Plans and Their Impact on Competition: Lessons from Dentistry," in M. Olsen, ed., *A New Approach to Economics of Health Care*, Washington, DC: American Enterprise Institute, 1981, pp. 359–374.

Greenlick, M. R., and Darsky, B. J., "A Comparison of General Drug Utilization in a Metropolitan Community with Utilization Under a Drug Prepayment Plan," *American Journal of Public Health* 1968; 58(11):2121–2136.

Grossman, M., *The Demand for Health: A Theoretical and Empirical Investigation,* New York: Columbia University Press (for the National Bureau for Economic Research), 1972a.

Grossman, M., "On the Concept of Health Capital and the Demand for Health," *Journal of Political Economy* 1972b; 80(2):223–255.

Hadley, J., *More Medical Care, Better Health? An Economic Analysis of Mortality Rates,* Washington, DC: The Urban Institute Press, 1982.

Hadley, J., Holohan, J., and Scanlon, W., "Can Fee for Service Co-Exist with Demand Creation?" *Inquiry* 1979; 16(3):247–258.

Hadley, J., "Physician Participation in Medicaid: Evidence from California," *Health Services Research,* 1979; 14:266–280.

Harberger, A. C., "Three Basic Postulates for Applied Welfare Economics: An Interpretive Essay," *Journal of Economic Literature* 1971; 9(3):785–797.

Harris, J., "The Internal Organization of Hospitals: Some Economic Implications," *Bell Journal of Economics* 1977; 8:467–482.

Harris, J., "Pricing Rules for Hospitals," *Bell Journal of Economics* 1979a; 10(1):224–243.

Harris, J. E., "Regulation and Internal Control in Hospitals," *Bulletin of the New York Academy of Medicine* 1979b; 55(1):88–103.

Harris, J., "Comment: Competition and Equilibrium as a Driving Force in the Health Services Sector," in R. P. Inman, ed., *Managing the Service Economy,* Cambridge: Cambridge University Press, 1985.

Harvard Medical Practice Study, *Patient, Doctors, Lawyers: Medical Injury, Malpractice Ligitation and Patient Compensation in New York,* Cambridge: Harvard University, 1990.

Hay, J., and Leahy, M., "Physician-Induced Demand: An Empirical Analysis of the Consumer Information Gap," *Journal of Health Economics* 1982; 3:231–244.

Health Care Financing Review, International Comparison of Health Care Financing and Delivery: Data and Perspectives 1989; 7 (annual supplement).

Health Insurance Association of America (HIAA), "Providing Employee Health Benefits: How Firms Differ," Washington, DC: The Health Insurance Association of America, 1990.

Held, P., "Access to Medical Care in Designated Physician Shortage Areas: An Economic Analysis." Princeton: Mathematica Policy Research, June 1976.

Herfindahl, O. C., "Concentration in the US Steel Industry," unpublished doctoral dissertation, Columbia University, 1950.

Hershey, J., Kunreuther, H., Schwartz, J. S., and Williams, S. V., "Health Insurance Under Competition: Would People Choose What Is Expected?" *Inquiry* 1984; 21(4):349–360.

Hickson, G. B., Altmeier, W. A., and Perrin, J. M., "Physician Reimbursement by Salary or Fee-for-Service: Effect on Physician Practice Behavior in a Randomized Prospective Study," *Pediatrics* 1987; 80(3):344–350.

Hirshman, A., *National Power and the Structure of Foreign Trade*, Berkeley: University of California Press, 1945.

Hollingsworth, J. W., and Bondy, P. K., "The Role of Veterans Affairs Hospitals in the Health Care System," *New England Journal of Medicine* June 28, 1990; 322(26):1851–1856.

Holmer, M., "Tax Policy and the Demand for Health Insurance," *Journal of Health Economics* 1984; 3:203–221.

Holzman, D., "Malpractice Crisis Therapies Vary," *Insight*, December 12, 1988.

Hotelling, H., "Stability in Competition," *Economic Journal* 1929; 39:41–57.

Hsiao, W. C., Braun, P., Kelly, P. L., and Becker, E. C., "Results, Potential Effects and Implementation Issues of the Resource-Based Relative Value System," *Journal of the American Medical Association* 1988; 260(16):2429–2438.

Hsiao, W., Braun, P., Yntema, D., and Becker, E., "Estimating Physicians' Work for a Resource-Based Relative Value System," *New England Journal of Medicine* 1988; 319(13):835–841.

Hughes, E. X. F., Fuchs, V. R., Jacoby, J. E., and Lewit, E. M., "Surgical Work Loads in a Community Practice," *Surgery* 1972; 71:315–327.

Hughes, R. G., Hunt, S. S., and Luft, H. S., "Effects of Surgeon Volume and Hospital Volume on Quality of Care in Hospitals," *Medical Care* 1987; 25(6):489–503.

Igelhart, J. K., "Health Policy Report: Japan's Medical Care System," *New England Journal of Medicine* 1988a; 319(12):807–812.

Igelhart, J. K., "Health Policy Report: Japan's Medical Care System—Part Two," *New England Journal of Medicine* 1988b; 319(17):1166–1171.

Igelhart, J. K., "Health Policy Report: Canada's Health Care System Faces Its Problems," *New England Journal of Medicine* 1990; 322(8):562–568.

Igelhart, J. K., "Health Policy Report: Germany's Health Care System" (first of two parts), *New England Journal of Medicine* 1991a; 324(7):503–508.

Igelhart, J. K., "Health Policy Report: Germany's Health Care System" (second of two parts), *New England Journal of Medicine* 1991b; 324(24):1750–1756.

Ikegami, N., "Health Technology Development in Japan," *International Journal of Technology Assessment in Health Care* 1988; 4:239–254.

Jackson-Beeck, M., and Kleinman, J. H., "Evidence for Self-Selection Among Health Maintenance Organization Enrollees," *Journal of the American Medical Association* 1983; 250(20): 2826–2829.

Janerich, D. T., Thompson, W. D., Varela, L. R., et al., "Lung Cancer and Exposure to Tobacco Smoke in the Household," *New England Journal of Medicine* 1990; 323(10):632–636.

Joskow, P., *Controlling Hospital Costs: The Role of Government Regulation,* Cambridge: MIT Press, 1981.

Joyce, T., Corman, H., and Grossman, M., "A Cost-Effectiveness Analysis of Strategies to Reduce Infant Mortality," *Medical Care* 1988; 26(4):348–360.

Kahneman, D., and Tversky, A., "Prospect Theory: An Analysis of Decision Under Risk," *Econometrica* 1979; 47:263–289.

Kass, D. I., and Paulter, P. A., "Physician Control of Blue Shield Plans," Washington: Federal Trade Commission, 1979.

Kasteler, J., Kane, R. L., Olsen, D. M., and Thetford, C., "Issues Underlying Prevalence of 'Doctor Shopping' Behavior," *Journal of Health and Social Behavior* 1976; 17:328–339.

Keeler, E. B., Buchanan, J. L., Rolph, J. E., et al., "The Demand for Episodes of Treatment in the Health Insurance Experiment," Santa Monica, CA: The RAND Corporation, Report R-3454-HHS, March 1988.

Keeler, E. B., Newhouse, J. P., and Phelps, C. E., "Deductibles and the Demand for Medical Care Services: The Theory of a Consumer Facing a Variable Price Schedule Under Uncertainty," *Econometrica* 1977; 45(3):641–655.

Kessel, R. A., "Price Discrimination in Medicine," *Journal of Law and Economics* 1958; 1(2):20–53.

Kessel, R. A., "Transfused Blood, Serum Hepatitis, and the Coase Theorem," *Journal of Law and Economics* 1974; 17:265–290.

Kitch, E. W., Isaac, M., and Kaspar, K., "The Regulation of Taxicabs in Chicago," *Journal of Law and Economics* 1971; 14(2):285–350.

Kleinman, E., "The Determinants of National Outlay on Health," in M. Perlman, ed., *The Economics of Health and Medical Care,* London: MacMillan, 1974.

Kuhn, T. S., *The Structure of Scientific Revolutions,* 2nd ed., Chicago: University of Chicago Press, 1970.

Kwoka, J. E., "Advertising and the Price and Quality of Optometric Services," *American Economic Review* 1984; 74(1):211–216.

Lancaster, K., "A New Approach to Consumer Demand Theory," *Journal of Political Economy* 1966; 74(2):132–157.

Lave, J. R., and Lave, L. B., "Hospital Cost Functions," *American Economic Review* 1970; 58:379–395.

Leffler, K., "Physician Licensure: Competition and Monopoly in American Medicine," *Journal of Law and Economics* 1978; 21(1):165.

Lerner, A. P., "The Concept of Monopoly and the Measurement of Monopoly Power," *Review of Economic Studies* 1934; 1:157–175.

Lewis, C. E., "Variations in the Incidence of Surgery," *New England Journal of Medicine* 1969; 281(16):880–884.

Long, J., and Scott, F., "The Income Tax and Nonwage Compensation," *Review of Economics and Statistics* 1982; 64(2):211–219.

Long, S. H., Settle, R. F., and Stuart, B. C., "Reimbursement and Access to Physicians' Services Under Medicaid," *Journal of Health Economics* 1986; 5:235–252.

Luft, H. S., "The Relationship Between Surgical Volume and Mortality: An Exploration of Causal Factors and Alternative Models," *Medical Care* 1980; 18:940–959.

Luft, H. S., *Health Maintenance Organizations: Dimensions of Performance,* New York: Wiley & Sons, 1981.

Luft, H. S., Bunker, J. P., and Enthoven, A. C., "Should Operations Be Regionalized? The Empirical Relation Between Surgical Volume and Mortality," *New England Journal of Medicine* 1979; 301:1364–1369.

Manning, W. G., Benjamin, B., Bailit, H. L., and Newhouse, J. P., "The Demand for Dental Care: Evidence from a Randomized Trial in Health Insurance," *Journal of the American Dental Association* 1985; 110:895–902.

Manning, W. G., Leibowitz, A., Goldberg, G. A., et al., "A Controlled Trial of the Effect of a Prepaid Group Practice on the Use of Services," *New England Journal of Medicine* 1984; 310(23):1505–1510.

Manning, W. G., Newhouse, J. P., Duan, N., et al., "Health Insurance and the Demand for Medical Care: Evidence from a Randomized Experiment," *American Economic Review* 1987; 77(3):251–277.

Marder, W. D., and Willke, R. J., "Comparison of the Value of Physician Time by Specialty," in H. E. Frech III, *Regulating Doctors' Fees: Competition, Benefits, and Controls Under Medicare*, Washington, DC: American Enterprise Institute, 1991, pp. 260–281.

Marquis, M. S., "Cost Sharing and Provider Choice," *Journal of Health Economics* 1985; 4:137–157.

Marquis, M. S., and Holmer, M., "Choice Under Uncertainty and the Demand for Health Insurance," Santa Monica, CA: The RAND Corporation, Note N2516-HHS, September 1986.

Marquis, M. S., and Phelps, C. E., "Demand for Supplemental Health Insurance," *Economic Inquiry* 1987; 25(2):299–313.

Mattison, N., "Pharmaceutical Innovation and Generic Drug Competition in the USA: Effects of the Drug Price Competition and Patent Term Restoration Act of 1984," *Pharmaceutical Medicine*, 1986, 1:177–185.

Maull, K. I., Kinning, L. S., and Hickman, J. K., "Culpability and Accountability of Hospital-Injured Alcohol-Impaired Drivers," *Journal of the American Medical Association* 1984; 252(14):1880–1883.

Maxwell, R. J., *Health and Wealth: An International Study of Health-Care Spending*, Lexington, MA: Lexington Books, 1981.

Mayhew, D. R., and Simpson, H. M., "Alcohol, Age, and Risk of Road Accident Involvement," *Alcohol, Drugs and Traffic Safety: Proceedings of the Ninth International Conference—1983*, Washington, DC: U.S. Dept. of Transportation, National Highway Traffic Safety Administration, 1983; 937–947.

McCarthy, T., "The Competitive Nature of the Primary-Care Physician Services Market," *Journal of Health Economics* 1985; 4(1):93–118.

McCombs, J. S., "Physician Treatment Decisions in a Multiple Treatment Model," *Journal of Health Economics* 1984; 3(2): 155–171.

McGuire, T., Nelson, R., and Spavins, T. " 'An Evaluation of Consumer Protection Legislation: The 1962 Drug Amendments': A Comment," *Journal of Political Economy* 1975; 83(3):655–662.

McKenzie, G. W., *Measuring Economic Welfare: New Methods*, Cambridge: Cambridge University Press, 1983.

McKenzie, G., and Pearce, I., "Exact Measures of Welfare and the Cost of Living," *Review of Economic Studies*, 1976; 43: 465–468.

McPherson, K., Strong, P. M., Epstein, A., and Jones, L., "Regional Variations in the Use of Common Surgical Procedures: Within and Between England and Wales, Canada, and the United States of America," *Social Science in Medicine* 1981, 15A:273–288.

McPherson, K., Wennberg, J. E., Hovind, O. B., and Clifford, P., "Small-Area Variations in the Use of Common Surgical Procedures: An International Comparison of New England, England, and Norway," *New England Journal of Medicine* 1982; 307(21):1310–1314.

Melnick, G. A., and Zwanziger, J., "Hospital Behavior Under Competition and Cost-Containment Policies," *Journal of the American Medical Association* 1988; 260(18):2669–2675.

Mills, D. H., Boyden, J. S., Rubsamen, D. S., and Engle, H. L., *Report on Medical Insurance Feasibility Study*, San Francisco: California Medical Association, 1977.

Mitchell, B. M., and Phelps, C. E., "National Health Insurance: Some Costs and Effects of Mandated Employee Coverage," *Journal of Political Economy* 1976; 84(3):553–571.

Moore, M. J., and Viscusi, W. K., "Doubling the Estimated Value of Life: Results Using New Occupational Fatality Data," *Journal of Policy Analysis and Management* 1988a; 7(3):476–490.

Moore, M. J., and Viscusi, W. K., "The Quantity-Adjusted Value of Life," *Economic Inquiry* 1988b; 31:369–388.

Moore, S. H., Martin, D. P., Richardson, W. C., "Does the Primary-Care Gatekeeper Control the Costs of Health Care? Lessons from the SAFECO Experience," *New England Journal of Medicine* 1983, 309(22):1400–1404.

Morrisey, M. A., Conrad, D. A., Shortell, S. M., and Cook, K. S., "Hospital Rate Review: A Theory and an Empirical Review," *Journal of Health Economics* 1984; 3(1):25–47.

Newhouse, J. P., "A Model of Physician Pricing," *Southern Economic Journal* 1970a; 37(2):174–183.

Newhouse, J. P., "Toward a Theory of Nonprofit Institutions: An Economic Model of a Hospital," *American Economic Review* 1970b; 60(1):64–74.

Newhouse, J. P., "The Economics of Group Practice," *Journal of Human Resources* 1973; 8(1):37–56.

Newhouse, J. P., "A Design for a Health Insurance Experiment," *Inquiry* 1974; 11(3):5–27.

Newhouse, J. P., "Medical Care Expenditure: A Cross-National Survey," *Journal of Human Resources* 1977; 12:115–125.

Newhouse, J. P., Manning, W. G., Morris, C. N., et al., "Some Interim Results from a Controlled Trial of Cost-Sharing in Health

Insurance," *New England Journal of Medicine* 1981; 305(25): 1501–1507.

Newhouse, J. P., Phelps, C. E., and Marquis, M. S., "On Having Your Cake and Eating It Too: Econometric Problems in Estimating the Demand for Health Services," *Journal of Econometrics* 1980; 13(3):365–390.

Newhouse, J. P., Williams, A. P., Bennett, B. W., and Schwartz, W. B., "Does the Geographical Distribution of Physicians Reflect Market Failure?" *Bell Journal of Economics* 1982a; 13:493–505.

Newhouse, J. P., Williams, A. P., Bennett, B. W., and Schwartz, W. B., "Where Have All the Doctors Gone?" *Journal of the American Medical Association* 1982b; 247(17):2392–2396.

Noether, M., "The Effect of Government Policy Changes on the Supply of Physicians: Expansion of a Competitive Fringe," *Journal of Law and Economics* 1986; 29(2):231–262.

Ohnuki-Tierney, E., *Illness and Culture in Contemporary Japan: An Anthropological View*, Cambridge: Cambridge University Press, 1984.

Olsen, D. M., Kane, R. L., and Kastler, J., "Medical Care as a Commodity: An Exploration of the Shopping Behavior of Patients," *Journal of Community Health* 1976; 2(2):85–91.

Parkin, D., McGuire, A., and Yule, B., "Aggregate Health Care Expenditures and National Income: Is Health Care a Luxury Good?" *Journal of Health Economics* 1987; 6(2):109–128.

Pauker, S. G., and Kassirer, J. P., "The Threshold Approach to Clinical Decision Making," *New England Journal of Medicine* 1980; 302:1109–1117.

Pauly, M. V., and Satterthwaite, M. A., "The Pricing of Primary Care Physicians' Services: A Test of the Role of Consumer Information," *Bell Journal of Economics* 1981; 12:488–506.

Pauly, M. V., "The Economics of Moral Hazard," *American Economic Review* 1968; 58(3):531–537.

Pauly, M. V., "Medical Staff Characteristics and Hospital Costs," *Journal of Human Resources* 1978; 13(Supp.):77–111.

Pauly, M. V., "The Ethics and Economics of Kickbacks and Fee Splitting," *Bell Journal of Economics* 1979; 10(1):344–352.

Pauly, M. V., *Doctors and Their Workshops*, Chicago: University of Chicago Press, 1980.

Pauly, M. V., "Taxation, Health Insurance, and Market Failure," *Journal of Economic Literature* 1986; 24(6):629–675.

Pauly, M. V., and Redisch, M., "The Not-for-Profit Hospital as a Physicians' Cooperative," *American Economic Review* 1973; 63(1):87–99.

Paxton, H. T., "Just How Heavy Is the Burden of Malpractice Premiums?" *Medical Economics* January 16, 1989: 168–185.

Peltzman, S., "An Evaluation of Consumer Protection Legislation: The 1962 Drug Amendments," *Journal of Political Economy* 1973; 81(5):1049–1091.

Perkins, N. K., Phelps, C. E., and Parente, S. T., "Age Discrimination in Resource Allocation Decisions: Evidence from Wrongful Death Awards," University of Rochester, Public Policy Analysis Program working paper, 1990.

Phelps, C. E., "The Demand for Health Insurance: A Theoretical and Empirical Investigation," Santa Monica, CA: The RAND Corporation Report R-1054-OEO, July 1973.

Phelps, C. E., "The Demand for Reimbursement Insurance," in R. N. Rosett, ed., *The Role of Health Insurance in the Health Services Sector*, New York: National Bureau for Economic Research, 1976.

Phelps, C. E., "Induced Demand—Can We Ever Know Its Extent?" *Journal of Health Economics* 1986a; 5:355–365.

Phelps, C. E., "Large-Scale Tax Reform: The Example of Employer-Paid Health Insurance Premiums," University of Rochester Working Paper No. 35, March 1986b.

Phelps, C. E., "Risk and Perceived Risk of Drunk Driving Among Young Drivers," *Journal of Policy Analysis and Management* 1987; 6(4):708–712.

Phelps, C. E., "Death and Taxes—An Opportunity for Substitution," *Journal of Health Economics* 1988; 7(1):1–24.

Phelps, C. E., "Bug-Drug Resistance: Sometimes Less is More," *Medical Care* 1989; 29(2):194–203.

Phelps, C. E., Hosek, S., Buchanan, J., et al., "Health Care in the Military: Feasibility and Desirability of a Closed Enrollment System," Santa Monica, CA: The RAND Corporation Report R-3145-HA, April 1984.

Phelps, C. E., and Mushlin, A. I., "Focusing Technology Assessment Using Medical Decision Theory," *Medical Decision Making* 1988; 8(3):279–289.

Phelps, C. E., and Newhouse, J. P., "Effects of Coinsurance: A Multivariate Analysis," *Social Security Bulletin* 1972; 35(6): 20–29.

Phelps, C. E., and Newhouse, J. P., "Coinsurance, the Price of Time, and the Demand for Medical Services," *Review of Economics and Statistics* 1974; 56(3):334–342.

Phelps, C. E., and Parente, S. T., "Priority Setting for Medical Technology and Medical Practice Assessment, *Medical Care* 1990; 28(8):703–723.

Phelps, C. E., and Reisinger, A. L., "Unresolved Risk in Medicare," in M. V. Pauly, ed., *Proceedings of University of Pennsylvania Conference on the Occasion of the 20th Anniversary of Medicare*, Philadelphia: University of Pennsylvania Press, 1988.

Phelps, C. E., and Sened, I., "Market Equilibrium with Not-for-Profit Firms," working paper, University of Rochester, 1990.

Phelps, C. E., Weber, L. A., and Parente, S. T., "Variations in Medical Practice Patterns: Doctor-Specific Differences in Length of Stay," Institute for Health Policy Studies, working paper, University of Rochester, 1990.

Poullier, J. P., "Health Data File: Overview and Methodology," *Health Care Financing Review* 1989; Annual Supplement, 111–118.

Pratt, J. W., "Risk Aversion in the Large and in the Small," *Econometrica* 1964; 32(1–2):122–136.

Pratt, J. W., Wise, D. A., and Zeckhauser, R., "Price Differences in Almost Competitive Markets," *Quarterly Journal of Economics* May 1979; 93:189–211.

Reinhardt, U., "A Production Function for Physician Services," *Review of Economics and Statistics* 1972; 54(1):55–66.

Reinhardt, U. E., "Manpower Substitution and Productivity in Medical Practices: Review of Research," *Health Services Research* 1973; 8(3):200–227.

Reinhardt, U. E., *Physician Productivity and Demand for Health Manpower*, Cambridge, MA: Ballinger Publishing Company, 1975.

Reinhardt, U., "The Theory of Physician-Induced Demand: Reflections After a Decade," *Journal of Health Economics* 1985; 4(2):187–193.

Rice, T. H., "Induced Demand—Can We Ever Know Its Extent?" *Journal of Health Economics* 1987; 6:375–376.

Rice, T. H., and Labelle, R. J., "Do Physicians Induce Demand for Medical Services?" *Journal of Health Politics, Policy and Law* 1989; 14(3):587–600.

Roemer, M. I., and Schwartz, J. L., "Doctor Slowdown: Effects on the Population of Los Angeles County," *Social Science in Medicine* 1979; 13C(4):213–218.

Roemer, M. I., "Bed Supply and Hospital Utilization: A Natural Experiment," *Hospitals* 1961; 35:36–42.

Rogers, J. F., and Muscaccio R. A., "Physician Acceptance of Medicare Patients on Assignment, *Journal of Health Economics* 1983; 2(1):55–73.

Rolph, J. E., "Some Statistical Evidence on Merit Rating in Medical Malpractice Insurance," *Journal of Risk and Insurance* 1981; 48:247–260.

Roos, N. P., Flowerdew, G., Wajda, A., and Tate, R. B., "Variations in Physician Hospital Practices: A Population-Based Study in Manitoba, Canada," *American Journal of Public Health* 1986; 76(1):45–51.

Rosa, J. J., ed., *Advances in Health Economics and Health Services Research—Comparative Health Systems: The Future of National Health Care Systems and Economic Analysis* Greenwich, CT: JAI Press, 1990 (Supplement).

Rosenthal, G., "Price Elasticity of Demand for General Hospital Services," in H. E. Klarman, ed., *Empirical Studies in Health Economics*, Baltimore: The Johns Hopkins University Press, 1970.

Rosett, R. N., and Huang, L. F., "The Effect of Health Insurance on the Demand for Medical Care," *Journal of Political Economy* 1973; 81:281–305.

Rossiter, L. F., and Wilensky, G. R., "A Reexamination of the Use of Physician Services: The Role of Physician-Initiated Demand," *Inquiry* 1983; 20:231–244.

Roueche, B., *Eleven Blue Men and Other Narratives of Medical Detection*, New York: Berkley Medallion Books, New Berkley Medallion Edition, 1965.

Russell, L. B., and Manning, C. L., "The Effect of Prospective Payment on Medicare Expenditures," *New England Journal of Medicine* 1989; 320:439–444.

Ruther, M., and Helbing, C., "Medicare Liability of Persons Using Reimbursed Physician Services: 1980," *Health Care Financing Notes*, December 1985.

Sadanand, A., and Wilde, L. L., "A Generalized Model of Pricing for Homogeneous Goods Under Imperfect Information," *Review of Economic Studies* 1982; 49:229–240.

Salkever, D. C., and Bice, T. W., "The Impact of Certificate of Need Controls on Hospital Investment," *Milbank Memorial Fund Quarterly* 1976 54:185–214.

Sandier, S., "Health Services Utilization and Income Trends," *Health Care Financing Review* December 1989; Annual Supplement, 33–48.

Sandler, D. P., Comstock, G. W., Helsing, K. J., and Shore, D. L., "Deaths from All Causes in Non-Smokers Who Lived with Smokers," *American Journal of Public Health* 1989; 79(2): 163–167.

Satterthwaite, M. A., "Consumer Information, Equilibrium, Industry Price, and the Number of Sellers," *Bell Journal of Economics* 1979; 10(2):483–502.

Satterthwaite, M. A., "Competition and Equilibrium as a Driving Force in the Health Services Sector," in R. P. Inman, ed., *Managing the Service Economy*, Cambridge: Cambridge University Press, 1985.

Scheiber, G. J., "Health Care Expenditures in Major Industrialized Countries, 1960–87" *Health Care Financing Review* 1990; 11(4):159–167.

Schwartz, W. B., Newhouse, J. P., Bennett, B. W., and Williams, A. P., "The Changing Geographic Distribution of Board-Certified Specialists," *New England Journal of Medicine* 1980; 303:1032–1038.

Schwartz, A., and Wilde, L. L., "Intervening in Markets on the Basis of Imperfect Information," *Pennsylvania Law Review* 1979; 127: 630–682.

Schwartz, A., Wilde, L. L., "Competitive Equilibria in Markets for Heterogeneous Goods Under Imperfect Information: A Theoretical Analysis with Policy Implications," *Bell Journal of Economics* 1982a; 13(1):181–193.

Schwartz, A., and Wilde, L. L., "Imperfect Information, Monopolistic Competition, and Public Policy," *American Economic Review* May 1982b; 72(2):18–23.

Scitovsky, A. A., and Snyder, N. M., "Effect of Coinsurance on the Demand for Physician Services," *Social Security Bulletin* 1972; 35(6):3–19.

Scitovsky, A. A., and McCall, N. M., "Coinsurance and the Demand for Physician Services: Four Years Later," *Social Security Bulletin* 1977; 40:19–27.

Shavell, S., "Strict Liability vs. Negligence," *Journal of Legal Studies* 1980; 9:1–25.

Simon, J. L., and Smith, D. B., "Change in Location of a Student Health Service: A Quasi-Experimental Evaluation of the Effects of Distance on Utilization," *Medical Care* 1973; 11(1):59–67.

Sindelar, J. L., "State Taxation of Health Insurance: Preferential Treatment," Yale University Working Paper, July 7, 1987.

Sloan, F. A., "Lifetime Earnings and Physicians' Choice of Specialty," *Industrial and Labor Relations Review* 1970; 24:47–56.

Sloan, F. A., "Physician Supply Behavior in the Short Run," *Industrial and Labor Relations Review* 1975; 28(4):549–569.

Sloan, F. A., "Rate Regulation as a Strategy for Hospital Cost Control: Evidence from the Last Decade," *Milbank Memorial Fund Quarterly/Health and Society* 1983; 61(2):195–221.

Sloan, F. A., and Feldman, R., "Competition Among Physicians," in W. Greenberg, ed., *Competition in the Health Care Sector: Past, Present, and Future,* Washington, DC: Federal Trade Commission, 1978.

Sloan, F. A., Blumstein, J. F., and Perrin, J. M., eds., *Uncompensated Hospital Care: Rights and Responsibilities,* Baltimore, The Johns Hopkins Press, 1986.

Sloan, F. A., Mergenhagen, P. M., Burfield, W. B., et al., "Medical Malpractice Experience of Physicians: Predictable or Haphazard?" *Journal of the American Medical Association* 1989; 262:3291–3297.

Sloan, F. A., Mitchell, J., and Cromwell, J., "Physician Participation in State Medicaid Programs," *Journal of Human Resources* 1978; 13(Supp.):211–245.

Sloan, F. A., and Steinwald, B., "Effects of Regulation on Hospital Costs and Input Use," *Journal of Law and Economics* 1980a; 23(1):81–110.

Sloan, F. A., and Steinwald, B., *Insurance, Regulation, and Hospital Costs,* Lexington, MA: Lexington Books, 1980b.

Sloan, F. A., Valvona, J., and Mullner, R., "Identifying the Issues: A Statistical Profile," in F. A. Sloan, J. F. Blumstein, and J. M. Perrin, eds., *Uncompensated Hospital Care: Rights and Responsibilities,* Baltimore: The Johns Hopkins University Press, 1986, pp. 16–53.

Sloan, F. A., and Vraciu, R. A., "Investor-Owned and Not-for-Profit Hospitals: Addressing Some Issues," *Health Affairs* Spring 1983; 25–34.

Sloss, E. M., Keeler, E. B., Brook, R. H., et al., "Effect of a Health Maintenance Organization on Physiologic Health: Results from a Randomized Trial," *Annals of Internal Medicine* May 1987; 1–9.

Smith, M. C., and Garner, D. D., "Effects of a Medicaid Program on Prescription Drug Availability and Acquisition," *Medical Care* 1974; 12(7):571–581.

Solow, R., "Blood and Thunder," *Yale Law Journal* 1971; 80:1711.

Stano, M., "An Analysis of the Evidence on Competition in the Physicians' Services Market," *Journal of Health Economics* 1985; 4:197–211.

Stano, M., "A Clarification of Theories and Evidence on Supplier-Induced Demand for Physicians' Services," *Journal of Human Resources* 1987a; 22:611–620.

Stano, M., "A Further Analysis of the Physician Inducement Controversy," *Journal of Health Economics* 1987b; 6:227–238.

Stano, M., and Folland, S., "Variations in the Use of Physician Services by Medicare Beneficiaries," *Health Care Financing Review* 1988; 9(3):51–57.

Steinwald, B., and Neuhauser, D., "The Role of the Proprietary Hospital," *Law and Contemporary Problems* 1970; 35:818.

Steinwald, B., and Sloan, F. A., "Determinants of Physicians' Fees," *Journal of Business* 1974; 47(4):493–511.

Swartz, K., "The Uninsured and Workers Without Employer-Group Health Insurance," monograph, Washington, DC: The Urban Institute, 3789-02, August 1988.

Swartz, K., "Strategies for Assisting the Medically Uninsured," monograph, Washington, DC: The Urban Institute, 1989.

Taylor, A. K., and Wilensky, G. R., "The Effect of Tax Policies on Expenditures for Private Health Insurance," in J. Meyer, ed., *Market Reforms in Health Care*, Washington, DC: American Enterprise Institute, 1983.

Thorpe, K. E., and Phelps, C. E., "Regulatory Intensity and Hospital Cost Growth," *Journal of Health Economics* 1990; 9:143–166.

Titmuss, R. M., *The Gift Relationship: From Human Blood to Social Policy*, New York: Vintage Books, 1972.

Torrance, G. W., "Measurement of Health State Utilities for Economic Appraisal," *Journal of Health Economics* 1986; 5(1):1–30.

Torrance, G. W., "Utility Approach to Measuring Health-Related Quality of Life," *Journal of Chronic Diseases* 1987; 40(6):593–600.

U.S. Department of Health and Human Services, *Health Care Financing Program Statistics: Medicare and Medicaid Data Book, 1988*, Baltimore, MD: Health Care Financing Administration Publication No. 03270, April 1989.

Viscusi, W. K., "Labor Market Valuations of Life and Limb: Empirical Evidence and Policy Implications," *Public Policy* 1978; 26(3): 359–386.

Vogel, R., "The Tax Treatment of Health Insurance Premiums as a Cause of Overinsurance," in M. V. Pauly, ed., *National Health Insurance: What Now, What Later, What Never?* Washington, DC: American Enterprise Institute, 1980.

Ware, J. E., Brook, R. H., Rogers, W. H., et al., "Comparison of Health Outcomes at a Health Maintenance Organization with Those of Fee-For-Service Care," *The Lancet* May 3, 1986; 1017–1022.

Warner, K. E., "Smoking and Health Implications of a Change in the Federal Cigarette Excise tax," *Journal of the American Medical Association* 1986; 255(6):1028–1032.

Warner, K. E., "Effects of the Antismoking Campaign: An Update," *American Journal of Public Health* 1989; 79(2):144–151.

Watt, J. M., Derzon, R. A., Renn, S. C., et al., "The Comparative Economic Performance of Investor-Owned Chain and Not-for-Profit Hospitals," *New England Journal of Medicine* 1986; 314(2):89–96.

Wennberg, J. E., "Dealing with Medical Practice Variations: A Proposal for Action," *Health Affairs* 1984; 3(2)6–31.

Wennberg, J. E., "Small Area Analysis and the Medical Care Outcome Problem," in L. Sechrest, E. Perrin, and J. Bunker, eds., *Research Methodology: Strengthening Causal Interpretation of Non-Experimental Data*, Rockville, MD: Department of Health and Human Services, PHS90-3454, 1990, pp. 177–213.

Wennberg, J. E., Freeman, J. L., and Culp, W. J., "Are Hospital Services Rationed in New Haven or Over-Utilised in Boston? *Lancet 1*, May 23, 1987, 1185–1188.

Wennberg, J. E., and Gittelsohn, A., "Health Care Delivery in Maine I: Patterns of Use of Common Surgical Procedures," *Journal of the Maine Med. Assn.* 1975; 66:123–130, 149.

Wennberg, J. E., McPherson, K., and Caper, P., "Will Payment Based on Diagnosis-related Groups Control Hospital Costs?" *New England Journal of Medicine* 1984; 311(5):295–330.

Wolinsky, F. D., and Corry, B. A., "Organizational Structure and Medical Practice in Health Maintenance Organizations," in *Profile of Medical Practice 1981*, Chicago: American Medical Assn., 1981.

Woodbury, S., "Substitution Between Wage and Nonwage Benefits," *American Economic Review* 1983; 73(1):166–182.

Woodward, R. S., and Warren-Boulton, F., "Considering the Effect of Financial Incentives and Professional Ethics on 'Appropriate' Medical Care," *Journal of Health Economics* 1984; 3(3):223–237.

Zeckhauser, R. J., "Medical Insurance: A Case Study of the Trade-Off Between Risk Spreading and Appropriate Incentives," *Journal of Economic Theory* 1970; 2(1):10–26.

Zuckerman, S., Bovbjerg, R. R., and Sloan, F. A., "Effects of Tort Reforms and Other Factors on Medical Malpractice Insurance Premiums," *Inquiry* 1990; 27:167–182.

Acknowledgments

Page 122, Table 5.1; page 123, Table 5.2; and page 126, Table 5.4, from "Health Insurance and the Demand for Medical Care: Evidence from a Randomized Experiment" by W. G. Manning, J. P. Newhouse, and N. Duan, *American Economic Review* 77(3): 251–277, 1987. Reprinted by permission of the American Economic Association, 2014 Broadway, Suite 305, Nashville, Tennessee 37203. Page 127, Table 5.5, from "The Demand for Dental Care: Evidence from a Randomized Trial on Health Insurance" by W. G. Manning, B. Benjamin, H. L. Bailit, and J. P. Newhouse, *Journal of the American Dental Association* 110:895–902, 1985. Reprinted by permission of the *Journal of the American Dental Association*. Page 135, Table 5.8, from "Effects of a Medicaid Program on Prescription Drug Availability and Acquisition" by M. C. Smith and D. D. Garner, *Medical Care* 12(7): 571–581, 1974. Reprinted by permission of J. B. Lippincott Company. Page 163, table in Box 6.2, from *Profile of Medical Practice, 1981*, table 38. Chicago: American Medical Association. Pages 189–190, Table 7.2, from "Does the Geographical Distribution of Physicians Reflect Market Failure?" by J. P. Newhouse, A. P. Williams, B. W. Bennett, and W. B. Schwartz, *Bell Journal of Economics*, 13: 493–505, 1982. Copyright 1982. Reprinted by permission of RAND. Pages 202–203, portions of Box 7.3, from "Consumer Information, Equilibrium, Industry Price, and the Number of Sellers" by M. A. Satterthwaite, *Bell Journal of Economics* 10(2): 483–502, 1979. Copyright 1979. Reprinted by permission of RAND. Page 204, Table 7.3, from "Medical Care as a Commodity: An Exploration of the Shopping Behavior of Patients" by D. M. Olsen, R. L. Kane, and D. Kastler, *Journal of Community Health* 2(2): 85–91, 1976. Reprinted by permission of Human Sciences Press, Inc. Page 212, portion of Box 7.4, from "The Role of Patient Time in the Pricing of Dental Services: The Fee-Provider Density Relation Explained" by A. S. DeVany, D. R. House, and T. R. Saving, *Southern Economic Journal* 49(3): 669–680, 1983. Reprinted by permission of Southern Economic Association. Pages 212–213, portion of Box 7.4, from "Comment" by V. R. Fuchs, *Journal of Health Economics* 5(3): 367, 1986. Reprinted by permission of Elsevier Science Publishers and V. R. Fuchs. Page 269, table in Box 9.3, from *AHA Hospital Statistics*, 1987. Reprinted by permission of American Hospital Association. Page 337, Table 12.1, from C. E. Phelps and A. L. Reisinger, "Unresolved Risk in Medicare" in *Lessons from the First 20 Years of Medicare* by Mark V. Pauly and William L. Kissick. Copyright 1988 by the University of Pennsylvania Press. Reprinted by permission of the University of Pennsylvania Press. Page 410, Table 14.1, from "Malpractice Awards in Single-Plaintiff Medical Malpractice Cases," April 28, 1989, *The Wall Street Journal*. Reprinted by permission of *The Wall Street Journal*, © 1989, Dow Jones & Company, Inc. All rights reserved worldwide.

Index